SCIENCE, INFORMATION, AND POLICY INTERFACE

for Effective Coastal and Ocean Management

SCIENCE, INFORMATION, AND POLICY INTERFACE
for Effective Coastal and Ocean Management

Edited by

Bertrum H. MacDonald

Suzuette S. Soomai

Elizabeth M. De Santo

Peter G. Wells

CRC Press
Taylor & Francis Group
Boca Raton London New York

CRC Press is an imprint of the
Taylor & Francis Group, an **informa** business

CRC Press
Taylor & Francis Group
6000 Broken Sound Parkway NW, Suite 300
Boca Raton, FL 33487-2742

Printed on acid-free paper
Version Date: 20160218

ISBN 13: 978-1-138-49083-3 (pbk)
ISBN 13: 978-1-4987-3170-6 (hbk)

Library of Congress Cataloging-in-Publication Data

Names: MacDonald, Bertrum H., editor.
Title: Science, information, and policy interface for effective coastal and ocean management / edited by Bertrum H. MacDonald, Suzuette S. Soomai, Elizabeth M. De Santo, and Peter G. Wells.
Description: Boca Raton, FL : CRC Press, 2016. | Includes bibliographical references and index.
Identifiers: LCCN 2015045134 | ISBN 9781498731706
Subjects: LCSH: Marine resources--Management. | Marine ecosystem management. | Coastal zone management.
Classification: LCC GC1018.5 .S35 2016 | DDC 333.91/64--dc23
LC record available at http://lccn.loc.gov/2015045134

Visit the Taylor & Francis Web site at
http://www.taylorandfrancis.com

and the CRC Press Web site at
http://www.crcpress.com

Contents

Section I Introduction

Section II Fundamental Concepts and Principles

Section III Case Studies

Section IV   The Way Forward

Foreword

Humans have been fascinated by oceans since the dawn of time. Living by the sea, they had little idea of the extent and depth of the vast expanses of water that lay before them. But they were keen to learn, soon discovering that fish, shellfish, and other coastal resources provided a plentiful supply of food and other valuable materials. Indeed, tools used by coastal dwellers have been found off the shores of Crete dating back 130,000 years. As the generations passed, so too did information about what each generation had learned. The thirst for greater understanding remained unquenched. By 4000 BC, the Egyptians had built boats capable of voyages of discovery across open seas, to be followed in subsequent centuries by the great ocean explorers and eventually international traders. Today, the extensive use of the oceans represents one of the key foundations of the global economy.

For those living inland, however, information about the oceans has often arrived second hand. Stories abounded about distant, exotic lands; terrible storms and tidal waves; and strange sea creatures encountered on various journeys. The challenge for the landlubbers must have been to distinguish between truth and myth. Imagine, for example, hearing about an eight-armed octopus for the first time, or sailors' tales of the leviathan, or of reaching the ends of the oceans on what was thought to be a flat Earth! Separating fact from fiction is still one of the greatest challenges for those dealing with information.

For most of human history, there have been less than half a billion people on Earth. But with the scientific and industrial revolutions, an unprecedented increase in the size of the population began, rising from ca. 1 billion in 1800 to around ca. 7.4 billion today. This rapid change demanded a concomitant increase in global trading, using the oceans to support the shipping of goods to meet needs. Furthermore, fishing pressures increased manyfold as the oceans became a major source of food. This intense increase in activity involving enormous numbers of humans has generated a plethora of unexpected consequences. Who could have imagined that so many fish and shellfish would be taken from the oceans that many fisheries would collapse? Or that the volume of our waste dumped in the sea would damage marine ecosystems and pose a threat to our health? Or that human diseases such as cholera would flourish as a result of exposure to a pathogen harbored in coastal waters?

Information about all aspects of the oceans has now been gathered in diverse forms by different actors over many centuries. Scientists in particular have made great progress in understanding how ocean ecosystems work. What we have learned is that careful management is the only way to sustain our oceans so that we can reap their myriad benefits for generations to come.

Tackling the challenges that we face requires in-depth knowledge and understanding. Understanding is based on accurate information that is scientifically robust, reproducible, and which covers all aspects and types of ocean environments. This information must be conveyed in a readily understandable form to those politicians and other decision makers who can take action through policy formulation and implementation.

It is this information challenge that the editors and authors of this book have addressed. It has been expertly drawn together to provide a lucid and illuminating account of the science–policy or science-information–policy interface relating to the oceans. The chapter authors have provided valuable, insightful analysis of the issues arising from and the barriers to more effective ocean management and sustainable resource use. They have considered how scientific advice is communicated and used, for example, in integrated coastal and ocean management. Also, they have delved into how various stakeholders can positively affect decision-making by using strategies for improved acquisition of information. The role of networks in information exchange is addressed and recommendations are offered to improve connectivity. How user engagement can ensure that scientific research is made more useful in the provision of relevant information is described, as well as the need for transparency regarding how information is gathered and used in public policy-making. Finally, those experienced in working at the science–policy interface are identified as an important resource that could be used much more effectively in future evidence-based and evidence-informed policy-making.

This volume will be of great use to the growing numbers of ocean practitioners from all sectors, dedicated to improving the health and sustainability of our oceans for generations to come.

Professor Michael H. Depledge, PhD, DSc, FRSA, FRSB, FRCP
Chair of Environment and Human Health,
University of Exeter Medical School, Exeter, Devon, UK
(Formerly chief scientist, Environment Agency UK, and member of the
Royal Commission on Environmental Pollution)

Preface

Despite the increase in scientific information and knowledge on a vast range of ocean topics and advances in the concept and practice of integrated coastal and ocean management (ICOM), solutions to many serious global coastal and ocean issues, for example, climate change, sustainability of fisheries, biodiversity and habitat loss, increased pollution, invasive species, and uncontrolled coastal development are not rapidly forthcoming. While this wave of new information—much of which is published as primary and gray literature—is available for evidence-based and evidence-informed policy-making, its use and influence are seldom explicitly recognized in ocean management arenas.

In this book, we examine the role of information in policy-making and decision-making for ICOM. Contributors have drawn on their expertise and experience in environmental and fisheries science, marine policy, public policy and administration, scientific advocacy, and information management. This range of disciplines exemplifies the dimensions of information and the science–policy interface in the policy-making process in ICOM.

Section I of the book provides the context for studying the role of scientific information in policy-making and presents a comprehensive overview of the characteristics of the science–policy interface. Section II describes fundamental concepts and principles germane to understanding the role of information in effective integrated coastal and ocean management. In Section III, national and international case studies reveal some of the factors that enable or inhibit awareness and use of information in policy-making contexts and the communication of information at the science–policy interface. Section IV presents highlights of the subject and future research challenges.

This book will be useful to all major groups in the policy-making process, including senior policy makers and decision makers, policy advisors, resource managers, information managers, scientists, and other practitioners in coastal and ocean management. We dedicate the book to the new generation of professionals involved in the challenging task of managing our ocean spaces and species now and for future generations.

Bertrum H. MacDonald, Suzuette S. Soomai,
Elizabeth M. De Santo, and Peter G. Wells

Acknowledgments

First, we thank all of our contributors and reviewers. Their special knowledge and commitment to the preparation of this book were pivotal to its success.

The genesis of this book was the Environmental Information: Use and Influence (EIUI) research program. EIUI has been funded by three research grants from the Social Sciences and Humanities Research Council of Canada (SSHRC) since 2007. EIUI is further supported by many in-kind contributions from its research partners: Environment Canada, Fisheries and Oceans Canada, Nova Scotia Department of Fisheries and Aquaculture, Northwest Atlantic Fisheries Organization, Food and Agriculture Organization of the United Nations, World Wildlife Fund Canada, Bay of Fundy Ecosystem Partnership, and the Canadian Parks and Wilderness Society.

Susan Rolston, of Seawinds Consulting Services, Hackett's Cove, Nova Scotia, provided copy-editing services for the final manuscript. Gail LeBlanc, of Dalhousie Design Services, designed the cover image. Lee Wilson assisted with the preparation of the final text. All are thanked for their contributions. We are also very grateful for the assistance and guidance offered to us by CRC Press while the book was in preparation.

Most importantly, we extend our utmost thanks to all of the graduate students who are and have been involved in the EIUI research program over the years. These students brought energy, new ideas, and commitment to the program. They represent a bright future for information management and the oceans.

The Editors

Editors

Bertrum H. MacDonald is professor of information management in the School of Information Management and dean of the Faculty of Management at Dalhousie University, Halifax, Nova Scotia. He holds degrees in science (BSc, Biology), history of science (MA), and information science (MLS, and PhD). He pursues research that investigates the dissemination and use of scientific information in historical and contemporary contexts. He is particularly interested in interdisciplinary research which led to the launch of the Environmental Information: Use and Influence research program at Dalhousie University. Since the mid-2000s, he and his research colleagues, along with a multidisciplinary team of students, have pursued research about information activities at the science–policy interface in marine management in collaboration with provincial, national, and international governmental and intergovernmental organizations. He has held a Fellowship at the Smithsonian Institution in Washington, DC, won the international Grey Net Award with his Dalhousie colleagues, and was awarded the Marie Tremaine Medal, the highest honor of the Bibliographical Society of Canada. In addition to administrative leadership at Dalhousie University, he is currently serving in executive positions with local, national, and international associations.

Suzuette S. Soomai is a postdoctoral fellow with the Environmental Information: Use and Influence (EIUI) research program at Dalhousie University, Halifax, Nova Scotia. Her research focuses on the role of scientific information in policy- and decision-making for marine fisheries management. She holds an interdisciplinary PhD and a master of marine management degree from Dalhousie University. She also holds a master of philosophy in zoology and a BSc (Hons) from the University of the West Indies. Prior to joining EIUI, she was a government fisheries scientist in Trinidad and Tobago, where she was primarily responsible for assessing groundfish resources, maintaining fisheries information systems, and for managing national technical programs under regional and global fisheries management projects. As a government scientist, she worked closely with the Caribbean Regional Fisheries Mechanism and the Food and Agriculture Organization of the United Nations. She has also published extensively in the technical report series of these organizations. She has worked closely with the fishing industry in a range of activities including tropical freshwater aquaculture farming and conducting at-sea trawl gear testing. Currently, her interests include understanding the role of scientific information in policy- and decision-making in fisheries management organizations.

Elizabeth M. De Santo is assistant professor of environmental studies in the Department of Earth and Environment at Franklin & Marshall College, and an adjunct professor at Dalhousie University. She holds a PhD in geography, cosupervised in laws, from University College London; master's degrees in the history of international relations (London School of Economics and Political Science) and environmental management (Duke University); and a BA in zoology from Connecticut College. Her research and teaching center on environmental governance, focusing on (1) the conservation and management of marine ecosystems and species, and (2) improving the science–policy interface in environmental decision-making. She is particularly interested in the challenges of effectively implementing marine protected areas and biodiversity conservation worldwide. Prior to joining Franklin & Marshall College, Elizabeth taught in the Marine Affairs Program and in the College of Sustainability at Dalhousie University in Halifax, Nova Scotia. She has an international and interdisciplinary background and has also worked in the nonprofit sector, including positions with the International Union for Conservation of Nature (IUCN), the World Environment Center, and the American Museum of Natural History. She is a member of the IUCN World Commission on Environmental Law, and has served on advisory boards for World Wildlife Fund Canada and the Canadian Parks and Wilderness Society. Originally from New York City, Elizabeth lived in the United Kingdom for six years and Canada for four, and currently resides in Lancaster, Pennsylvania.

Peter G. Wells is an adjunct professor at Dalhousie University, Halifax, Nova Scotia, and a senior research fellow, International Ocean Institute, Halifax, Nova Scotia. He holds a BSc in biology (McGill University, 1967), an MS in zoology (University of Toronto, 1969), and a PhD in zoology (University of Guelph, 1976). He worked as a marine scientist and aquatic toxicologist for the Canadian Federal Government for 34 years, retiring from Environment Canada in June 2006. His academic interests are marine ecotoxicology, marine information management, and community service on issues affecting the Bay of Fundy, Gulf of Maine and North-West Atlantic. Since 1970, he has written, contributed to, or edited over 300 primary and technical publications in the fields of water pollution and marine environmental science, including several books. He is the current editor of the *Proceedings of the Nova Scotian Institute of Science*. He served on various national and international technical committees, including with the National Academy of Sciences (USA), the Organisation for Economic Co-operation and Development, and the United Nations. He was a long-serving member of the United Nations Joint Group of Experts on the Scientific Aspects of Marine Environmental Protection. His honors and awards include: Fellow of the AAAS (2000); the SETAC Presidential Citation for Exemplary Service (2002); Dalhousie University's Award of Excellence for Teaching (2003); Bay of Fundy Ecosystem Partnership, Environmental Stewardship Award (2006); and the Susan Snow-Cotter Leadership Award, Gulf of Maine Council on the Marine Environment (2013). When not hiking and climbing mountains, he can be found working at Dalhousie as co-lead of the Environmental Information: Use and Influence research program, which initiated this book.

Coordinator

 **James D. Ross** is a member of the Environmental Information: Use and Influence research program and a lecturer at Dalhousie University's Faculty of Engineering. He holds an MA in English and an MLIS. His MLIS research focused on state of the environment reporting via a case study of the State of the Scotian Shelf Report which formed the basis for a chapter in this volume.

Authors

Lahsen Ababouch is director of the Fisheries and Aquaculture Policy and Economics Division at the Food and Agriculture Organization of the United Nations. He was previously professor at the King Hassan II Institute of Agronomy and Veterinary Science, Morocco, where he held senior positions in teaching, research, and outreach. He has published and communicated extensively in the fields of agriculture and fisheries.

Kristiann Allen is the chief of staff in the Office of the Chief Science Advisor to the Prime Minister of New Zealand. Her professional experience has spanned multiple levels at the nexus of science, government, and civil society. She has previously advised research agenda setters and government decision makers in academic institutions and in the Canadian health science funding system. She has also facilitated the use of scientific evidence within United Nations agencies. She continues to explore the multiple roles, tensions, and opportunities in national and international science systems at the University of Auckland.

Shannon Arnold has worked in marine conservation and small-scale fisheries research and advocacy since 2007 in Canada and internationally. She is currently the fisheries program manager for a 30,000-member-strong Filipino confederation of fishers and farmers focused on organizing and empowering the sector to fully participate in policy and resource management. Shannon led the elasmobranch conservation work at Canada's Ecology Action Centre. She continues research on sharks and mobula rays in the Philippines and is also investigating the international trafficking of marine wildlife.

Diana L. Ascher is a doctoral candidate in University of California, Los Angeles's Department of Information Studies. She focuses on the evaluation, classification, organization, communication, interpretation, and prioritization of information, drawing from more than two decades of academic and corporate experience in behavioral science, education, finance, journalism, law, leadership, management, medicine, and public policy.

William Ascher is the Donald C. McKenna Professor of Government and Economics at Claremont McKenna College, where he directs the Roberts Environmental Center. He also directs the Pacific Basin Research Center at Soka University of America. His research is on natural resource and environmental policy, development policy, and political psychology.

Alexi Baccardax Westcott is the project officer of the Atlantic Coastal Zone Information Steering Committee Secretariat and has held this position since 2010. She holds a BSc in environmental science and a Master of Environmental Studies from Dalhousie University. A focus of her graduate studies was coastal management.

Angela Bednarek works at The Pew Charitable Trusts that connects science and policy-making about ocean ecosystems. Dr. Bednarek was an AAAS Diplomacy Fellow at the U.S. Department of State, where she negotiated U.S. positions on Global Environment Facility and World Bank projects, Organisation for Economic Co-operation and Development initiatives, and chemicals agreements and was the U.S. representative to the United Nations Dams and Development Project. She received fellowships at the Earth Institute at Columbia University and a Udall Fellowship in Environmental Public Policy. She has a PhD in biology from the University of Pennsylvania and a BS in biology and art from the University of Notre Dame.

Heather Breeze is a project leader with the Canada Department of Fisheries and Oceans in Dartmouth, Nova Scotia. Over the past 10 years, she has worked on integrated oceans and coastal management projects, marine protected areas, cold-water coral conservation, and state of the oceans reporting with the Gulf of Maine Council on the Marine Environment.

Brian Coffey is a lecturer (sustainability and urban planning) at the Royal Melbourne Institute of Technology University (Victoria, Australia). His research interests center on the policy and governance dimensions of sustainability and science-policy relations, and he has published in a variety of public policy and environmental policy, planning, and management journals.

Elizabeth R. DeSombre is the Camilla Chandler Frost Professor of Environmental Studies at Wellesley College, where she directs the Environmental Studies Program. Her research has focused primarily on environmental problems of the global commons, with a special focus on ocean issues.

Susanna D. Fuller is a marine biologist with a background in benthic ecology. Susanna has been involved in policy development at the national and international scale to protect deep sea vulnerable ecosystems from the impacts of fishing activities. She has also worked on sustainable fisheries markets based projects. She is the marine conservation coordinator at the Ecology Action Centre, a board member of the Deep Sea Conservation Coalition, and sits on the Steering Committee of the High Seas Alliance. Susanna has a PhD in biology from Dalhousie University and a BSc from McGill University.

Luca Garibaldi, a biologist, has worked with the Food and Agriculture Organization since 1990. In 2000, he was appointed fishery statistician in charge of the Food and Agriculture Organization global capture database. He has published several studies related to the database and seven chapters on capture fisheries trends in *The State of World Fisheries and Aquaculture*.

Peter Gluckman is chief science advisor to the Prime Minister of New Zealand, a post he has held since it was established in 2009. Sir Peter is recipient of New Zealand's highest civilian and scientific honors. He is internationally respected for his work promoting the use of evidence in public policy formation and the translation of scientific knowledge into better social, economic, and environmental outcomes. He is the founding chair of both the Asia-Pacific Economic Cooperation Economies' Chief Scientists and Equivalents Network and the International Network of Government Science Advice.

Janice E. Graham is a professor in medical anthropology and infectious diseases, and former Canada Research Chair in bioethics (2002–2012) in the Faculty of Medicine at Dalhousie University. Working in the area of emerging biotherapeutics and vaccines in Canada, Europe, and Africa, Graham studies safety, efficacy, and trust in the construction and legitimization of clinical research and regulatory practice. She has presented evidence to the Science Policy Directorate, Office of Legislative and Regulatory Modernization, Parliament of Canada, World Health Organization, and United Nations on regulatory risks of industry capture to independent research, open data, and public health.

Heather J. Grant has been working at the Ecology Action Centre since 2012 as marine communications campaigner. She specializes in communicating science and policy in a way that is accessible to the public in order to further campaign goals and raise awareness of marine conservation and fisheries issues in Canada. Her concern about the status of large pelagic species like sharks and tuna began at a young age. Heather has a BSc in marine biology from Dalhousie University, with a minor in environmental science.

Anatoliy Gruzd is a Canada Research Chair in social media data stewardship, director of the Social Media Lab, and associate professor in the Ted Rogers School of Management at Ryerson University, Canada. He is also a coeditor of a multidisciplinary journal on big data and society. His research initiatives explore how social media are changing the ways in which people communicate, collaborate, and disseminate information and how these changes impact the social, economic, and political norms and structures of modern society. He is also actively developing new approaches and tools to support social media data analytics.

Troy W. Hartley is a research associate professor of Marine Science and Policy and director of the Virginia Sea Grant College Program. He received a PhD in environmental and natural resource policy (University of Michigan), MAIS in policy and dispute resolution (George Mason University) and a BS in zoology (University of Vermont). His research focuses on governance network structures, multi-stakeholder collaboration, and marine policy, particularly on fisheries, land-use, water resource, and climate adaptation management.

Mavis Jones holds a doctorate in politics from the University of East Anglia and has studied citizen and expert policy deliberations in human genetics, drug regulation, xenotransplantation, agricultural biotechnology, climate change, and other matters. For the last several years, she has been making a living handling scientific evidence.

Elizabeth C. McNie is a research scientist at the Western Water Assessment, Cooperative Institute for Research in Environmental Sciences, University of Colorado at Boulder. Her research explores how to produce usable science and the role of boundary organizations in linking knowledge with action. She was an assistant professor of political science and earth and atmospheric sciences at Purdue University. She received her PhD in environmental studies and certificate in science and technology policy from the University of Colorado, Boulder, and a BS in marine transportation from the California Maritime Academy.

Ryan Meyer is senior scientist at the California Ocean Science Trust, an independent, nonprofit organization that works across government, science, and communities to build trust and understanding and to promote a constructive role for science in natural resource management. He leads the organization's Citizen Science Initiative, focused on expanding the ways that citizen science programs can link with coastal and marine policy and management. Ryan earned a PhD at Arizona State University, studying climate science funding in the United States and Australia. He is a Fulbright Scholar, a University Fellow in the Research Institute for Environment and Livelihoods at Charles Darwin University, and an affiliate of the Consortium for Science, Policy, and Outcomes.

Arthur P. J. Mol is vice-chancellor of Wageningen University and Research Centre, the Netherlands, and professor of environmental policy at the same institution. He is joint editor of the journal *Environmental Politics* and editor of the book series *New Horizons in Environmental Politics* (Edward Elgar). He has published widely on environmental governance, international environmental policy, China and environment, and social theory and environment.

Kevin O'Toole is an adjunct associate professor at Deakin University (Victoria, Australia). He publishes in the area of sustainability, including knowledge systems in coastal zone management; catchment to regional scale indicators; integrating environmental, social, and economic measures; community governance; demand management for domestic water; and comparative land use analysis.

Adam Parris is an interdisciplinary scientist, working on social and environmental change in U.S. coastal zones. Adam serves as the executive director of the Science and Resilience Institute at Jamaica Bay, New York. He is the lead author on the "Global Mean Sea Level Rise Scenarios for the U.S. National Climate Assessment" and provided technical guidance and leadership on the Sea Level Rise Tool for Sandy Recovery. He also served as program manager for the National Oceanic and Atmospheric Administration's Regional Integrated Sciences and Assessments program. He holds a BA in environmental geology and English literature from Bucknell University and an MS in geology from the University of Vermont.

Julian Plummer is currently a publications coordinator in the Fisheries and Aquaculture Department of the Food and Agriculture Organization. He has also worked with the Food and Agriculture Organization as an English language editor since 1995. He has coordinated and edited many high-profile publications such as the past four editions of *The State of World Fisheries and Aquaculture*.

Kate Porter is a research coordinator and analyst with the MacEachen Institute for Public Policy and Governance, Dalhousie University. Before joining the MacEachen Institute, she studied human-wildlife conflict biology and non-invasive wildlife tracking techniques. She obtained a master of environmental studies from Dalhousie University, and a BA in human ecology from College of the Atlantic.

Kevin Quigley is director of the MacEachen Institute for Public Policy and Governance and an associate professor at the School of Public Administration at Dalhousie University, Halifax, Canada. He specializes in public sector risk and crisis management, risk governance, security and critical infrastructure protection understood in broad social context. He is particularly interested in research methods that employ interdisciplinary and comparative approaches.

Jake Rice recently retired as chief scientist for the Canada Department of Fisheries and Oceans (DFO). He has published extensively in the fields of fisheries, marine science, and conservation of marine biodiversity. He spent 10 years as director of Peer Review and Scientific Advice for DFO, and has served as scientific advisor for many Canadian and international policy and management groups.

Kathryn E. Schleit is the marine campaign coordinator on the Ecology Action Centre's Marine Team. Before joining the Ecology Action Centre, she worked at the Pew Environment Group in Washington, DC, doing international marine policy on improving fisheries management and combating illegal fishing. She also spent time with the Peace Corps working with women fishworker cooperatives in a coastal town in the Philippines and developing coastal education programs. She has a master of marine affairs degree from the University of Washington, focused on community-based coastal management and marine protected areas, and a bachelor's degree from Franklin & Marshall College.

Andrew G. Sherin was a physical scientist for the Geological Survey of Canada for 35 years, holding several positions related to the management of marine data. He completed his career organizing the semiannual performance review of science programs. He has held the position of director of the Atlantic Coastal Zone Information Steering Committee Secretariat since 2010.

Marc Taconet is chief, Statistics and Information Branch, Fisheries and Aquaculture Department, Food and Agriculture Organization, and secretary of the Fisheries and Resources Monitoring System partnership. With the Food and Agriculture Organization since 1987, he has led the development of the Fisheries Global Information System, played a leading user-community role in several European Union–funded projects, and chaired the iMarine board.

Hilde M. Toonen is a postdoctoral research fellow and lecturer, based at the Environmental Policy group of Wageningen University, the Netherlands. Her research interests lie in the field of (marine) environmental governance, with emphasis on informational modes of decision-making and processes of inclusion and exclusion.

Stefania Vannuccini is the Food and Agriculture Organization (FAO) fishery statistician responsible for statistics and analysis of utilization, production, trade, and consumption of fish and fishery products. In addition, she is responsible for fish outlook models developed jointly by the FAO with the Organization for Economic Co-operation and Development and with the World Bank, and International Food Policy Research Institute.

Reviewers

The editors would like to extend their utmost gratitude to our team of peer reviewers, whose contributions were invaluable to the development of this volume. The following list is not exhaustive, and the editors also thank those reviewers who preferred to remain anonymous.

Julian Barbière
Head, Marine Policy and Regional
   Coordination Section
Intergovernmental Oceanographic
   Commission of United Nations
   Educational, Scientific and
   Cultural Organisation

Alex Bielak
Adjunct Professor
Department of Communication
   Studies and Multimedia
McMaster University

Samantha Burgess
Acting Chief Scientific Advisor
UK Marine Management Office

Jacquie Burkell
Assistant Dean, Research
Associate Professor
Faculty of Information and Media
   Studies
University of Western Ontario

Michael J. A. Butler
Director, International Ocean
   Institute – Canada, and Adjunct
   Research Associate
Marine Affairs Program
Dalhousie University

Ratana Chuenpagdee
Canada Research Chair in Natural
   Resource Sustainability and
   Community Development
Professor
Department of Geography
Memorial University of
   Newfoundland

Kevern Cochrane
Director
Food and Aquaculture Department
Fisheries and Agriculture
   Organization of the United
   Nations
(Retired)

Ruth Cordes
Editorial Consultant
Halifax, Nova Scotia

Stephen Decker
Environmental Studies Program
   Chair
Lecturer in Environmental Studies
   and Geography
Grenfell Campus, Memorial
   University of Newfoundland

Susanna Eden
Assistant Director
Water Resources Research Center
University of Arizona

Philip Enros
Science Policy
Environment Canada
(Retired)

Lucia Fanning
Professor
Marine Affairs Program
Dalhousie University

Tim Hall
Regional Manager, Ocean and
    Coastal Management
Canada Department of Fisheries
    and Oceans
(Retired)

Milton Haughton
Executive Director
Caribbean Regional Fisheries
    Mechanism

Richard Haworth
Adjunct Professor
School of Public Administration
Dalhousie University

Patrick Helm
Department of the Prime Minister
    and Cabinet
Government of New Zealand

Dick Hodgson
Adjunct Professor
Marine Affairs Program
Dalhousie University

Petter Holm
Professor
Department of Social and
    Marketing Studies
MaReMa Centre
Norwegian College of Fishery
    Science
University of Tromsø

John Karau
Senior Policy Advisor
Canada Department of Fisheries
    and Oceans
(Retired)

David Keeley
The Keeley Group
Providence, Rhode Island

Donald Macrae
Advisor
United Nations Economic
    Commission for Europe

Robin Mahon
Professor Emeritus
Centre for Resource Management
    and Environmental Studies
University of the West Indies,
    Barbados

Ronald Pelot
Professor
Marine Affairs Program
Department of Industrial
    Engineering
Dalhousie University

Susan Russell-Robinson
Associate Program Coordinator,
Coastal and Marine Geology
U.S. Geological Survey

Carol Tenopir
Chancellor's Professor
School of Information Sciences
University of Tennessee, Knoxville

Juliette Young
Biodiversity Policy Researcher
Center for Ecology and Hydrology
Natural Environment Research
    Council, United Kingdom

List of Figures

List of Tables

List of Boxes

List of Acronyms

AAAS	American Association for the Advancement of Science
ABNJ	areas beyond national jurisdiction
ACIP	Atlantic Coastal Information Portal
ACOA	Atlantic Canada Opportunities Agency
ACZISC	Atlantic Coastal Zone Information Steering Committee
AGLINK	Worldwide Agribusiness Linkage Program
AHWGW	Ad Hoc Working Group of the Whole (United Nations)
ALPAC	Atlantic Large Pelagic Advisory Committee
AMOP	Arctic and Marine Oilspill Program
AoA	assessment of assessments (United Nations)
AOI	area of interest
AP	advisory panel
API	application program interface
ASMFC	Atlantic States Marine Fisheries Commission
BBNJ	biodiversity beyond national jurisdiction
BG	blue growth
BGI	Blue Growth Initiative
BIO	Bedford Institute of Oceanography
BSE	bovine spongiform encephalopathy
CASS	collaborative acoustic stock survey
CBD	Convention on Biological Diversity
CCME	Canadian Council of Ministers of the Environment
CDAST	COINAtlantic Data Accessibility Self-Assessment Tool
CFIA	Canadian Food and Inspection Agency
CGDI	Canadian Geospatial Data Infrastructure
CGG	COINAtlantic Geocontent Generator
CITES	Convention on International Trade in Endangered Species of Wild Fauna and Flora
COBRA	cabinet office briefing room A
COFI	Committee on Fisheries (FAO)
COIN	Coastal and Ocean Information Network
COP	conference of parties
CoP	community of practice
COSEWIC	Committee for the Status of Endangered Wildlife in Canada
COSIMO	Commodity Simulation Model
CSIRO	Commonwealth Scientific and Industrial Research Organisation (Australia)
CSU	COINAtlantic Search Utility
CWP	Coordinating Working Party on Fishery Statistics (FAO)

CWS	Canadian Wildlife Service (Environment Canada)
CZM	coastal zone management
DDT	dichloro-diphenyl-trichloroethane
DFO	Department of Fisheries and Oceans (Canada)
DGAC	Dietary Guidelines Advisory Committee
DPSIR	Driving forces-Pressures-State-Impacts-Responses (framework)
DWFN	distant-water fishing nation
EAC	Ecology Action Centre
EAF	ecosystem approach to fisheries
EBFM	ecosystem-based fisheries management
EBM (1)	ecosystem-based management
EBM (2)	evidence-based medicine
EBSA	ecologically and biologically significant area
EC	Environment Canada
ECNASAP	East Coast of North America Strategic Assessment Project
EEZ	exclusive economic zone
EIA	environmental impact assessment
EIUI	Environmental Information: Use and Influence (research program)
EMSA	European Maritime Safety Agency
ENGO	environmental nongovernmental organization
EPS	Environmental Protection Service (Environment Canada)
ESSIM	Eastern Scotian Shelf Integrated Management Initiative
EU	European Union
FAO	Food and Agriculture Organization of the United Nations
FBS	food balance sheets
FEP	fisheries ecosystem plans
FIRMS	Fisheries and Resources Monitoring System
FMP	fisheries management plan
FOC	flag of convenience
FoE	Friends of the Earth
FSRS	Fishermen and Scientists Research Society Research Society
GAN	Global Action Network
GDP	gross domestic product
GESAMP	Joint Group of Experts on the Scientific Aspects of Marine Environmental Protection (United Nations)
GIS	geographic information system
GOM	Gulf of Maine
GOMC	Gulf of Maine Council on the Marine Environment
GR	genetic resources
GSK	GlaxoSmithKline
HLPE	high-level panel of experts

HOTO	Health of the Oceans (DFO, United Nations)
HPFB	Health Products and Food Branch (Health Canada)
HPV	human papilloma virus
IACS	International Association of Classification Societies
ICAM	integrated coastal area management
ICAN	International Coastal Atlas Network
ICCAT	International Commission for the Conservation of Atlantic Tunas
ICES	International Council for the Exploration of the Sea
ICM	integrated coastal management
ICMA	integrated coastal marine areas
ICOIN	Inland Waters Coastal and Ocean Information Network
ICOM	integrated coastal and ocean management
ICZM	integrated coastal zone management
IFMP	integrated fisheries management plan
IFPRI	International Food Policy Research Institute
IGH	interest group hypothesis
IGO	intergovernmental organization
IM	integrated management
IMO	International Maritime Organization
IMPACT	International Model for Policy Analysis of Agricultural Commodities
IOC	Intergovernmental Oceanographic Commission (UNESCO)
IODE	International Ocean Data Exchange
IPCC	Intergovernmental Panel on Climate Change
IRGC	International Risk Governance Council
ISA	International Seabed Authority
IUCN	International Union for Conservation of Nature
IUU	illegal, unreported, and unregulated fishing
IWD	Inland Waters Directorate (Environment Canada)
KML	keyhole markup language
KTE	knowledge transfer and exchange
LBA	land-based activities
LiDAR	laser induced detection and ranging
LME	large marine ecosystem
LOMA	large ocean management area
LRIS	Land Registration and Information Service
LTC	Legal and Technical Commission (ISA)
MARPOL	International Convention for the Prevention of Pollution from Ships (1973/78)
MBA	Monterey Bay Aquarium
MCI	Marine Conservation Institute (US)
MDSG	Maryland Sea Grant
MEA	Millennium Ecosystem Assessment

MEOPAR	Marine Environmental Observation Prediction and Response Network
MEPC	Marine Environmental Protection Committee (IMO)
MEQ	marine environmental quality
MFH	market failure hypothesis
MFish	Ministry of Fisheries (New Zealand)
MGDI	Marine Geospatial Data Infrastructure
MGR	marine genetic resource
MOU	memorandum of understanding
MPA	marine protected area
MPI	Ministry for Private Industries (New Zealand)
MSC	Marine Stewardship Council
NAFO	Northwest Atlantic Fisheries Organization
NAS	National Academy of Sciences (US)
NGO	nongovernment organization
NICE	National Institute for Health and Clinical Excellence (UK)
NMFS	National Marine Fisheries Service (US)
NOAA	National Oceanic and Atmospheric Administration (United States)
NPA	National Program of Action
NRC	National Research Council (US)
NRDC	National Resources Defense Council
NSB	National Science Board (US)
NSERC	Natural Sciences and Engineering Research Council of Canada
NSF	National Science Foundation (US)
OBIS	Ocean Biogeographic Information System
OECD	Organisation for Economic Co-operation and Development
OPEC	Organization of the Petroleum Exporting Countries
ORH	opinion responsive hypothesis
PAH	polycyclic aromatic hydrocarbons
PATH	Program in Appropriate Technologies
PCB	polychlorinated biphenyls
PCRD	Public Comment Draft Report
PDT	plan development teams
PSC	port state control
PWB	Programme of Work and Budget (FAO)
QA/QC	quality assurance/quality control
RAP	rational actor paradigm
RCCOM	Regional Committee for Coastal and Ocean Management (DFO)
RCT	randomized control trial
RFB	regional fishery body
RMFO/A	regional fisheries management organization/arrangement

RPA	recovery population assessment
SAC	Stakeholder Advisory Committee (DFO)
SAFE	stock assessment and fisheries evaluation
SAGE	UK Scientific Advisory Group for Emergencies
SARA	Species at Risk Act (Canada)
SBSTTA	Subsidiary Body on Scientific, Technical, and Technological Advice (CBD)
SBT	segregated ballast tanks
SCOPE	Scientific Committee on Problems of the Environment
SDG	sustainable development goals
SEA	strategic environmental assessment
SEAFO	South East Atlantic Fisheries Organization
SEEA	System of Environmental-Economic Accounting (United Nations)
SOE	state of the environment
SOFIA	State of World Fisheries and Aquaculture (FAO)
SoSS	State of the Scotian Shelf
SPI	science–policy interface
STEM	science, technology, engineering, mathematics
Strategy–STF	Strategy for Improving Information on Status and Trends of Capture Fisheries
TAC	total allowable catch
TRAC	Transboundary Resources Assessment Committee (Canada/US)
UN	United Nations
UNCLOS	UN Convention on the Law of the Sea
UNEP	United Nations Environment Programme
UNGA	United Nations General Assembly
UNWTO	United Nations World Tourism Organization
US	United States
USEPA	United States Environmental Protection Agency
VISIT	Voluntary Initiatives for Sustainability in Tourism (EU)
VME	vulnerable marine ecosystem
VME-DB	Vulnerable Marine Ecosystems Database
VRE	virtual research environment
WB	World Bank
WMO	World Meteorological Organization
WMS	web mapping services
WOA	World Ocean Assessment
WSSD	World Summit on Sustainable Development
WWF	World Wide Fund for Nature (also World Wildlife Fund)

Section I

Introduction

1

Introduction

Bertrum H. MacDonald, Suzuette S. Soomai,
Elizabeth M. De Santo, and Peter G. Wells

CONTENTS

1.1 The Coastal and Ocean Management Challenge

Integrated coastal and ocean management (ICOM) "is a dynamic, multidisciplinary, iterative and participatory process to promote sustainable management of coastal and ocean areas balancing environmental, economic, social, cultural and recreational objectives over the long term" (UNESCO 2006, p. 6). ICOM is a concept and process that has evolved in overlapping stages over the past few decades, encompassing shoreline management, defining the *coastal zone*, and integrating coastal/shoreline management with the marine environment. Coastal management began on the terrestrial side of the coastal zone, focusing on particular challenges posed by this dynamic environment, such as shoreline erosion, wetland protection, coastal development, and public access (Clark 1995; Cicin-Sain and Knecht 1998; Sorensen 1997; among others). For example, the U.S. approach to coastal zone management was first formalized with the Coastal Zone Management Act of 1972, which focused more on the management of land use at the shore than on coastal water-related issues. In the decades since, efforts to protect and manage the coastal zone in the United States and beyond (e.g., Canada and the United Kingdom) have expanded to include integrated coastal management (ICM), coastal zone management (CZM), integrated coastal zone management (ICZM), and perhaps the broadest approach, ICOM. Canada incorporated ICOM into its

Oceans Act (1996, section 30(b)), recognizing the need for "integrated man-agement of activities in estuaries, coastal waters and marine waters" and adopting the term integrated management. Hence, the terms are often used interchangeably in legislation and in the literature, reflecting a broadening geography of what is included in the coastal zone, from land use to coastal to ocean waters.

Beyond biophysical components, ICOM also takes into account the socio-economic and political aspects of managing the range of competing uses and jurisdictions found in the coastal zone. In their seminal text on ICOM, Cicin-Sain and Knecht (1998, p. 461) define ICM as "a continuous and dynamic process by which decisions are made for the sustainable, use, development, and protection of coastal and marine areas and resources" aimed at over-coming the jurisdictional and management fragmentation inherent in the coastal zone. The integration aspect of ICOM aims to bring together several dimensions, ranging from different economic sectors, levels of government, and the terrestrial/sea divide, to integrating science with management, and bringing different countries together in cooperative programs (Cicin-Sain and Knecht 1998).

The ICOM dimension that focuses on integrating science with manage-ment is the starting point for this volume. Twenty years after the publication of Cicin-Sain and Knecht's (1998) seminal text on ICOM, the world continues to struggle with implementing this management approach and ocean health continues to deteriorate (IPSO 2013; UNEP 2006; Hoegh-Guldberg et al. 2015; among others). Cicin-Sain and Knecht (1998, p. 171) argue that "the most fun-damental tenet underlying the ICOM concept is that ICOM decision-making is based on the use of the best information and the best science available," integrating the natural, social, and economic sciences. Coastal systems can be viewed as shared systems (Parkes and Manning 1998), requiring mul-tiple disciplines and practitioners in coastal management to work together for their effective management. Also paramount are the roles of stakeholder engagement and the generation, use, and influence of relevant information for decision-making. Our book reemphasizes the role of information in ICOM and augments the efforts of Cicin-Sain and Knecht (1998), GESAMP (1996), and many others in their efforts to strengthen coastal science and management around the world.

1.1.1 Production, Communication, and Use of Scientific Information in Policy-Making

Since the 1972 Stockholm Conference on the Human Environment, the United Nations (UN) system has been the focal point for addressing global environmental issues at the international level (Chasek et al. 2013). Within the UN system, considerable quantities of gray literature on the marine envi-ronment are produced by intergovernmental organizations, for example, the United Nations Environmental Programme (UNEP) and its many advisory bodies, including the Joint Group of Experts on the Scientific Aspects of Marine Environmental Protection (GESAMP) and the Scientific Committee on Problems of the Environment (SCOPE). National governmental and nongovernmental organizations have also produced thousands of scientific publications on marine environments aimed at guiding public policy on aspects of sustainable development. Given the increase in scientific information addressing a vast range of marine environmental topics, for example, climate change, biodiversity and habitat loss, increased pollution, invasive species, and uncontrolled coastal development (e.g., DFO 2010; Halpern et al. 2008; GESAMP 2001; IPCC 2013, 2014a, 2014b; MEA 2005; Rogers and Laffoley 2011), it follows that, in theory, solutions to these serious global coastal and ocean issues should also be increasing—but they are not, and coastal and ocean problems continue to exist.

New information continues to be developed from wide-ranging scientific research and synthesis by governmental, intergovernmental, and nongovernmental organizations, universities, industry, and other independent bodies, for example, consultancies and think tanks. Much of this information is referred to as gray literature, meaning that it is not controlled by commercial publishers (GreyNet 2015). Although this gray literature is generally created through rigorous peer review, its credibility is often questioned, especially when compared to the primary literature, that is, literature produced by commercial publishers. Nonetheless, numerous studies have pointed to the increasing generation of information by government departments, international intergovernmental bodies, and nongovernmental organizations, resulting in greater reliance on the publishing practice of gray literature for disseminating information to inform policy decisions (e.g., Luzi 2000; Schöpfel and Farace 2010; Thelwall et al. 2010; Webster and Collins 2005). In most policy settings, gray literature may also be of great importance because of its rapid production—compared with much of the primary literature—which can facilitate knowledge diffusion where decisions are based on complex competing factors and pressures in political processes (Bremer and Glavovic 2013; Shanley and López 2009; Pielke 2007). An example of the importance of gray literature produced by governmental organizations is also seen in the public attention to climate change reports, for example, the Stern Review on the economics of climate change released for the British government (Stern 2006) and technical assessment reports produced by the Intergovernmental Panel on Climate Change (IPCC) (IPCC 2014a,b).

Scientific information is available in different formats, for example, print and digital, and is communicated to decision-making audiences and the public though formal and informal information dissemination strategies, such as libraries, websites, meetings, and personal communication. Over the past 20 years, communication patterns have rapidly changed due to the prominence of ever advancing information technologies, including the Internet and social media (Cossins 2014). Formal and informal networks are also becoming increasingly important in communicating information. Groups outside government, for example, nongovernmental organizations such as the World Wildlife Fund (or World Wide Fund for Nature [WWF]) and the International Union for Conservation of Nature (IUCN), among others, have become more active in disseminating scientific information to policy communities at various levels of government.

Senior decision makers in government are increasingly expected to rely on evidence-based approaches to policy-making and to program implementation based on the best available evidence from research (Gluckman 2014; Nutley et al. 2007). Decision makers, or their advisors, are expected to choose from among the wide range of available information, often with competing views advocated by diverse stakeholders. Decision-making is further challenged by the current financial austerity measures that limit the time and resources available to make decisions. These challenges heighten the importance and relevance of rapid access to directly pertinent information, much of it in the gray literature of government and interest groups.

Evidence-based policy-making is also being applied within the context of ICOM and resource management, a modern approach to ensuring sustainable development of marine areas (Bremer and Glavovic 2013; Coffey and O'Toole 2012; Hiscock et al. 2003; Holmes and Lock 2010; Levin et al. 2009). The ICOM approach.

> implies a conscious management process that acknowledges interrelationships among coastal and ocean uses and the environments … the process is designed to overcome the fragmentation inherent in single-sector management approaches, e.g., fishing operations, oil and gas development, etc. (Cicin-Sain and Knecht 1998, p. 1)

It follows that the information needs in ICOM are quite complex, depending on the issue to be addressed As a result, ICOM requires input from a range of sectors including natural and social sciences, as well as local and traditional knowledge. Appropriate governance mechanisms are also needed to achieve an integrated information flow for decision-making relating to coastal and ocean management, for example, including mechanisms to facilitate consultation with a wide range of stakeholders to build consensus and implement management measures, as well as communication or information pathways between stakeholders, scientists, and decision makers.

Ready access to scientific information by users (policy makers, senior decision makers, resource managers, and other practitioners) is often assumed, given all the available websites and search engines, but key information can also remain undetected. Although information is only one part of policy formulation behavior, identifying solutions to marine environmental issues depends as much on efficiently finding and understanding existing information and applying it in policy solutions as on the creation of new relevant information. For example, although it is known that access to information occurs—politicians and their staff use the Library of Parliament in Ottawa and the Library of Congress in Washington DC—exactly how policy makers access and use this large body of information, or not, is still poorly understood (Ascher et al. 2010; Briggs and Knight 2011; Holmes and Lock 2010; Likens 2010; McNie 2007; Mitchell 2010; Mitchell et al. 2006; Stojanovic et al. 2009). Furthermore, few organizations have undertaken an analysis of the use of their information products (publications), and information pathways in decision-making contexts are still being elucidated (Economic Commission for Europe 2003; MacDonald et al. 2004; Soomai 2013; Soomai et al. 2011a, 2011b, 2013; Wells 2003; Wells et al. 2002).

Use of research is not unidimensional. Use can mean reading the findings from research as general background briefing, examining research in making a decision even if the research findings are rejected, or having a direct impact on policy choices (Weiss 1979). The literature describes a spectrum of types of use ranging from conceptual (simple awareness) to instrumental, for example, application in a decision or policy, and behavior (Weiss 1979; Nutley et al. 2007). In fact, the stages of information production and use in policy-making exist as an iterative process or a type of continuum (Nutley et al. 2007). Similarly, the generation, transmission, and use of environmental information is highly complex; its use depends on many factors, such as access, availability, and transparency, among others (Ascher et al. 2010). These important points should be kept in mind as we attempt to unravel the challenge of the science-policy interface in ICOM.

1.1.2 The Role of Information at the Science–Policy Interface

The pivotal role of scientific information in the search for solutions to environmental problems is clearly demonstrated.

> [I]t is the production, the processing, the use and the flow of, as well as the access to and the control over, information that is increasingly becoming vital in environmental governance practices and institutions. … Information and knowledge are … key resources in environmental politics … and the motivations and sources for changing unsustainable behaviour are increasingly informational. (Mol 2008, p. 277)

Yet, as the previous quotation points out, in spite of increasing knowledge about stresses on the world's oceans (published as both primary and gray literature) and evidence-based policy-making, many problems persist. This suggests a gap or disconnect between the information *produced* and the information *used* in decision-making, which often limits its role in policy formulation and environmental management (Koetz et al. 2012; Nursey-Bray et al. 2014; Spruijt et al. 2014; Stoutenborough and Vedlitz 2014; Wells 2003).

The apparent disconnect at the science–policy interface between the information and knowledge produced by scientists and that used by policy makers has been given considerable attention over the years (e.g., Moksness et al. 2013; NRC 2002, 2012). Scholars have ascribed this to "inherent differences between the fundamental structures and traditions of science and policy" (Francis et al. 2005, p. 35) contributing to a far from optimal "flow of knowledge between researchers, policy makers, and resource managers" (Roux et al. 2006, p. 5). Numerous studies have highlighted that most models of communication also ignore the use of scientific information in public policy-making, where information use is different from its use in pure research contexts (e.g., Doern 2001; Duff 1997; Dunn 2005; Søndergaard et al. 2003; Van de Veer Martens and Goodrum 2005).

The science–policy interface operates on several scales: geographic, institutional, political, and temporal (Cash and Moser 2000; Bremer and Glavovic 2013; Young 2014). The interface is also more inclusive than the label implies, as it involves social processes and encompasses different types of knowledge. In fact, multiple interfaces exist due to many decision-making contexts, each comprising complex informational connections and networks, for example, scientific and local traditional knowledge that are influenced by societal factors, some of which are unique to individual policies, decision makers, and the environmental issues. The *information universe* is multidimensional, and information flow may not be linear or unidirectional, which also accounts for the complexity of activity at the science–policy interface.

Developing a clearer understanding of the processes at the science–policy interface is critical for all of the actors involved in identifying, managing, and solving the many complex marine environmental problems, from pollution to fisheries to the vulnerability of coastal cities. Chapter 2 describes the science–policy interface in more detail, as do all the chapters from the perspectives of the contributors to this volume.

1.2 Origins of the Book

Since 2007, the Environmental Information: Use and Influence (EIUI) research program at Dalhousie University, Halifax, Canada, has been addressing the role in policy-making of marine environmental and resource information produced by governmental and nongovernmental organizations. The EIUI team is interdisciplinary, capitalizing on its diverse expertise to build a greater understanding of the production, use, and influence of marine scientific information in public management settings. The guiding framework of our research focuses on the interface between production of scientific information, especially gray literature, and its use in policy-making and decision-making contexts, that is, the science–policy interface. (Figure 1.1).

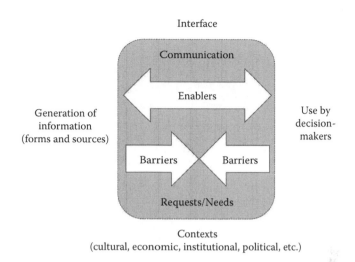

Contexts
(cultural, economic, institutional, political, etc.)

FIGURE 1.1
Key elements of information flow at the multidimensional science–policy interface(s).

Our research assumes a prominent role for such information in policy formation and problem resolution, an assumption examined closely in many case studies. Using this framework, we are developing techniques to measure information use and influence, to identify and mitigate communication barriers, and to understand the many processes at work at the interface between information producers and users.

Our research to date has been conducted primarily through case studies of governmental, intergovernmental, and nongovernmental organizations representing national, regional, and global settings, which give extensive attention to coastal and ocean environmental matters, and are involved in public policy development. In particular, we are examining the ways in which information impacts the development of legislation and policy, and how it affects environmental decision-making and management programs within ICOM. Our research also addresses stakeholder perceptions of the policy process and the role of information, acknowledging the wider influence of socioeconomic and political contexts on all of these relationships.

We are involved in ongoing research partnerships with many governmental organizations and other partners. Our findings have prompted recommendations about the communication of scientific information that are being considered by the case study organizations to increase the use and influence of their information in policy-making and decision-making important to the marine environment (Avdić 2013; Cano Chacón 2013; Chamberlain 2015; Cordes 2004; Cossarini 2010; Cossarini et al. 2014; Hutton 2009, 2010; MacDonald et al. 2004, 2007; MacDonald et al. 2010, 2013; McLean 2014; Ross 2014; Soomai 2009, 2013, 2015; Soomai et al. 2011a, 2011b, 2013; Soomai et al. 2011; Wilson 2015). Integration of the findings from all of the data sets is informing the development of theoretical models of the life cycles of marine environmental information and general measures of the influence of marine environmental information within the ICOM process.

In 2013, the EIUI program hosted a 2-day workshop, "Marine Information Matters," for discussion between EIUI researchers and its partners. The meeting discussed current approaches to the research, methods to measure use and influence of information in light of ongoing advances in information technologies, and subject areas for future studies (EIUI 2013).

The workshop led to a special session, "Does Information Matter? A Critical Question for the Future of Coastal Zone Management," at the 2014 Coastal Zone Canada Conference, in Halifax, Nova Scotia. This session brought together scholars and practitioners to address questions such as: Where does evidence come from? How does it make its way out of the laboratories and offices of scientists and other researchers into the minds and documents of policy makers and decision makers? Who is responsible for bridging the communication gap between these different groups? What enables information mobilization? What inhibits it? How can the science–policy communication gap be bridged effectively? Following the conference, the EIUI team initiated this book, the contributing authors being panelists from the conference session and other invited experts.

1.3 Objectives of the Book

This book deals explicitly with the role of scientific information in the policy-making process critical to ICOM. It assumes a belief in evidence-based policy-making, recognizing that such an approach is not universally accepted. The fundamental concepts and principles and the case studies explored in this book aim to increase understanding of the multidimensional processes at the science–policy interface by which scientific information is incorporated into policy decisions. The case studies of coastal and ocean reports and organizations producing information examined in this book provide empirical evidence of the benefits and challenges of communication of information, and its use in policy contexts. The use of multiple methods in the case studies demonstrates the complexity of studying information flows at the many manifestations/occurrences of the science–policy interface—for example, within and among organizations—and elucidates enablers and barriers to that activity.

The contributors are experts in various disciplines: information management, marine environmental science, resource management, health sciences, environmental governance, risk management, public policy, and coastal and ocean management. All of these disciplines and others play a role in

ICOM, and all are information based. They include leading researchers in the emerging research area on information/knowledge utilization and the science–policy interface. These experts are among the senior professionals from governmental, intergovernmental, and nongovernmental organizations with coastal and ocean management mandates. Similarly, the reviewers of the book chapters were selected from the leading researchers and experts in the fields of information management, environmental governance, public policy, and resource management, with national, regional, and global experience.

Through this collective effort, we have begun to attain a comprehensive understanding of how coastal and ocean scientific information is currently used in related policy formulation and management decisions. Above all, the content of this book contributes to increased awareness of the critical role of scientific information in the policy-making process among policy makers, senior decision makers, resource managers, scientists, students, and other interested groups, especially those working on coastal and ocean issues.

1.4 Topics Covered in the Book

Chapter 2 provides an overview of the science–policy interface in ICOM, drawing on relevant scholarship. The next section of the book (Section II) addresses fundamental concepts and principles, focusing on the role of information in the following contexts: the science–policy interface in ICOM (Coffey and O'Toole, Chapter 3); global fisheries governance (Rice, Chapter 4); risk regulation and governance in coastal areas (Quigley and Porter, Chapter 5); fisheries certification, global shipping, and tourism (Toonen and Mol, Chapter 6); strategies for stakeholders seeking information on ICOM (Ascher and Ascher, Chapter 7); network analysis and trust in science (Hartley, Chapter 8); coastal and ocean decision-making in the United States (McNie et al., Chapter 9); science in public policy (Gluckman and Allen, Chapter 10); and an analysis of methods for analyzing information awareness, use, and influence (Soomai et al., Chapter 11).

The third section of the book, Section III, analyzes the role of information in the science–policy interface via the following case studies: Canada's *State of the Scotian Shelf Report* (Ross and Breeze, Chapter 12); global ocean shipping (DeSombre, Chapter 13); processes of regulatory decision-making in the health sciences (Graham and Jones, Chapter 14); the Atlantic Coastal Zone Information Steering Committee and ICOM (Sherin and Baccardax Westcott, Chapter 15); a Canadian federal government department, Environment Canada (Wells, Chapter 16); the Food and Agriculture Organization of the United Nations (Ababouch et al., Chapter 17), and the Ecology Action Centre, an environmental nongovernmental organization (Fuller et al., Chapter 18).

The final section of the book (Section IV, Chapter 19) synthesizes key points from the preceding chapters, drawing lessons for the production, use, and influence of information at the science–policy interface in ICOM. Additional challenges and areas of further research are also set out.

1.5 Wider Contributions of the Book

This volume advances our knowledge of the role(s) of information in policy decision-making on priority coastal and ocean environmental matters. This new knowledge contributes to the disciplines of information management, marine environmental studies, resource management, and public policy. In addition to building theoretical understanding in this understudied area, insights provided, particularly in the case studies, will inform information diffusion and knowledge management practices for the development and implementation of effective coastal and ocean policies. Specifically, this book provides (1) solution-oriented benefits related to the marine environment, (2) an increased understanding of the science–policy interface, and (3) an examination of current methods for studying information flows and knowledge utilization in coastal and ocean management.

The book contributes to increased awareness of the critical role of information in solving environmental problems. Many of the case studies examined in this volume examine information management dimensions of environmental issues within government departments. Lessons drawn from the fundamental concepts and principles and the case studies described in the book can enhance best practices for more effective communication and use of marine environmental information, particularly at the science–policy interface.

This volume explores the different occurrences and types of science–policy interfaces existing within and between organizations. It also examines the varying roles of different types of organizations—apart from government—in information production and dissemination. For example, reports by nongovernmental organizations and scientific academies and professional associations are increasingly important. Understanding the difference between use and influence of information and assessing impact or influence is a major challenge. There is often no easy way of differentiating between use and influence or to measure the impact of one individual or organization against the influence of a similar organization.

Research aimed at understanding the role of information at the science–policy interface needs an interdisciplinary approach encompassing the disciplines of information management, marine environmental studies, resource management, public policy, and governance. Our research highlights the fact that a range of methods is needed to study communication of information and use in policy-making. No single research method can provide a total picture of information flows. The questions posed by our research about the communication process help to increase understanding of the enablers and barriers to communicating research findings, as well as set out directions for further research.

References

Ascher, W., T. Steelman, and R. Healy. 2010. *Knowledge and Environmental Policy: Re-imagining the Boundaries of Science and Politics.* Cambridge, MA: MIT Press.

Avdić, V. 2013. Measuring use and influence: An assessment of the FAO's flagship report, The State of World Fisheries and Aquaculture. Master's Project Report. Dalhousie University.

Bremer, S. and B. Glavovic. 2013. Mobilizing knowledge for coastal governance: Re-framing the science–policy interface for integrated coastal management. *Coastal Management* 41: 39–56.

Briggs, S. V. and A. T. Knight. 2011. Science–policy interface: Scientific input limited. *Science* 333: 696–697.

Cano Chacón, M. 2013. The role of the information of the Marine Stewardship Council Certification process in developing countries: A case study of two MSC certified fisheries in Mexico. Master's Project Report. Dalhousie University.

Cash, D. W. and S. C. Moser. 2000. Linking global and local scales: Designing dynamic assessments and management processes. *Global Environmental Change* 10: 109–120.

Chamberlain, S. 2015. Developing and implementing a research framework to determine the overall use and influence of a long-term marine environmental monitoring program: A case study on Gulfwatch in Nova Scotia. Master's Project Report. Dalhousie University.

Chasek, P., D. Downie, and J. W. Brown. 2013. *Global Environmental Politics.* 6th ed. Boulder, CO: Westview.

Cicin-Sain, B. and R. W. Knecht. 1998. *Integrated Coastal and Ocean Management: Concepts and Practices.* Washington, DC: Island Press.

Clark, J. R. 1995. *Coastal Zone Management Handbook.* Boca Raton: CRC.

Coastal Zone Management Act of 1972. 1972, 16 U.S.C. §§1451–1465.

Coffey, B. and K. O'Toole. 2012. Towards an improved understanding of knowledge dynamics in integrated coastal zone management: A knowledge systems framework. *Conservation and Society* 10 (4): 318–329.

Cordes, R. E. 2004. Is grey literature ever used? Using citation analysis to measure the impact of GESAMP, an international marine scientific advisory body. *Canadian Journal of Information and Library Science* 28: 45–65.

Cossarini, D. M. 2010. Marine environmental grey literature: A case study of the Gulf of Maine Council on the Marine Environment. Master's thesis. Dalhousie University.

Cossarini, D. M., B. H. MacDonald, and P. G. Wells. 2014. Communicating marine environmental information to decision makers: Enablers and barriers to use of publications (grey literature) of the Gulf of Maine Council on the Marine Environment. *Ocean & Coastal Management* 96: 163–172.

Cossins, D. 2014. Setting the record straight. *The Scientist,* October 1. http://www.the-scientist.com/?articles.view/articleNo/41056/title/Setting-the-Record-Straight/.

DFO (Department of Fisheries and Oceans). 2010. 2010 Canadian marine ecosystem status and trends report. DFO Canada Science Advisory Secretariat Science Advisory Report 2010/030. Ottawa: DFO.

Doern, G. B. 2001. Science and scientists in regulatory governance: A mezzo-level framework for analysis. *Science and Public Policy* 28: 195–205.

Duff, A. S. 1997. Some post-war models of the information chain. *Journal of Librarianship and Information Science* 29: 179–187.

Dunn, K. 2005. Impact of the inclusion of grey literature on the scholarly communication patterns of an interdisciplinary specialty. Paper presented at the Sixth International Greynet Conference, New York Academy of Medicine, December 2004. Amsterdam: TextRelease.

Economic Commission for Europe. 2003. *Environmental Policy in Transition: Ten Years of UNECE Environmental Performance Reviews.* New York: United Nations.

EIUI (Environmental Information: Use and Influence). 2013. Marine information matters: Probing its use and influence in policy and decision making. Report of the EIUI-Partnership Workshop, 20–21 September, 2013. Halifax: Dalhousie University. http://eiui.ca/?p=1843.

Francis, T. B., K. A. Whittaker, V. Shandas, et al. 2005. Incorporating science into the environmental policy process: A case study from Washington State. *Ecology and Society* 10: 35. http://www.ecologyandsociety.org/vol10/iss1/art35.

GESAMP (Joint Group of Experts on the Scientific Aspects of Marine Environmental Protection). 1996. The contributions of science to integrated coastal management. GESAMP Reports and Studies No. 61. Rome: FAO.

GESAMP (Joint Group of Experts on the Scientific Aspects of Marine Environmental Protection). 2001. Sea of troubles. GESAMP Reports and Studies No. 70. Nairobi: UNEP.

Gluckman, P. 2014. Policy: The art of science advice to government. *Nature* 207 (7491): 163–165.

GreyNet. 2015. Grey literature network service. http://www.greynet.org/greynethome/aboutgreynet.html.

Halpern, B. S., S. Walbridge, K. A. Selkoe, et al. 2008. A global map of human impact on marine ecosystems. *Science* 319 (5865): 948–952.

Hiscock, K., M. Elliot, D. Laffoley, et al. 2003. Data use and information creation: Challenges for marine scientists and for managers. *Marine Pollution Bulletin* 46 (5): 534–541.

Hoegh-Guldberg, O., D. Beal, T. Chaudry, et al. 2015. *Reviving the Ocean Economy. The Case for Action—2015.* Gland: WWF International.

Holmes, J. and J. Lock. 2010. Generating the evidence for marine fisheries policy and management. *Marine Policy* 34: 29–35.

Hutton, G. R. G. 2009. Developing an inclusive measure of influence for marine environmental grey literature. Master's thesis. Dalhousie University.

Hutton, G. R. G. 2010. Understanding influence of scientific information in the digital age: A study of the grey literature of a United Nations advisory group. *Proceedings of the Nova Scotian Institute of Science* 45 (2): 91–101.

IPCC (Intergovernmental Panel on Climate Change). 2013. *Climate Change 2013: The Physical Science Basis.* Geneva: IPCC.

IPCC (Intergovernmental Panel on Climate Change). 2014a. Summary for policy-makers. In Climate change 2014: Mitigation of climate change. Contribution of Working Group III to the Fifth Assessment Report of the Intergovernmental Panel on Climate Change, edited by O. Edenhofer, R. Pichs-Madruga, Y. Sokona, E. Farahani, S. Kadner, K. Seyboth, A. Adler, et al. Cambridge, MA: Cambridge University Press.

IPCC (Intergovernmental Panel on Climate Change). 2014b. Climate change 2014: Mitigation of climate change. Contribution of Working Group III to the Fifth Assessment Report of the Intergovernmental Panel on Climate Change, edited by O. Edenhofer, R. Pichs-Madruga, Y. Sokona, E. Farahani, S. Kadner, K. Seyboth, A. Adler, et al. Cambridge, MA: Cambridge University Press.

IPSO (International Programme on the State of the Ocean). 2013. The global state of the ocean: Interactions between stresses, impacts and some possible solutions. *Marine Pollution Bulletin* 74 (2): 491–552.

Koetz, T., K. N. Farrell, and P. Bridgewater. 2012. Building better science–policy inter-faces for international environmental governance: Assessing potential within the Intergovernmental Platform for Biodiversity and Ecosystem Services. *International Environmental Agreements* 12 (1): 1–21.

Levin, P. S., M. J. Fogarty, S. A. Murawski, et al. 2009. Integrated ecosystem assess-ments: Developing the scientific basis for ecosystem-based management of the ocean. *PloS Biology* 7 (1): 23–28.

Likens, G. E. 2010. The role of science in decision making: Does evidence-based sci-ence drive environmental policy? *Frontiers of Ecology and the Environment* 8: 1–9.

Luzi, D. 2000. Trends and evolution in the development of grey literature: A review. *International Journal on Grey Literature* 1: 105–116.

MacDonald, B. H., R. E. Cordes, and P. G. Wells. 2004. Grey literature in the life of GESAMP: An international marine scientific advisory body. *Publishing Research Quarterly* 20: 26–41.

MacDonald, B. H., R. E. Cordes, and P. G. Wells. 2007. Assessing the diffusion and impact of grey literature published by international intergovernmental scien-tific groups: The case of the Gulf of Maine Council on the Marine Environment. *Publishing Research Quarterly* 23: 30–46.

MacDonald, B. H., E. M. De Santo, K. Quigley, et al. 2013. Tracking the influence of grey literature in public policy contexts: The necessity and benefit of interdisci-plinary research. *The Grey Journal* 9: 61–69.

MacDonald, B. H., P. G. Wells, R. E. Cordes, et al. 2010. The use and influence of infor-mation produced as grey literature by international, intergovernmental marine organizations: Overview of current research. In *Grey Literature in Library and Information Studies*, edited by D. J. Farace and J. Schöpfel, 167–180. Berlin: De Gruyter Saur.

McLean, S. 2014. A study of the use of data provided by coastal atlases in coastal policy and decision-making. Master's Project Report. Dalhousie University.

McNie, E. C. 2007. Reconciling the supply of scientific information with user demands: An analysis of the problem and review of the literature. *Environmental Science & Policy* 10: 17–38.

MEA (Millennium Ecosystem Assessment). 2005. *Ecosystems and Human Well-being: Synthesis*. Washington, DC: Island Press. http://www.millenniumassessment.org/documents/document.356.aspx.pdf.

Mitchell, R. B. 2010. *International Politics and the Environment*. London: Sage.

Mitchell, R. B., W. C. Clark, and D. W. Cash. 2006. Information and influences. In *Global Environmental Assessments. Information and Influence*, edited by R. B. Mitchell, W. C. Clark, D. W. Cash, and N. M. Dickson, 307–338. Cambridge, MA: MIT Press.

Moksness, E., E. Dahl, and J. Stottrup. 2013. *Global Challenges in Integrated Coastal Zone Management*. Chichester, UK: John Wiley.

Mol, A. P. J. 2008. *Environmental Reform in the Information Age: The Contours of Informational Governance*. New York: Cambridge University Press.

NRC (National Research Council). 2002. *Science and its Role in the National Marine Fisheries Service*. Washington, DC: National Academies Press.

NRC (National Research Council). 2012. *Using Science as Evidence in Public Policy*. Washington, DC: National Academies Press.

Nursey-Bray, M. J., J. Vince, M. Scott, et al. 2014. Science into policy? Discourse, coastal management and knowledge. *Environmental Science & Policy* 38: 107–119.

Nutley, S. M., I. Walter, and H. T. O. Davies. 2007. *Using Evidence: How Research Can Inform Public Services*. Bristol: Policy Press.

Oceans Act, 1996 S.C., ch. 31.

Parkes, J. G. M. and E. W. Manning. 1998. An historical perspective on coastal zone management in Canada. Oceans Conservation Report Series, Canadian Technical Report. Fisheries and Aquatic Science No. 2213. Ottawa: Department of Fisheries and Oceans.

Pielke, R. A. 2007. *The Honest Broker: Making Sense of Science in Policy and Politics*. New York: Cambridge University Press.

Rogers, A. D. and D. d'A. Laffoley. 2011. International Earth System Expert Workshop on Ocean Stresses and Impacts: Summary Report. Oxford: International Program on the State of the Ocean.

Ross, J. D. 2014. What do users want from a state of the environment report? A study of awareness and use of the State of the Scotian Shelf Report. Master's thesis. Dalhousie University.

Roux, D. J., K. H. Rogers, H. C. Briggs, et al. 2006. Bridging the science–management divide: Moving from unidirectional knowledge transfer to knowledge interfacing and sharing. *Ecology and Society* 11 (1): 4. http://www.ecologyandsociety.org/vol1/iss1/art/.

Schöpfel, J. and D. J. Farace. 2010. Grey literature. In *Encyclopedia of Library and Information Science*, 3rd ed., 2029–2039. London: Taylor & Francis.

Shanley, P. and C. López. 2009. Out of the loop: Why research rarely reaches policy makers and the public and what can be done. *Biotropica* 41: 535–544.

Søndergaard, T. F., J. Andersen, and B. Hjørland, B. 2003. Documents and the communication of scientific and scholarly information: Revising and updating the UNISIST model. *Journal of Documentation* 59 (3): 278–320.

Soomai, S. S. 2009. Information and influence in fisheries management: A preliminary study of the shrimp and groundfish resources in the Brazil-Guianas continental shelf. Master's Project Report. Dalhousie University.

Soomai, S. S. 2013. Understanding the science–policy interface: Measuring use and influence of information in policy making. In *Building Bridges in Ocean Management: Connecting the Policy, Science and Public Spheres*, edited by J. MacIntosh and J. Stoner, Marine Affairs Technical Report No. 11. Halifax: Marine Affairs Program, Dalhousie University.

Soomai, S. S. 2015. Elucidating the role of scientific information in decision-making for fisheries management. PhD diss. Dalhousie University.

Soomai, S. S., B. H. MacDonald, and P. G. Wells. 2011a. The 2009 State of Nova Scotia's Coast Report: An initial study of its use and influence. Halifax: Environmental Information: Use and Influence, Dalhousie University.

Soomai, S. S., B. H. MacDonald, and P. G. Wells. 2011b. The State of the Gulf of Maine Report: An initial study on awareness, use, and influence of the theme papers. Halifax: Environmental Information: Use and Influence, Dalhousie University.

Soomai, S. S., B. H. MacDonald, and P. G. Wells. 2013. Communicating environmental information to the stakeholders in coastal and marine policy-making. Case studies from Nova Scotia and the Gulf of Maine/Bay of Fundy Region. *Marine Policy* 40: 176–186.

Soomai, S. S., P. G. Wells, and B. H. MacDonald. 2011. Multi-stakeholder perspectives on the use and influence of "grey" scientific information in fisheries management. *Marine Policy* 35 (1): 50–62.

Sorensen, J. 1997. National and international efforts at integrated coastal management: Definitions, achievements, and lessons. *Coastal Management* 25 (1): 3–41.

Spruijt, P., A. B. Knol, E. Vasileiadou, et al. 2014. Roles of scientists as policy advisers on complex issues: A literature review. *Environmental Science & Policy* 40: 16–25.

Stern, N. 2006. *Stern Review: The Economics of Climate Change.* London: HM Treasury. http://webarchive.nationalarchives.gov.uk/+/http:/www.hm-treasury.gov.uk/sternreview_index.htm.

Stojanovic, T. A., I. Ball, R. C. Ballinger, et al. 2009. The role of research networks for science–policy collaboration in coastal areas. *Marine Policy* 33: 901–911.

Stoutenborough, J. W., and A. Vedlitz. 2014. The effect of perceived and assessed knowledge of climate change on public policy concerns: An empirical comparison. *Environmental Science & Policy* 37: 23–33.

Thelwall, M., A. Klitkou, A. Verbeek, et al. 2010. Policy-relevant webometrics for individual scientific fields. *Journal of the American Society for Information Science and Technology* 61: 1464–1475.

UNEP (United Nations Environment Programme). 2006. *The State of the Marine Environment: Trends and Processes.* The Hague: UNEP/GPA Coordination Office.

UNESCO (United Nations Educational, Scientific and Cultural Organization). 2006. A handbook for measuring the progress and outcomes of integrated coastal and ocean management. IOC Manuals and Guides No. 46; ICAM Dossier No. 2. Paris: UNESCO.

Van der Veer Martens, B. and A. A. Goodrum. 2005. The diffusion of theories: A functional approach. *Journal of the American Society for Information Science and Technology* 57: 330–341.

Webster, J. G. and J. Collins. 2005. *Fisheries Information in Developing Countries: Support to the Implementation of the 1995 FAO Code of Conduct for Responsible Fisheries.* Rome: FAO.

Weiss, C. H. 1979. The many meanings of research utilization. *Public Administration Review* 39 (5): 426–431.

Wells, P. G. 2003. State of the marine environment (SOME) reports—A need to evaluate their role in marine environmental protection and conservation. *Marine Pollution Bulletin* 46: 1219–1223.

Wells, P. G., R. A. Duce, and M. E. Huber. 2002. Caring for the sea—Accomplishments, activities and future of the United Nations GESAMP (the Joint Group of Experts on the Scientific Aspects of Marine Environmental Protection). *Ocean & Coastal Management* 45 (1): 77–89.

Wilson, L. T. 2015. The communication of information in multi-sectoral networks: A case study of tidal power network(s) in the Bay of Fundy region of Atlantic Canada. Master's thesis. Dalhousie University.

Young, S. P., ed. 2014. *Evidence-Based Policy-Making in Canada: A Multidisciplinary Look at How Evidence and Knowledge Shape Canadian Public Policy.* Don Mills: Oxford University Press.

2

Understanding the Science–Policy Interface in Integrated Coastal and Ocean Management

Bertrum H. MacDonald, Suzuette S. Soomai,
Elizabeth M. De Santo, and Peter G. Wells

CONTENTS

2.1 Introduction

In February 2013, the Halifax *Chronicle Herald* reported that the Nova Scotia minister of fisheries, Sterling Belliveau, had suggested to members of the Canadian federal Senate Committee on Fisheries and Oceans that "Nova Scotia fishermen should be allowed to fish several 'undeveloped' species provided scientific data shows that it's feasible." Belliveau went further to state, "Several Nova Scotia fishermen have applied to the department ... to fish stone crab and other undeveloped species and have been denied due to a lack of scientific information or old data.... Well let's get some," he emphasized (McLeod and Medel 2013). Whether the minister received the evidence that he requested is now a moot point (the government fell in an election eight months later), but the call for additional scientific information highlights evidence-based policy-making in action, a concept that has received attention from researchers and practitioners for over two decades. This topic also figures in public discourse. A year after *The Chronicle Herald* reported the minister of fisheries' statements, an editorial on the storage of fracking waste water in that newspaper expressed the view that "governments ... must base their decisions on the best objective criteria available" ("Respecting the Evidence on Fracking Waste Water," *The Chronicle Herald*, January 31, 2014). Although these news media accounts may suggest that *evidence-based policy-making* is common and straightforward in application, in fact the practice is more complex and multidimensional than might be expected, as many intersecting factors and actors play a part in the picture, that is, for governments, using the best science in the public interest in a sea of competing interests and political considerations.

As the growing body of literature on the science–policy interface (or, more broadly, the research–policy interface) shows, a gap frequently exists between scientific evidence and decisions that could be informed by and benefit from that evidence. The presence of the gap, which has been noted by scientists, policy makers, journalists, and other stakeholders alike, has prompted numerous calls to address the problem. In 2010, for example, the journal *Nature* declared that "scientists ... can and must continue to inform policy makers about the underlying science and the potential consequences of policy decisions" (*Nature* 2010, p. 141). *Nature's* editors have been by no means alone in underscoring the importance of communication between scientists and policy makers so that credible, timely scientific information is considered in decisions about global issues (see, e.g., Bremer and Glavovic 2013; Eden 2011; Elfner et al. 2011; Schenkel 2010). The American Association for the Advancement of Science (AAAS) has also been particularly active for many years on this issue of communication through its policy forum column in the journal *Science*. Although scientific facts and understanding related to an issue/problem are only one dimension of decision-making and policy-making, the importance of the facts should not be underestimated.

The deteriorating health of the world's oceans also led Dr. Jane Lubchenco (a distinguished marine ecologist, former AAAS president and former undersecretary of commerce for Oceans and Atmosphere and administrator of the National Oceanic and Atmospheric Administration) and Nancy Sutley (former chair of the White House Council on Environmental Quality and the Interagency Ocean Policy Task Force) to state that "the need for science-based solutions and forward-thinking, holistic approaches to management has never been greater" (Lubchenco and Sutley 2010, p. 1485). These views were repeated by Lubchenco at a high-profile ocean health seminar at the AAAS conference in Boston in 2013 (AAAS 2013).

At the 2010 Coastal Zone Canada Conference held in Charlottetown, Prince Edward Island, participants in a panel on Information Management for Integrated Coastal and Ocean Management (ICOM) debated why the use of information and its management seemed to be an invisible part of the ICOM process. Every issue in coastal and ocean management and every stage of ICOM involve the creation, distribution, awareness, and use of information, although clearly to varying degrees. The speakers wondered what accounts for blind spots regarding the role(s) and values of information in ocean and coastal management contexts. Can the challenges be attributed to the ubiquity of information? Is information so common that everyone takes it for granted? Indeed, is the very question of information's role naïve and unnecessary? We are immersed in information from birth and perhaps assume that everyone knows how information "functions" for individuals, groups, and society; as a consequence, we do not recognize or appreciate the complexity of information behavior and systems that operate all around us and their significance in decision-making and management activities. Information is simply taken for granted, much like air in our daily lives. As well, other challenges abound, including issues of communication, hidden information (deliberately or not), the overwhelming volume of information, limited awareness or understanding of coastal and ocean management issues, the complex relationships between sectors and disciplines, trust or lack of trust among practitioners in various fields, and political considerations. In short, the role of information in integrated ocean and coastal management seems to be the "elephant in the room"—a factor of unrecognized importance and hence, understudied.

The number of research studies and the resulting literature on marine subjects are growing substantially, yet the health of the oceans is deteriorating rapidly (UNEP 2006; WWF 2015). This observation initiated the research program of the interdisciplinary Environmental Information: Use and Influence group at Dalhousie University more than 10 years ago (Wells 2003; Cordes 2004; MacDonald et al. 2004). Researchers have estimated that global scientific output doubles about every nine years (Bornmann and Mutz 2014; Van Noorden 2014), and other recent research emphasizes that the acceleration of scientific publishing is resulting in a new paper being published "roughly every 20 seconds" (Munroe 2013). Two other scholars have estimated that at least 114 million English-language scholarly documents are available on the web (Khabsa and Giles 2014). The production and availability of information are not uniform throughout the world, however, as some regions face greater challenges than others as a result of limited capacity in financial resources and available information technology (Holmgren and Schnitzer 2004).

Great effort and resources have been poured into producing research publications, including many major scientific reports on the state of the marine environments, but as Wells (2003) asked, are the reports being noticed and read, and are they influencing decisions where it matters most, that is, in protecting the oceans? At a superficial level, one might conclude that the research has limited or no effect, because marine environments are continuing to degrade. Such a conclusion is, of course, far too simplistic. More information is available, but this coincides with more people living along the world's coastlines, contributing to increased pressures on the system. As well, direct application of research information in public policy or decisions may not occur, the pathway of information to a policy or a decision is not always obvious, and often a significant time lag exists between the identification of a problem and its solution. In addition, researchers and other stakeholders "often despair that clear findings are sometimes not heeded when decisions are made about the direction and delivery of public services. Indeed, policy decisions sometimes seem to fly in the face of what is considered to be the best available evidence [or information] about 'what works'" (Nutley et al. 2007, p. 1) or what is needed, for example, slowing climate change and limiting the discharge of pollutants.

This chapter introduces the topic of *evidence-based policy-making*, characteristics of the science–policy interface, and methods for enhancing communication of information at the interface, in the context of ICOM. As the chapter progresses, the role of information and the need to understand these concepts, processes, and methods will emerge.

2.2 Information Use in Policy-Making for Integrated Ocean and Coastal Management

2.2.1 Information Needs

The need for a solid scientific basis for decision-making in ICOM is evident (Cicin-Sain and Knecht 1998), given the complexity of interactions between the environment, resource users, economics, and social well-being of communities. Specific information needs are related to particular coastal and ocean management issues. While this book does not focus on data per se, for example, data contained in geographic information systems, environmental trend analysis, and fisheries statistics, it explicitly examines how advice for management available in research literature, including technical and summary reports, is used in decision-making.

As mentioned in Section 2.1, the oceans face many problems related to the sustainability of living marine resources, the protection and conservation of habitats and biodiversity, and the protection of ocean and human health. The leading environmental issue recognized globally is anthropogenic climate change, which has been receiving particular attention over the past two decades and is the focus of interest of many scientists, policy makers, and international institutions. The work of the Intergovernmental Panel on Climate Change (IPCC) is an informative example of how scientists and policy makers can work together, bridging the so-called science–policy interface, to tackle an urgent global issue in marine environmental management. The IPCC's five incremental assessment reports to date illustrate how the best available science about a given problem can be evaluated, summarized, and then presented to policy makers, with their involvement and for their use (IPCC 2014). The policy summaries are the product of many discussions and reviews and are written specifically for the decision maker, policy maker, and non-expert audience (including politicians).

Since the 1990s, as recognized by Cicin-Sain and Knecht, a "continuing challenge in the management of coastal resources generally centers on the science-policy interface. Improvements in resource management usually depend on improvements in our understanding of the processes involved" (Cicin-Sain and Knecht 1998, p. 173). Insights from a study on the science–policy interface conducted by the US National Academy of Science (NAS), cochaired by Cicin-Sain, included suggestions for improving communications across the interface (NRC 1995). More than a decade later, a study by the US NAS added that understanding how science is used in policy requires an investigation into what makes reliable, valid, and compelling policy arguments from the perspective of policy makers (NRC 2012). Often, due to the focus and demands of their research, scientists see their primary job as contributing to the body of scientific knowledge, while producing information that is directly useful in decision-making is sometimes regarded as less important. National academies such as the National Research Council (United States), the Council of Canadian Academies, and the Royal Society (United Kingdom) have undertaken research to bridge this gap while addressing key societal issues. Along with these advances, numerous researchers have probed the issue of the attributes of information per se and consider "useful" information in decision-making to be salient, that is, relevant to the needs of decision makers and other stakeholders; credible, that is, scientifically adequate in the eyes of the stakeholders; and legitimate, reflecting the perception that its production was unbiased (Clark et al. 2006; Delaney and Hastie 2007; McNie 2007; see also Chapters 3, 9, and 10 in this volume).

Despite this progress, understanding information use and influence in policy-making on major issues of societal concern remains a relatively new area of investigation. The case studies in Section III of this book describe how marine information is produced, communicated, and used in ICOM and associated areas critical to the oceans.

2.2.2 Evidence-Based Policy-Making

Evidence-based policy-making can be traced back to at least the 1960s, when large-scale planning processes took place in the United States in areas such as defense, urban redevelopment, and budgeting (Howlett 2009). The term *evidence-based*, or the increasingly common *evidence-informed policy-making*, encompasses efforts to guide policy processes by giving priority to evidentiary decision-making criteria. The objective is to avoid or minimize policy failures resulting from a mismatch between expectations of governments and "on-the-ground" conditions (Howlett 2009, p. 154). While there has been no fundamental challenge to the basic concept of evidence-based policy-making, UK researchers such as Nutley et al. (2007) found that the evidence confirming any benefits or dysfunctions of the process to be rather thin. Notwithstanding this finding, enthusiasm for the insights offered by research gained ground over a decade ago, "epitomised in the UK by the rhetoric of Tony Blair's Labour government" at the end of the 1990s, which adopted the slogan "what matters is what works" (Nutley et al. 2007, p. 10). This perspective "was intended to signal an end to ideologically-based decision-making in favour of evidence-based thinking" (Nutley et al. 2007, p. 10). Governments in other jurisdictions embraced evidence-based decision-making practices to varying degrees (e.g., Denmark and New Zealand) or not at all, to the consternation of researchers and the informed public, for example, recently in Canada (see Turner 2013; Harris 2014; Winfield 2013).

Concerns about wholesale adoption of the evidence-based approach have been raised (e.g., see Cherney and Head 2010), chief among these being suggestions that research evidence is only one factor involved in decision-making. Anne Glover, former chief scientific advisor to the European Commission, noted at the AAAS conference in Boston in 2013: "I know that our world is not just based on evidence, nor should it be, and that scientists are not responsible for making [policy] decisions. Where I feel we're in the wrong place is that at the moment we are very relaxed and quite cavalier sometimes about the evidence" (AAAS 2013a,b). The conference report further stated that "[e]ven when heads of state make policies that deviate from the available scientific evidence, Glover would like them to nonetheless state that 'we accept the evidence, but for other reasons—political, social, economic reasons—we go in this direction'" (AAAS 2013a,b). In Glover's view, this will "stimulate much better dialogue with citizens in enabling the use of new technologies." Glover's views were echoed to a large degree by Sir Peter Gluckman, chief science advisor to the New Zealand prime minister, in a lecture given at the University of Sussex on January 21, 2014 (Gluckman 2014). At a subsequent conference on Science Advice to Governments in Auckland, New Zealand, in 2014, both Glover and Gluckman were among the senior advisors participating in this first international summit on science advice.

The official media report of the conference stated that "science advice to governments has emerged as a discipline in its own right, which is both art and science" (International Network for Government Science Advice 2014).

Concerns about the adoption and application of evidence-based policy, while meriting attention because they highlight the complexity of factors characterizing a gap between science and policy, do not override the relevance and importance of drawing on scientific information in efforts to find solutions to pressing problems facing society at every jurisdictional level. The growing body of research literature on the science–policy interface is one indicator that this matter is gaining greater recognition among academics and practitioners. Since 2014 alone, more than 125 research papers have discussed aspects of the science–policy interface (based on a Web of Science, EBSCO, and ProQuest search in September 2015). Research aimed at understanding what contributes to the gap between the availability of information and knowledge arising from research and the processes of policy-making and decision-making has increased as both researchers and policy makers lament the distance between the two (e.g., Coffey and O'Toole 2012; Jasanoff 1994; Lalor and Hickey 2013; Mitchell 2010; Nursey-Bray et al. 2014; Pielke 2007; Shanley and Löpez 2009; Schenkel 2010; Spruijt et al. 2014; Stoutenborough and Vedlitz 2014).

2.3 Role(s) of Information at the Science–Policy Interface

The role(s) that scientific information plays in decision-making is either understudied or underemphasized. However, it is entwined in decision-making processes and is often in the foreground of the rhetoric about the gap between science and policy, as noted in Section 2.2.2 and in comments from the Science Advice to Governments Conference in New Zealand (International Network for Government Science Advice 2014). Researchers have highlighted that most models of scientific communication ignore the use of research information in public policy-making, where information assimilation is a quite different process than in pure research contexts (Duff 1997; Doern 2001; Dunn 2005; Søndergaard et al. 2003; Van der Veer Martens and Goodrum 2005).

Some of the dimensions of the science–policy interface are illustrated in Figure 2.1, which shows a canvas composed of many actors and factors operating under the realms of science and policy (with a focus on public policy and decision-making in government). Conceptually, the bridge encompasses numerous features regarding infrastructure that facilitate communication channels across the gap between the science and policy realms. Information arising from research can be communicated in the direction of policy, that is,

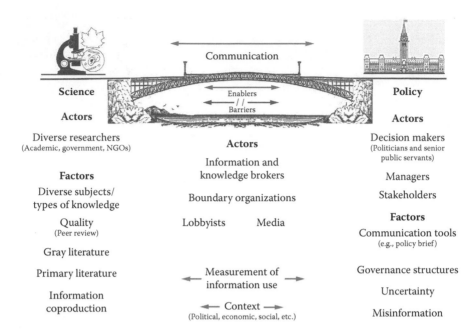

FIGURE 2.1
Bridging the science–policy interface: some of the actors and factors to consider.

science advice can be given, and in turn, policy can communicate questions or problems that researchers can tackle and then forward new evidence back to the policy community. Some have argued that the information to policy challenge is largely a communications issue, whereby inadequate or ineffective communication results in a barrier to information or knowledge transfer (e.g., Kahan 2010). If a message lacks clarity or fails to take account of the perspective or needs of an audience, communication will falter. While on the surface this assessment may seem entirely intuitive, the matter may be considerably more complicated, because many factors can be wrapped up in the communication challenge. A range of actors in the science and policy realms and at the bridge facilitate communication. Enablers and barriers characterize the activities in this dynamic process, and the context of these activities brings numerous additional factors to bear on what must be accepted as a complex and nontrivial phenomenon (MacDonald et al. 2010; Soma et al. 2016).

The sections that follow draw on recent studies reported in the science–policy interface literature to examine how researchers are attempting to build understanding of the activities at the interface, especially what often causes or creates the apparent gap between the two realms. Further discussion of these themes is presented in the chapters in this volume, particularly Chapter 19.

2.4 Characteristics of the Science–Policy Interface

The science–policy interface operates at several scales: geographic, institutional, political, and temporal (Nutley et al. 2007; Young 2014). However, the interface is more inclusive than the label implies, as it encompasses many social processes and may include traditional or local knowledge. In fact, multiple interfaces exist due to many information, decision-making, and policy contexts. Furthermore, the information universe is multidimensional, and information flow may be nonlinear, which also accounts for the complexity of activity at the science–policy interface. While information can follow a direct pathway from published research to a decision-making context, more often information arising from research moves concurrently through multiple channels at varying paces (sometimes rapidly, as can be the case with social media) and involves a variety of actors, for example, nongovernmental organizations (NGOs), journalists, and the interested public.

The remainder of this section examines some of the main characteristics of the interface: actors involved in the policy- and decision-making processes, diverse subjects and available knowledge, available information products and framing of the issues, politicizing of science, uncertainty, and organizational aspects. Many of these characteristics can act as enablers or barriers to information flow at the science–policy interface (e.g., Cossarini et al. 2014; Oltmann 2015; Suhay and Druckman 2015).

2.4.1 Actors

The primary actors in the production of scientific information are researchers working in a range of organizations: national governmental agencies, international intergovernmental agencies, universities and colleges, public and private research institutes, business and industry, and NGOs. The primary actors in the policy processes are science advisors, policy analysts, policy makers, and decision makers in provincial/state, regional, and national governmental departments and agencies and in intergovernmental organizations. Science advisors distill the information and prepare options and recommendations outlined in briefing notes and documents. Policy analysts support other civil servants and are primarily responsible for providing informational inputs and policy advice to the policy makers. The latter are senior civil servants and politicians, while senior decision makers at the policy level are considered to have political powers and are often politicians (Bardach 2004; Ouimet et al. 2009).

Apart from researchers and policy makers, the interface may involve multiple additional stakeholder groups with external influences on the process, for example, NGOs and industry. Within these groups, the "interested public" has been identified as playing a role, since interested and informed individuals are often engaged in ICOM activities (Soomai et al. 2011, 2013).

Furthermore, journalists and the media, who have been described as "competent outsiders to science" (Polman et al. 2014, p. 766), can exert direct or indirect influence in policy contexts by defining issues, swaying national policies, highlighting conferences, and generally providing information to the public (Brüggemann and Engesser 2014; Clarke et al. 2015; Cooper 2011; Dikou and Dionysopoulou 2011; Ford and King 2015; Luokkanen et al. 2014; Voyer et al. 2013; Weitkamp and Eidsvaag 2014). Nisbet and Fahy (2015, p. 38) assert that "journalists and their news organizations can contextualize and critically evaluate expert knowledge, facilitate discussion that bridges entrenched ideological divisions, and promote consideration of a broader menu of policy options and technologies."

2.4.2 Diverse Subjects and Available Knowledge

As noted in Section 2.1, the publication of scientific research is accelerating, and our understanding of coastal and marine (ocean) environments is growing. For ICOM to work effectively—given ecosystem approaches to management—integrating information and knowledge from social, economic, and political domains, as well as local and traditional knowledge, is needed to resolve the problems that are the focus of coastal and ocean management (Ascher et al. 2010; Raymond et al. 2010; Taylor and de Loë 2012; Bremer 2014; Bremer and Glavovic 2013; Singh et al. 2014). Governmental agencies can act as nodes of networks, and may lever the resources of many actors, including industry, scientists, government bodies, NGOs, and interest groups, to communicate information. However, integrating natural, social, economic, and local perspectives is often challenging, as gaps or uncertainties are inherent in data sets, for example, in long time series of ecological data. The case study on awareness and use of state of the environment reports in this volume (see Chapter 12) elaborates on this characteristic.

2.4.3 Information Products and Framing of Issues

A policy brief (or briefing note, as mentioned in Section 2.2.1) is a common format by which information reaches senior managers, policy makers, and politicians in many government organizations. Despite their frequency as a means of disseminating the results of research, few studies have explored the effectiveness of this form of communication (Beynon et al. 2012; Masset et al. 2013). In one recent study, little evidence was found that policy briefs led to a change in prior beliefs (Rajabi 2012). The results of research can be reported in a multitude of information products and formats, which can enable information dissemination, but at the same time can confuse potential readers as to which source is reliable. The case studies in this volume shed further light on this topic.

The framing of a problem or issue in a policy brief or another format of advice defines its primary elements and scope, and communicates the options for decision makers. Therefore, the way in which a problem or issue is framed can determine how important and worthy of attention it becomes, and how well it can influence how political and societal actors, for example, government, NGOs, industry, professional communities, and the general public, view the complexity and uncertainty related to the issue (Lakoff 2010; Kahan 2015; Rudd 2015). Morton et al. (2011), for example, showed that individual intentions to behave environmentally decreased when climate change predictions with high uncertainty were negatively framed, that is, highlighting possible losses. However, when the same predictions were presented with a positive frame, that is, underlining the possibility of losses not materializing, this produced stronger intentions to act. In another recent study, McComas et al. (2015, p. 50) found that framing a marine health issue (the health of oysters) as a public health matter resonated more with readers than messages focusing specifically on marine health, which led the authors to suggest that "linking marine disease to public health could increase concern and support for marine policy that can protect not only public health but also reduce risks to marine organisms." This topic is explored further in the chapters on science information and governance (Chapter 4), risk (Chapter 5), and participatory approaches (Chapter 6).

2.4.4 Politicizing of Science

Policy-neutral science is characterized by transparency, reproducibility, and independence (Heazle 2004). Science advice is also expected to be unbiased, objective, impartial, and policy neutral (Rice 2011). This neutrality is a characteristic of "best available scientific advice" for evidence-based policymaking (Nutley et al. 2007). However, science and politics often do not exist as completely independent entities; they can be inseparable, with a blurred boundary between policy advocacy and science, whereby scientific advice can carry a political bias even if it is not intentional (Lackey 2007; Rice 2011; see also Chapters 4 and 10 in this volume).

Misinformation or misuse of information may be unintentional, that is, honest errors are made, or intentional, that is, incorrect information is presented to purposely mislead. Misinformation can be a common practice when there is an absence of information (Deeming 2013) or where there are high economic stakes for industry and government, for example, funding anti–climate change groups and studies (e.g., Wogan 2013). The experiential judgment of politicians and other key decision makers may also be considered to be just as important as data, information, and analysis by trained policy analysts. This topic is explored further in Chapter 10.

2.4.5 Uncertainty

Scientific uncertainty is largely due to the inherent variability in natural systems and to statistical uncertainty arising from assessment methods.

How uncertainty is understood and dealt with by scientists and policy makers can determine whether and how information is used in decision-making. Uncertainty can be exaggerated, underemphasized, or even ignored, and is most often not understood by scientifically untrained personnel and the general public.

Policy makers and decision makers can delay making policy decisions in the face of uncertainty in scientific recommendations (Wardekkera et al. 2008). As another means to postpone action, managers can demand more information to address uncertainty. Scientific uncertainty about a problem or measurement can also be overplayed by scientists and policy makers, resulting in more time and resources being focused on acquiring data and providing advice to describe the uncertainties. Furthermore, the requirements for very low uncertainty expected in some policy contexts, for example, healthcare, where evidence-based practice is championed today, may not be directly applicable in other policy areas, where it may be very difficult, and probably not necessary, to obtain research evidence to the same high degree of certainty.

Underemphasizing uncertainty can be even more dangerous than overplaying uncertainties and can do lasting damage to the credibility of the science (Maxim and van der Sluijs 2011; van der Sluijs 2012; Wilson 2009). Examples from the literature on climate change show that exaggerating certainty can create a false confidence in the credibility and legitimacy of scientific advice, which can then be exploited to obstruct and delay policy intervention (Russell 2010; van der Sluijs et al. 2008). In the climate data *Climategate* controversy and the IPCC, problems arose when the scientific advice did not reveal all sources of uncertainty in spite of the rigorous peer review that was applied. The research community believed that some uncertainties in the input parameters were benign, as the outcome of the assessment would not change whether or not the uncertainties were accounted for (Netherlands Environmental Assessment Agency 2010; Ravindranath 2010). However, a lack of transparency in the assessment process was used as a point of contention and an attempt to publicly discredit the scientific advice. Similar dangers from overselling certainty were noted in other scientific advisory bodies, for example, some International Council for the Exploration of the Sea (ICES) fisheries scientists identified with Climategate, as they regularly "simplify" uncertainties in stock assessments when they communicate advice to fisheries managers (Dankel et al. 2012; Wilson 2009). This topic is further explored in Chapter 9, where the role of communication networks in addressing uncertainty and building trust is discussed.

2.4.6 Organizational Aspects

The divergent professional motivations and different timescales of the output associated with science and policy communities affect information flow across the interface (Tribbia and Moser 2008). Scientists are interested in and focus on understanding natural phenomena, making discoveries, and solving problems, using scientific procedures that often require producing an extensive time series of data. Policy makers and program managers, on the other hand, typically need advice on solutions to immediate problems. Compromise between the two ways of addressing problems and information needs is often weak or impossible. For example, the demand for technical audits and program evaluations in government-funded science programs is often based on the fiscal year (Doern and Reed 2000), and some individuals' priorities are driven by the electoral cycle; scientific research and related reports for policy-making most often cannot be completed within similar time frames. The traditional hierarchical structure of government bureaucracies also creates departmentalization and centralization, which can limit communication of information and information needs and potentially cause conflict within the public service (Yang and Maxwell 2011).

Political decisions are typically publically visible, sometimes the outcome of intense partisan debate, and can be contentious. However, the processes by which the decisions are made are generally not transparent, even though the decisions often have far-reaching implications, that is, through new or revised legislation. Gaining access to decision makers to build understanding of how scientific information is used is difficult and often unachievable. Consequently, many studies have focused on the need for and use of information by environmental managers (e.g., Delaney and Hastie 2007; Jacobson et al. 2013; Kirchhoff 2013; Tribbia and Moser 2008; White et al. 2008). Recently, two researchers at McGill University were able to gain some understanding of decision-making at the highest levels of governance in Canada and Australia by interviewing former environment ministers (senior politicians) and deputy ministers/department secretaries (senior public servants) (Lalor and Hickey 2013). The politicians and public servants "generally believed in the value of science as a foundation for decision-making," but they called for "more inclusive and contextualized knowledge" to inform decisions (Lalor and Hickey 2013, p. 774). The different motivations of scientists, managers, and senior decision makers to act in policy contexts are explored in more detail in the chapters on governance (see Chapters 4 through 6 and 10) and in the case studies in Section III.

2.4.7 Quality Control (Peer Review) and Attributes of Information

Numerous studies have pointed to the growing production of gray literature by government departments, international intergovernmental organizations, and NGOs, resulting in greater reliance on gray literature for disseminating information to inform decisions and policies (e.g., MacDonald et al. 2004, 2007; Luzi 2000; Schöpfel and Farace 2010; Webster and Collins 2005). Although the use of gray literature in policy development is deemed very important, and while these publications are generally created through rigorous peer review, their credibility is often questioned compared with the primary literature, that is, the journals produced by commercial publishers. Moreover, questions about the credibility of some open access journals contribute to distrust of the quality of scientific information (Bohannon 2013).

Given an environmental issue at hand, the best available information can be determined based on its attributes, that is, credibility, salience, and legitimacy, as described in Cash et al. (2003) and Mitchell et al. (2006). These attributes can be used to filter out information in decision-making processes. The credibility of the available information refers to the perceived validity of the information used, relevance reflects the extent to which the work carried out is relevant to the policy process, and legitimacy reflects the perceived fairness and political acceptability of the information used in a decision. This subject is given further attention in Chapters 10 and 14.

2.5 Enhancing Communication of Information at the Science–Policy Interface

Different interpretations of the science–policy interface in integrated coastal and ocean management exist (e.g., Boelens 1992; Bremer and Glavovic 2013; Koetz et al. 2011). In a science-based interface that follows a linear model, the inherent uncertainty in science is attributed to a lack of available information, warranting the creation of new science to fill gaps in the information delivered to decision makers. Alternatively, in a participatory approach or a more collaborative model of a science–policy interface, uncertainty is considered to be inevitable, necessitating the integration of existing knowledge, including scientific knowledge and local knowledge, in an interdisciplinary approach. Knowledge mobilization is seen to be dependent on the characteristics of each science–policy interface, and a variety of factors can contribute to the use of information in these contexts.

2.5.1 Boundary Organizations

Boundary work facilitated by boundary organizations extends peer communities by sometimes including increased public participation and changes to traditional management processes, for example, adaptive management, in an effort to bridge the divide between science and policy-making by facilitating greater interaction between the producers and the users of information. The overall impact of these initiatives is increased perception of salience or relevance of information to the policy process, increased legitimacy related to inputs from multiple sources, and credibility related to the validity of findings. The case studies in Section III of this volume provide more detail on efforts to enhance communication of information at the science–policy interface.

Boundary work seeks to bridge the divide between science and policy-making and to connect two or more otherwise disparate groups, that is, scientists and policy makers, allowing the policy network to share information efficiently and quickly (in scale and time) and often leading to more productive policy-making (Guston 1999, 2001). Boundary organizations employ specialists, known as *interpreters*, *bridgers*, or *mediators*, from both sides of the science–policy interface to broker links between advisors or policy makers and scientists (Guston 2001; Pielke 2007; Holmes and Clark 2008; Van der Sluijs et al. 2008). These specialists are often co-opted from external groups into the decision-making structure of the boundary organization and play the role of *science arbiter* and *honest broker* by offering insights from the scientific advice or the policy requirements. They may assist in the selection of policy options, facilitate the development of researchable questions to meet policy needs and communicate these to researchers. They also provide an up-to-date balanced overview and synthesis of what is known, and what the key uncertainties are, in relation to a policy issue (Cossarini et al. 2014; Drimie and Quinlan 2011; Godfrey et al. 2010; Grainger 2013; Huitema and Turnhout 2009; Mitchell et al. 2006).

2.5.2 Extended Peer Communities and Increased Public Participation

Extended peer communities and increased public participation facilitate coproduction of information, which enhances communication, for example, across multiple organizations, jurisdictions, and stakeholder groups. Participation of stakeholders in knowledge production can also help to increase the quality of the information product; for example, including more viewpoints in a risk assessment can improve the usefulness of the assessment. The extended peer community becomes the foundation for salient, credible, and legitimate science for policy advice (Cash et al. 2003). Coproduction of information increases its salience, credibility, and legitimacy (McNie 2007; Hegger et al. 2012) and has been addressed in the context of global environmental assessments, an important component of ICOM (Mitchell et al. 2006). Given the number of actors, organizations, and stakeholders that can contribute to policy development, the importance of networks in communicating information at the interface is directly related to this topic (see Bodin et al. 2006; Hartley 2010; Hartley and Glass 2010). Further details about the relevance of networks in understanding information activities at the science–policy interface are described in Section II, particularly Chapter 8, and case studies in Section III, for example, Chapter 15.

2.5.3 Changes to Traditional Management Processes

Adaptive management is a characteristic of ICOM that promotes iterative relationships between producers and users of information to facilitate two-way communication, thereby increasing the usability of information

(Forst 2009; Linkov et al. 2006; Katsanevakis et al. 2011; Sarkki et al. 2015). Adaptive management—involving monitoring, evaluation, and modification of management actions—facilitates decision-making in the face of inherent uncertainty associated with social and ecological systems. The importance of the iterative relationship between scientists and policy makers is seen in studies involving climate change data (e.g., Dilling and Lemos 2011; Sarewitz and Pielke 2007). In conceptualizing a connection between science, decision-making, and societal outcomes, science is viewed in terms of the "supply" of information, while societal outcomes are seen as the "demand" function that seeks to apply information to achieve specific societal goals. Decision-making is then conceptualized as reconciling the dynamic relationship between supply and demand. Sarewitz and Pielke (2007) showed that a poor reconciliation between supply and demand of climate information occurred when users could not make efficient use of relevant available information, and as a result, decision-making was affected.

2.6 Discussion and Conclusion

This chapter, as part of the introduction to this book, identified some of the characteristics of the science–policy interface (or interfaces) that must be understood to achieve effective ICOM. The chapters that follow will treat these and other characteristics in more detail. The science–policy interface is dynamic, and its characteristics are expected to change with time and with regard to the issue at hand.

This discussion of the topics described our current understanding of some of the characteristics of the science–policy interface and highlighted gaps in the knowledge:

- The complexity of information behavior and systems involving many different actors in decision-making and management activities is frequently encountered. Policy networks now include a wide range of stakeholders, among them scientists, managers, policy makers (including politicians), the international community, NGOs, industry, journalists and the news media, think tanks, and the interested and general public.
- ICOM is characterized by a collaborative model of science–policy interrelations, likely due to the public demand for increased transparency in science and policy-making. Government and intergovernmental organizations that inform the public about environmental risks increasingly recognize that uncertainty must be dealt with in a transparent and effective manner involving collaboration among the scientific (research) community, policy makers, and interested

stakeholders. This collaboration includes the framing of the problem, choice of assessment methods, data collection strategies, interpretation of results, individual roles in knowledge production, open discourse on uncertainty, and the use of information in policy-making circles to represent the diversity of perspectives from which a policy problem and results of assessments can be viewed.

- Questions remain about what accounts for blind spots regarding the role(s) and values of information in ocean and coastal management contexts. For example, can the challenges be attributed to the ubiquity of information, that is, is information so common that everyone takes it for granted? How do the different actors or stakeholder groups define "information"?

- The need for further studies to understand the characteristics of the science–policy gap is evident. For instance, the relationship between science and politics is complex, and understanding decision-making in contexts where value judgments appear to be unavoidable should be a priority area for research. Furthermore, the motivations of policy makers related to the use and influence of scientific information in policy-making are still unclear. In evidence-based decision-making, do decision makers, such as managers and senior policy makers, understand the scientific information? As policy-making circles expand, who are the new actors, and what are their roles in the process? Are there other characteristics to consider in addition to those described in this chapter?

- The major attributes of information—credibility, relevance, and legitimacy—influence its uptake in decision-making. The provision of information resulting from interdisciplinary work and the formation of diverse peer review communities that adhere to ecosystem-based approaches to management are the cornerstone of ICOM. Given the vast and growing volume of such scientific information, what standards or filters do decision makers employ to determine usable information?

- The need to develop methods to measure the use and influence of information in policy contexts warrants being addressed. This would offer direction for the creation of useable information and information products, as well as clarifying which channels to use to ensure that the information reaches decision makers in a timely manner.

- Boundary organizations and specialists, such as knowledge brokers, were identified as organizational components that may have the ability to bridge the science–policy gap. However, such organizations and individuals cannot be expected to solve all of the problems inherent in the interface. Further understanding on how boundary organizations and knowledge brokers link the science and policy communities is needed.

It is expected that the concepts treated in this book will advance the discussion of how new information and knowledge about the problems facing coasts and oceans (produced largely by natural and social scientists and coupled with local or traditional knowledge) will move more effectively to the primary users of that information and knowledge: ocean managers, science advisors, policy makers, and decision makers.

That this subject is vitally important is becoming increasingly evident, as problems appear faster than solutions, or persistent problems prove to be almost intractable, despite the huge repository of potentially useful information. Hence, the goal of the book—to advance understanding of the interface between information produced by researchers and its use and influence in critical decision-making and policy-making, all focused on enhancing ICOM and sustainability of marine ecosystems, economies, and communities—is timely.

References

AAAS (American Association for the Advancement of Science). 2013a. European science policy on the move. Report from the American Association for the Advancement of Science Conference, Boston, February 14–18, 2013. http://news.aaas.org/2013_annual_meeting/0216european-science-policy-on-the-move.shtml.

AAAS (American Association for the Advancement of Science). 2013b. Program of the Annual Meeting, February. http://www.aaas.org/AM2013.

Ascher, W., T. Steelman, and R. Healy. 2010. *Knowledge and Environmental Policy: Re-imagining the Boundaries of Science and Politics*. Cambridge, MA: MIT Press.

Bardach, E. 2004. *A Practical Guide for Policy Analysis: The Eightfold Path to More Effective Problem Solving.* Washington, DC: CQ.

Beynon, P., M. Gaarder, C. Chapoy, et al. 2012. Passing on the hot potato: Lessons from a policy brief experiment. *IDS Bulletin* 43 (5): 68–75.

Bodin, Ö., B. Crona, and H. Ernstson. 2006. Social networks in natural resource management: What is there to learn from a structural perspective? *Ecology and Society* 11 (2): r2. http://www.ecologyandsociety.org/vol11/iss2/resp2/.

Boelens, R. G. V. 1992. From policies to science: Strategies for marine environmental protection. *Marine Pollution Bulletin* 24 (1–4): 14–17.

Bohannon, J. 2013. Who's afraid of peer review? *Science* 342 (6154): 60–65.

Bornmann, L. and R. Mutz. 2014. Growth rates of modern science: A bibliometric analysis based on the number of publications and cited references. *Journal of the Association for Information Science and Technology.* http://arxiv.org/abs/1402.4578v2.

Bornmann, L. and R. Mutz. 2014. Growth rates of modern science: A bibliometric analysis based on the number of publications and cited references. *Journal of the Association for Information Science* 66 (11).

Bremer, S. 2014. "No right to rubbish": Mobilising post-normal science for planning Gisborne's wastewater outfall. *Marine Policy* 46: 22–30.

Bremer, S. and B. Glavovic. 2013. Mobilising knowledge for coastal governance: Re-framing the science–policy interface for integrated coastal management. *Coastal Management* 41: 39–56.

Brüggemann, M. and S. Engesser. 2014. Between consensus and denial: Climate journalists as interpretive community. *Science Communication* 36 (4): 399–427.

Cash, D. W., W. C. Clark, F. Alcock, et al. 2003. Science and technology for sustainable development special feature: Knowledge systems for sustainable development. *Proceedings of the National Academy of Science* 100 (14): 8086–8091.

Cherney, A. and B. Head. 2010. Evidence-based policy and practice: Key challenges for improvement. *Australian Journal of Social Issues* 45 (4): 509–526.

Cicin-Sain, B. and R. W. Knecht. 1998. *Integrated Coastal and Ocean Management: Concepts and Practices*. Washington, DC: Island Press.

Clark, W. C., R. B. Mitchell, and D. W. Cash. 2006. Evaluating the influence of global environmental assessments. In *Global Environmental Assessments: Information and Influence*, edited by W. C. Clark, R. B. Mitchell, D. W. Cash, and N. M. Dickson, 1–28. Cambridge, MA: MIT Press.

Clarke, C. E., G. N. Dixon, A. Holton, et al. 2015. Including "evidentiary balance" in news media coverage of vaccine risk. *Health Communication* 30 (5): 461–472.

Coffey, B. and K. O'Toole. 2012. Towards an improved understanding of knowledge dynamics in integrated coastal zone management: A knowledge systems framework. *Conservation and Society* 19 (4): 318–329.

Cooper, C. B. 2011. Media literacy as a key strategy toward improving public acceptance of climate change science. *Bioscience* 61 (3): 231–237.

Cordes, R. E. 2004. Is grey literature ever used? Using citation analysis to measure the impact of GESAMP, an international marine scientific advisory body. *Canadian Journal of Information and Library Science* 28 (1): 45–65.

Cossarini, D. M., B. H. MacDonald, and P. G. Wells. 2014. Communicating marine environmental information to decision makers: Enablers and barriers to use of publications (grey literature) of the Gulf of Maine council on the marine environment. *Ocean & Coastal Management* 96: 163–172.

Dankel, D. J., R. Aps, G. Padda, et al. 2012. Advice under uncertainty in the marine system. *ICES Journal of Marine Science* 69 (1): 3–7.

Deeming, C. 2013. Trials and tribulations: The "use" (and "misuse") of evidence in public policy. *Social Policy & Administration* 47 (4): 359–381.

Delaney, A. E. and J. E. Hastie. 2007. Lost in translation: Differences in role identities between fisheries scientists and managers. *Ocean & Coastal Management* 50: 661–682.

Dikou, A. and N. Dionysopoulou. 2011. Communicating a marine protected area through the local press: The case of the national marine park of Alonissos, Northern Sporades, Greece. *Environmental Management* 47 (5): 777–788.

Dilling, L. and M. C. Lemos. 2011. Creating usable science: Opportunities and constraints for climate knowledge use and their implications for science policy. *Global Environmental Change* 21 (2): 680–689.

Doern, G. B. 2001. Science and scientists in regulatory governance: A mezzo-level framework for analysis. *Science and Public Policy* 28: 195–205.

Doern, G. B. and T. Reed. 2000. *Risky Business: Canada's Changing Science-Based Policy and Regulatory Regime*. Toronto: University of Toronto Press.

Drimie, S. and T. Quinlan. 2011. Playing the role of a "boundary organization": Getting smarter with networking. *Health Research Policy and Systems* 9 (Supplement 1): S11.

Duff, A. S. 1997. Some post-war models of the information chain. *Journal of Librarianship and Information Science* 29: 179–187.

Dunn, K. 2005. Impact of the inclusion of grey literature on the scholarly communication patterns of an interdisciplinary specialty. In *Work on Grey in Progress: Sixth International Conference on Grey Literature*, New York Academy of Medicine, NYAM Conference Center, December 6–7, 2004, New York, NY, 50–54. Amsterdam: TextRelease.

Eden, S. 2011. Lessons on the generation of usable science from an assessment of decision support systems. *Environmental Science & Policy* 14: 11–19.

Elfner, L. E., H. J. Falk-Krzesinski, K. Sullivan, et al. 2011. Team science. Heaving walls & melding silos. *American Scientist* 99: A1–A8.

Ford, J. D. and D. King. 2015. Coverage and framing of climate change adaptation in the media: A review of influential North American newspapers during 1993–2013. *Environmental Science & Policy* 48: 137–146.

Forst, M. F. 2009. The convergence of integrated coastal zone management and the ecosystem approach. *Ocean & Coastal Management* 52 (6): 294–306.

Gluckman, P. 2014. Evidence based policy: A quixotic challenge? Address Presented at the Science Policy Research Unit, University of Sussex, Brighton, UK, January 21, 2014. http://www.pmcsa.org.nz/wp-content/uploads/Sussex_Jan-21_2014_Evidence-in-Policy_SPRU.pdf.

Godfrey, L., N. Funke, and C. Mbizvo. 2010. Bridging the science–policy interface: A new era for South African research and the role of knowledge brokering. *South African Journal of Science* 106: 5–6.

Grainger, R. 2013. How does information influence policy? The role of fishery organizations in policy-making for fisheries. Lecture, Information Management Public Lecture Series, Dalhousie School of Information Management, September 20.

Guston, D. H. 1999. Stabilizing the boundary between US politics and science: The role of the Office of Technology Transfer as a boundary organization. *Social Studies of Science* 29: 87–111.

Guston, D. H. 2001. Boundary organizations in environmental policy and science: An introduction. *Science, Technology, and Human Values* 29: 87–112.

Harris, M. 2014. *Party of One*. Canada: Viking.

Hartley, T. W. 2010. Fishery management as a governance network: Examples from the Gulf of Maine and the potential for communication network analysis research in fisheries. *Marine Policy* 34: 1060–1067.

Hartley, T. W. and C. Glass. 2010. Science-to-management pathways in US Atlantic herring management: Using governance network structure and function to track information flow and potential influence. *ICES Journal of Marine Science* 67: 1154–1163.

Heazle, M. 2004. Scientific uncertainty and the International Whaling Commission: An alternative perspective on the use of science in policy making. *Marine Policy* 28 (5): 361–374.

Hegger, D., M. Lamers, A. Van Zeijl-Rozema, et al. 2012. Conceptualising joint knowledge production of regional climate change adaptation projects: Success conditions and levers for action. *Environmental Science & Policy* 18: 52–65.

Holmes, J. and R. Clark. 2008. Enhancing the use of science in environmental policy-making and regulation. *Environmental Science & Policy* 11: 702–711.

Holmgren, M. and S. A. Schnitzer. 2004. Science on the rise in developing countries. *PLOS Biology* 2 (1): 0010–0013.

Howlett, M., 2009. Policy analytical capacity and evidence-based policy-making. Lessons from Canada. *Canadian Public Administration* 52 (2): 153–175.

Huitema, D. and E. Turnhout. 2009. Working at the science–policy interface: A discursive analysis of boundary work at the Netherlands environmental assessment agency. *Environmental Politics* 18 (4): 576–594.

International Network for Government Science Advice. 2014. Science advice to governments comes of age at Auckland conference, Media Release, August 29. http://www.globalscienceadvice.org/media/.

IPCC (Intergovernmental Panel on Climate Change). 2014. Assessment reports. http://www.ipcc.ch/publications_and_data/publications_and_data_reports.shtml.

Jacobson, C., A. Lisel, R. W. Carter, et al. 2013. Improving technical information use: What can be learnt from a manager's perspective? *Environmental Management* 52: 221–233.

Jasanoff, S. 1994. *The Fifth Branch: Science Advisers as Policymakers*. Cambridge, MA: Harvard University Press.

Kahan, D. 2010. Fixing the communications failure. *Nature* 463: 296–297.

Kahan, D. M. 2015. Climate-science communication and the measurement problem. *Advances in Political Psychology* 36: 1–43.

Katsanevakis, S., V. Stelzenmüller, A. South, et al. 2011. Ecosystem-based marine spatial management: Review of concepts, policies, tools, and critical issues. *Ocean & Coastal Management* 54 (11): 807–820.

Khabsa, M. and C. L. Giles. 2014. The number of scholarly documents on the public web. *PloS ONE* 9. doi:10.1371/journal.pone.0093949.

Kirchhoff, C. J. 2013. Understanding and enhancing climate information use in water management. *Climatic Change* 119 (2): 495–509.

Koetz, T., K. N. Farrell, and P. Bridgewater. 2011. Building better science–policy interfaces for international environmental governance: Assessing potential within the Intergovernmental Platform for Biodiversity and Ecosystem Services. *International Environmental Agreements: Politics, Law and Economics* 12 (1): 1–21.

Lackey, R. T. 2007. Science, scientists, and policy advocacy. *Conservation Biology* 21 (1): 12–17.

Lakoff, G. 2010. Why it matters how we frame the environment. *Environmental Communication* 4 (1): 70–81.

Lalor, B. M. and G. M. Hickey. 2013. Environmental science and public policy in executive government: Insights from Australia and Canada. *Science and Public Policy* 40: 767–778.

Linkov, I., F. K. Satterstrom, G. Kiker, et al. 2006. From comparative risk assessment to multi-criteria decision analysis and adaptive management: Recent developments and implications. *Environment International* 32 (8): 1072–1093.

Lubchenco, J. and N. Sutley. 2010. Proposed U.S. policy for ocean, coast, and great lakes stewardship. *Science* 328: 1485–1486.

Luokkanen, M., S. Huttunen, and M. Hilden. 2014. Geoengineering, news media and metaphors: Framing the controversial. *Public Understanding of Science* 23 (8): 966–981.

Luzi, D. 2000. Trends and evolution in the development of grey literature: A review. *International Journal on Grey Literature* 1: 105–116.

MacDonald, B. H., R. E. Cordes, and P. G. Wells. 2004. Grey literature in the life of GESAMP: An international marine scientific advisory body. *Publishing Research Quarterly* 20: 26–41.

MacDonald, B. H., R. E. Cordes, and P. G. Wells. 2007. Assessing the diffusion and impact of grey literature published by international intergovernmental scientific groups: The case of the Gulf of Maine Council on the Marine Environment. *Publishing Research Quarterly* 23 (1): 30–46.

MacDonald, B. H., P. G. Wells, R. E. Cordes, et al. 2010. The use and influence of information produced as grey literature by international, intergovernmental marine organizations: Overview of current research. In *Grey Literature in Library and Information Studies*, edited by D. J. Farace and J. Schöpfel, 167–180. Berlin: De Gruyter Saur.

Masset, E., M. Gaarder, P. Beynon, et al. 2013. What is the impact of a policy brief? Results of an experiment in research dissemination. *Journal of Development Effectiveness* 5 (1): 50–63.

Maxim, L. and J. P. van der Sluijs. 2011. Quality in environmental science for policy: Assessing uncertainty as a component of policy analysis. *Environmental Science & Policy* 14 (4): 482–492.

McComas, K. A., J. P. Schuldt, C. A. Burge, et al. 2015. Communicating about marine disease: The effects of message frames on policy support. *Marine Policy* 57: 45–52.

McLeod, P. and B. Medel. 2013. N.S. casting for new fisheries. *The Chronicle Herald* (Halifax, NS), February 12. Accessed September 21, 2015. http://thechronicle-herald.ca/novascotia/685778-ns-casting-for-new-fisheries.

McNie, E. 2007. Reconciling the supply of scientific information with user demands: An analysis of the problem and review of the literature. *Environmental Science & Policy* 10: 17–38.

Mitchell, R. B. 2010. *International Politics and the Environment*. London: SAGE.

Mitchell, R. B., W. C. Clark, and D. W. Cash. 2006. Information and influence. In *Global Environmental Assessments*, edited by R. B. Mitchell, W. C. Clark, D. W. Cash, and N. M. Dickson, 307–338. Cambridge, MA: MIT Press.

Morton, T. A., A. Rabinovich, D. Marshall, et al. 2011. The future that may (or may not) come: How framing changes responses to uncertainty in climate change communications. *Global Environmental Change* 21 (1): 103–109.

Munroe, R. 2013. How much science is there? *Science* 342 (6154): 58–59.

Nature. 2010. Climate of fear. *Nature* 464 (7286): 141.

Netherlands Environmental Assessment Agency. 2010. Assessing an IPCC assessment: An analysis of statements on project regional impacts in the 2007 report. The Hague: Netherlands Environmental Assessment Agency (PBL).

Nisbet, M. C. and D. Fahy. 2015. The need for knowledge-based journalism in politicized science debates. *Annals of the American Academy of Political and Social Science* 658 (1): 223–234.

NRC (National Research Council). 1995. *Science, Policy and the Coast*. Washington, DC: National Academy Press.

NRC (National Research Council). 2012. *Using Science as Evidence in Public Policy*. Washington, DC: National Academies Press.

Nursey-Bray, M. J., J. Vince, M. Scott, et al. 2014. Science into policy? Discourse, coastal management and knowledge. *Environmental Science & Policy* 38: 107–119.

Nutley, S. M., I. Walter, and H. T. O. Davies. 2007. *Using Evidence: How Research Can Inform Public Services*. Bristol: Policy Press.

Oltmann, S. M. 2015. Data, censorship, and politics: Analyzing the restricted flow of information in federal scientific policy development. *Journal of the Association for Information Science and Technology* 66 (1): 144–161.

Ouimet, M., R. Landry, S. Ziam, et al. 2009. The absorption of research knowledge by public civil servants. *Evidence and Policy* 5 (4): 331–350.

Pielke, R. A., Jr. 2007. *The Honest Broker. Making Sense of Science in Policy and Politics.* Cambridge: Cambridge University Press.

Polman, J. L., A. Newman, E. W. Saul, et al. 2014. Adapting practices of science journalism to foster science literacy. *Science Education* 98 (5): 766–791.

Rajabi, F. 2012. Evidence-informed health policy making: The role of policy brief. *International Journal of Preventive Medicine* 3 (9): 596–598.

Ravindranath, N. H. 2010. IPCC: Accomplishments, controversies and challenges. *Current Science* 99 (1): 26–35.

Raymond, C. M., I. Fazey, M. S. Reed, et al. 2010. Integrating local and scientific knowledge for environmental management. *Journal of Environmental Management* 91: 1755–1777.

Rice, J. C. 2011. Advocacy science and fisheries decision-making. *ICES Journal of Marine Science* 68 (10): 2007–2012.

Rudd, M. A. 2015. Scientists' framing of the ocean science–policy interface. *Global Environmental Change* 33: 44–60.

Russell, M. 2010. The independent climate change e-mails review. Norwich: University of East Anglia. http://www.cce-review. org/pdf/FINAL%20REPORT.pdf.

Sarewitz, D. and R. Pielke. 2007. The neglected heart of science policy: Reconciling supply of and demand for science. *Environmental Science & Policy* 10: 5–16.

Sarkki, S., R. Tinch, J. Niemela, et al. 2015. Adding "iterativity" to the credibility, relevance, legitimacy: A novel scheme to highlight dynamic aspects of science–policy interfaces. *Environmental Science & Policy* 55: 505–512.

Schenkel, R. 2010. The challenge of feeding scientific advice into policy-making. *Science* 330: 1749–1751.

Schöpfel, J. and D. J. Farace. 2010. Grey literature. In *Encyclopedia of Library and Information Science*, edited by Marcia J. Bates and Mary Niles Maack, 3rd ed., 2029–2039. Boca Raton: CRC.

Shanley, P. and C. López. 2009. Out of the loop: Why research rarely reaches policy makers and the public and what can be done. *Biotropica* 41 (5): 535–544.

Singh, G. G., J. Tam, T. D. Sisk, et al. 2014. A more social science: Barriers and incentives for scientists engaging in policy. *Frontiers in Ecology and the Environment* 12 (3): 161–166.

Soma, K., B. H. MacDonald, K. Termeer, et al. 2016. Informational governance and environmental sustainability. *Current Opinion in Environmental Sustainability,* 18.

Søndergaard, T. F., J. Andersen, and B. Hjørland. 2003. Documents and the communication of scientific and scholarly information: Revising and updating the UNISIST model. *Journal of Documentation* 59 (3): 278–320.

Soomai, S. S., B. H. MacDonald, and P. G. Wells. 2013. Communicating environmental information to the stakeholders in coastal and marine policy making: Case studies from Nova Scotia and the Gulf of Maine/Bay of Fundy Region. *Marine Policy* 40: 176–186.

Soomai, S. S., P. G. Wells, and B. H. MacDonald. 2011. Multi-stakeholder perspectives on the use and influence of "grey" scientific information in fisheries management. *Marine Policy* 35 (1): 50–62.

Spruijt, P., A. B. Knol, E. Vasileiadou, et al. 2014. Roles of scientists as policy advisers on complex issues: A literature review. *Environmental Science & Policy* 40: 16–25.

Stoutenborough, J. W. and A. Vedlitz. 2014. The effect of perceived and assessed knowledge of climate change on public policy concerns: An empirical comparison. *Environmental Science & Policy* 37: 23–33.

Suhay, E. and J. N. Druckman. 2015. The politics of science: Political values and the production, communication, and reception of scientific knowledge. *Annals of the American Academy of Political and Social Science* 658: 6–15.

Taylor, B. and R. C. de Loë. 2012. Conceptualizations of local knowledge in collaborative environmental governance. *Geoforum* 42: 1207–1217.

Tribbia, J. and S. C. Moser. 2008. More than information: What coastal managers need to plan for climate change. *Environmental Science & Policy* 11: 315–328.

Turner, C. 2013. *The War on Science: Muzzled Scientists and Wilful Blindness in Stephen Harper's Canada*. Vancouver: Greystone Books.

UNEP (United Nations Environment Programme). 2006. *The State of the Marine Environment: Trends and Processes*. The Hague: UNEP/GPA Coordination Office.

UNESCO (United Nations Educational, Scientific and Cultural Organization). 2010. UNESCO Science Report 2010. The current status of science around the world. Paris: UNESCO.

Van der Sluijs, J. P. 2012. Uncertainty and dissent in climate risk assessment: A post-normal perspective. *Nature and Culture* 7 (2): 174–195.

Van der Sluijs, J. P., A. C. Petersen, P. H. M. Janssen, et al. 2008. Exploring the quality of evidence for complex and contested policy decisions. *Environmental Research Letters* 3 (2): 1–9.

Van der Veer Martens, B. and A. A. Goodrum. 2005. The diffusion of theories: A functional approach. *Journal of the American Society for Information Science and Technology* 57: 330–341.

Van Noorden, R. 2014. Global scientific output doubles every nine years. *Nature.* May 7. http//blogs.nature.com/news/2014/05/global-scientific-output-doubles-every-nine-years.html.

Voyer, M., T. Dreher, W. Gladstone, et al. 2013. Who cares wins: The role of local news and news sources in influencing community responses to marine protected areas. *Ocean & Coastal Management* 85: 29–38.

Wardekkera, J. A., J. P. van der Sluijs, P. H. M. Janssen, et al. 2008. Uncertainty communication in environmental assessments: Views from the Dutch science–policy interface. *Environmental Science & Policy* 11 (7): 627–641.

Webster, J. G. and J. Collins. 2005. Fisheries information in developing countries. Support for the implementation of the 1995 Code of Conduct for Responsible Fisheries. Rome: FAO.

Weitkamp, E. and T. Eidsvaag. 2014. Agenda building in media coverage of food research: Superfoods coverage in UK national newspapers. *Journalism Practice* 8 (6): 871–886.

Wells, P. G. 2003. State of the marine environment (SOME) reports: A need to evaluate their role in marine environmental protection and conservation. *Marine Pollution Bulletin* 46: 1219–1223.

White, D. D., E. A. Corley, and M. S. White. 2008. Water managers' perceptions of the science–policy interface in Phoenix, Arizona: Implications for an emerging boundary organization. *Science and Natural Resources* 21 (3): 230–243.

Wilson, D. C. 2009. *The Paradoxes of Transparency: Science and the Ecosystem Approach to Fisheries Management in Europe.* MARE Publication Series, 5. Amsterdam: Amsterdam University Press.

Winfield, M. S. 2013. The environment, "Responsible Resource Development," and evidence-based policy-making in Canada. In *Evidence-Based Policy-Making in Canada*, edited by S. P. Young, 196–221. Don Mills, ON: Oxford University Press.

Wogan, D. 2013. The well-funded and organized campaigns that influence climate change science online. *Scientific American*, April 9. http://blogs.scientificameri-can.com/plugged-in/the-well-funded-and-organized-campaigns-that-influ-ence-climate-change-science-online/.

WWF (World Wide Fund for Nature). 2015. *Living Blue Planet Report. Species, Habitats and Human Well-Being*, edited by J. Tanzer, C. Phua, A. Lawrence, A. Gonzales, T. Roxburgh, and P. Gamblin. Gland: WWF.

Yang, T-M. and T. A. Maxwell. 2011. Information-sharing in public organizations: A literature review of interpersonal, intra-organizational and inter-organizational success factors. *Government Information Quarterly* 28 (2): 164–175.

Young, S. P., ed. 2014. *Evidence-Based Policy-Making in Canada: A Multidisciplinary Look at How Evidence and Knowledge Shape Canadian Public Policy.* Don Mills, ON: Oxford University Press.

Section II

Fundamental Concepts and Principles

3

Exploring the Role of Science in Coastal and Ocean Management: A Review

Brian Coffey and Kevin O'Toole

CONTENTS

When scientists add their findings to the mix, they do not put an end to politics; they add new ingredients to the collective process.

(Latour 1998, p. 208)

3.1 Introduction: Situating the Challenge of Science-Informed Coastal and Ocean Management

Sustainable governance of coastal and marine environments is necessary given that most of the world's population lives adjacent to the coast and many people rely on coastal and ocean resources for their livelihoods

(Harvey and Caton 2003). Just as importantly, human activities impact on other species and the processes that sustain them. Available evidence also demonstrates that coastal and marine environments are not being managed sustainably and that these ecosystems are "amongst the most productive yet highly threatened systems in the world" (Agardy and Alder 2005, p. 515).

Sustainably managing such systems is challenging. Coastal and marine environments are created, sustained, and transformed by diverse and complex biophysical processes occurring over multiple time frames and spatial scales (Harvey and Caton 2003; Kay and Alder 2005). Further, human management of coastal and marine environments is administratively complex, often involving multiple authorities and jurisdictions (Cicin-Sain and Knecht 1998; Sorenson 1997), as well as being subject to diverse viewpoints and values (Thompson 2007). Berkes and Folke (1998) use the terms *social ecological systems* and *complex adaptive systems* to describe these mutually dynamic interactions between human and biophysical processes. In a similar vein, interactive governance theory (Kooiman 2008) characterizes governance as involving three interacting subsystems—a governing system, a system to be governed, and a system of governing interactions (links between the other two systems)—where governability is complicated by the presence of multiple wicked problems and varying levels of complexity in each subsystem (Jentoft and Chuenpadgee 2009, p. 553).

Considerable attention needs to be given to improving the ways in which the diverse challenges and inputs to coastal and ocean management are understood, investigated, mediated, and managed (Stepanova 2015). In some respects, the concept of integrated coastal zone management (ICZM, see Chapter 1) provides an overall model for responding to such challenges. Sorenson (1997, p. 9) defines ICZM as

> [t]he integrated planning and management of coastal resources in a manner that is based on the physical, socioeconomic and political interconnections both within, and among, the dynamic coastal systems, which when aggregated together define a coastal zone.

However, the literature on ICZM provides relatively limited guidance on the contribution of science and other forms of knowledge in pursing more integrated management. This is problematic given that scientific evidence is recognized as an important, if inadequate, element of sustainable coastal and ocean management. For example, Cahoon and Dumas (2011, p. 224) argue that "the role of science will be critical in informing the coastal policy process and supporting and defending better policy choices." McFadden (2007) laments "the case of disappearing science in coastal management." In making this assessment, she argues that coastal zone management has come to focus on the mediation of stakeholder conflict, with the consequence that "ICZM is becoming divorced from

progress on the scientific underpinning of integrated coastal management" (McFadden 2007, p. 435). For McFadden, the implications of this
are that "due to an absence of knowledge on integrated coastal behaviour,
stakeholders may be making decisions now that are damaging the system
in the long term" (McFadden 2007, p. 438). And yet Hulme, writing about
the need to embrace other ways of understanding the challenges associated with climate change, states:

> The contemporary political orthodoxy is that investment in science, tech-
> nology, engineering and maths (the STEM disciplines) provides the most
> assured basis for securing future economic vibrancy, social well-being
> and environmental protection. Yet the STEM disciplines by themselves
> carry a hubris that they seemingly cannot shake off. On their own they
> are inadequate for tackling "wicked" problems such as climate change.
> (Hulme 2001, p. 178)

Arguably, such an assessment is equally applicable in coastal and marine
settings. For example, Nursey-Bray et al. (2014, p. 107) raise related, and
potentially broader, concerns in arguing that the "arbitrary separation [of
coastal knowledge] into a binary discursive landscape mitigates against science–policy integration in practice." They say that

> to better understand how to build scientific research outputs into policy,
> decision-makers and researchers need to understand how knowledge
> works in practice, overcome [the] dichotomous construction of knowl-
> edge and, specifically, reconstruct or transition the notion of "science as
> knowledge" into "all knowledge types" into policy. (Nursey-Bray et al.
> 2014, p. 107)

In addition, it can be argued that some forms of scientific knowledge
have had too great an impact on decision-making, with neoliberal (Harvey
2005) ideas recognized as having shaped environmental policy and governance (Coffey and Marston 2013; Heynen and Robbins 2005; McCarthy and
Prudham 2004). For example, Prudham (2004) highlights the contribution of
neoliberal regulatory regimes to the contamination of the municipal water
supply in Walkerton, Ontario, which resulted in the deaths of seven people.

Given the importance of the issues at stake, this chapter considers the role
of science in coastal and ocean management decision-making. We make the
case that the role of science in coastal and ocean management is more complex
than is often appreciated and that science is necessary, but inadequate on its
own, for sustainable coastal and ocean management. To explore these issues,
we first explore what is meant by "science" and "decision-making" as a way
of introducing some of the complexity associated with coastal and ocean
management. We then introduce the concept of *knowledge systems* as a way of
characterizing the relationship between science and other forms of knowledge in decision-making. Finally, we consider how better understanding the

operating environment, within which science takes place, can contribute to the development of more productive, science–policy interfaces.

3.2 Unpacking Science and Decision-Making

The terms *science* and *decision-making* are seemingly straightforward and uncontroversial. Providing unambiguous definitions of such terms turns out, however, to be no easy task. It quickly becomes clear that the terms, and the activities and practices to which they refer, are complex and nuanced.

3.2.1 Unpacking Science

Rather than attempting to provide a definitive definition of science, examples drawn from three aspects of science, namely, its philosophy, types, and the purposes to which it can be oriented, are used to provide ways of understanding the complexity and diversity of science.

It is useful to start with the question posed by *What Is This Thing Called Science?* (Chalmers 2010). Chalmers' informative exploration of the history and philosophy of the physical sciences introduces and assesses different philosophies of science that have been advocated at different points in time, and which often continue to be advocated under different circumstances. For example, Chalmers discusses the problems and limitations associated with relying on uncritical observations (science as the knowledge derived from the facts of experience), experimentation, induction, deduction, falsification, and paradigms. Based on this analysis, he concludes that "there is no general account of science and scientific method to be had that applies to all sciences at all historical stages in their development" (Chalmers 2010, p. 247). Put simply, it is not possible to provide an unambiguous definition of science. Further, while rejecting Feyerabend's post-structurally inspired definition of science, Chalmers nonetheless accepts some of the challenging insights raised by Feyerabend's (1975) analysis. Chalmers claims that Feyerabend argues that as "matter[s] of historical fact, classic instances of scientific progress in science do not conform to the theories of science which they are taken to exemplify" (Chalmers 2010, p. 150). Broadly speaking, Feyerabend's approach to the history and philosophy of science focuses on truth as historically specific rather than objective and timeless.

The second way in which science is more complex than appreciated is in the differences that are apparent between the physical and social sciences. Firstly, they are qualitatively different undertakings (Flyvberg 2001), with the knowledge produced from the disciplines of history, politics, sociology, and anthropology being different to that produced by chemistry, physics, and mathematics. Secondly, Smith argues that

> the social sciences have a crucial role to play in ocean and coastal management: through their direct academic contribution; through professional practice within management organisations; and in promoting integrated decision-making at different levels ranging from simple communication among individuals and organisations to full structural integration of organisations. (Smith 2002, p. 581)

Finally, as clearly demonstrated by Funtowicz and Ravetz (1993), particular kinds of scientific research are appropriate for answering certain types of questions in some situations, but not others. For example, in discussing the emergence of post-normal science, Funtowicz and Ravetz (1993) distinguish between applied science (where there is technical uncertainty and low decision stakes), professional consultancy (where there is methodological uncertainty and moderate decision stakes), and post-normal science (where there is uncertainty and high decision stakes). When the decision stakes are high and system uncertainty is great, it is less likely that applied science or technical consultancy will prove useful. Instead, research that draws on wide participation, local knowledge, and recognition of values is more likely to be effective. With regard to integrated coastal zone management, Cummins and McKenna (2010) argue that *sustainability science* is needed. They propose six principles to guide the implementation of sustainability science in coastal settings: (1) resolve sustainable development issues by a problem-driven agenda; (2) coproduce knowledge in collaboration with stakeholder groups; (3) implement an interdisciplinary approach; (4) address earth system complexity; (5) focus communication and research activities at the local level; and (6) provide a process of social learning rather than providing definitive answers.

Different types of science can also be distinguished in terms of their orientations to decision-making. For example, in policy studies, distinctions have been made between the analysis *of* policy and analysis *for* policy (Hogwood and Gunn 1984, Table 3.1). Notwithstanding that such a view imposes an "either or" view of the purpose of different kinds of policy research, it nonetheless provides a sense of the orientations that research can take in relation to decision-making. It is possible to conduct research about policy while at the same time hoping that it also informs policy.

3.2.2 Unpacking Decision-Making

Decision-making also turns out to be more complex than might be imagined in the frequently expressed statement that "science should inform decision-making." In analyzing this statement, it is useful to focus on what is meant by "utilization" and what is meant by "decision-making". In two classic papers, Carol Weiss (1979; 1980) provides useful insights into these two areas and the notion of the *stages model* of the policy process. Three key points emerge from her analysis: research may be utilized in many ways, what constitutes a "decision" is far from clear, and research may be more or less useful at different parts of the policy process.

TABLE 3.1

Different Kinds of Policy Analysis

Analysis *of* Policy	
Studies of policy content	Studies describing and explaining the genesis and development of policies
Studies of policy outputs	Studies explaining why levels of expenditure or service provision vary over time
Studies of policy process	Studies focusing on how policy decisions are made and how policies are shaped in action
Analysis *for* Policy	
Evaluation	Studies focusing on the impact that policies have
Information for policy-making	Studies that marshal data in order to assist policy makers to reach decisions
Process advocacy	Studies seeking to improve the nature of policy-making systems through reallocation of functions and tasks
Policy advocacy	Studies where analysts promote specific options and ideas

Source: Data from Hogwood, B. and L. Gunn. 1984. *Policy Analysis for the Real World.* Oxford: Oxford University.

Weiss considers that if research utilization is to be encouraged, "it is essential to understand what 'using research' actually means" (Weiss 1979, p. 426). While Weiss is concerned with social science, her insights are broadly relevant across both the physical (i.e., natural) and social sciences, and arguably other bodies of knowledge. She identifies seven meanings of research utilization (summarized in Table 3.2). In terms of what constitutes a "decision" it is difficult to overlook Weiss' (1980) article "Knowledge Creep and Decision Accretion," where the title makes the point that knowledge is not often used in any direct and instrumental fashion to make a decision. Further, for Weiss (1980, p. 381), "many policy actions, even those of fateful order, are not 'decided' in brisk and clear cut style." Instead, Weiss (1980, p. 399) argues that three conditions make it difficult to pin down "decisions"—namely,

1. The dispersal of responsibility over many offices and the participation of many actors in decision-making [means that] no one individual feels that he or she has a major say

2. The division of authority among federal, state, and local levels in federal systems

3. The series of gradual and amorphous steps through which many decisions take shape.

Decision-making can also be viewed as a process, as is assumed within the stages model of the policy process (Howlett and Ramesh 2003; Althaus et al. 2012). In this approach, policy-making occurs within a relatively rational

TABLE 3.2

Types of Research Utilization

Type of Utilization	Explanation
Research-driven model	This model is based on the assumed sequence of events: basic research informs applied research, which informs development, which then finds application. For Weiss, this model is based on "the notion that basic research discloses some opportunity that may have relevance for public policy: applied research is conducted to define and test the finding of basic research for practical action: if all goes well, appropriate technologies are developed to implement the findings, whereupon application occurs" (p. 427).
Problem-solving model	This model assumes the direct application of results of a specific [social science] study to a pending decision and that "a problem exists and a decision has to be made" and "research provides the missing knowledge" (p. 427). Weiss sees two variations to this model: (1) policy makers search for information from pre-existing research; and, (2) policy makers engage scientists to fill an existing knowledge gap. In this model the sequence of events is: definition of pending decision > identification of missing knowledge > acquisition of social science research > interpretation of the research for the decision context > policy choice (p. 428).
Interactive model	This model assumes that those engaged in developing policy seek information from a variety of sources—administrators, practitioners, politicians, planners, journalists, clients, interest groups, aides, friends, and social scientists. The process is not one of linear order from research to decision but a disorderly set of interconnections and back-and-forthness that defies neat diagrams (p. 428).
Political model	This model is where "the constellation of interests around a policy issue predetermines the positions that decision makers take," such that research "becomes ammunition for the side that finds its conclusions congenial and supportive. Partisans flourish the evidence in an attempt to neutralize opponents, convince waverers, and bolster supporters" (p. 429).
Tactical model	This model assumes occasions where research is used for purposes that have little relation to the substance of the research. In these circumstances, research findings are not invoked, but the sheer fact that research is being done, is i.e., the fact that research has been commissioned, shows that the issue is taken seriously.
Enlightenment model	Under this model research utilization is not based on the findings of a single study nor even of a body of related studies that directly affect policy. Rather, the concepts and theoretical perspectives that social science research has engendered permeates the policy-making process (p. 429).
Research as part of the intellectual enterprise of society	This model assumes that like policy, social science research responds to the currents of thought, the fads, and fancies, of the period. Social science and policy interact, influence each other and are influenced by the larger fashions of social thought (p. 430).

Source: Data from Weiss, C. 1979. *Public Administration Review* 39: 426–431.

TABLE 3.3

Five Stages of the Policy Cycle and Their Relationship to Applied Problem Solving

Steps in Applied Problem Solving	Stages in the Policy Cycle
Problem recognition	Agenda-setting: The process by which problems come to the attention of governments
Proposal of solution	Policy formulation: How policy options are formulated within government
Choice of solution	Decision-making: The process by which governments adopt a particular course of action or non-action
Putting solution into effect	Policy implementation: How governments put policies into effect
Monitoring results	Policy evaluation: Processes by which the results of policies are monitored and assessed

Source: Data from Howlett, M. and M. Ramesh. 2003. *Studying Public Policy: Policy Cycles and Policy Subsystems.* Oxford: Oxford University Press.

cycle of decision-making, comprising separate but interrelated steps, which broadly align with generic problem-solving criteria, as summarized in Table 3.3 (Howlett and Ramesh 2003).

A modified version of this approach is the eight-stage policy cycle model presented by Althaus et al. (2012): (1) identification of issue; (2) analysis of policy; (3) consideration of policy instruments; (4) consultation with external stakeholders; (5) coordination across government agencies; (6) decision-making by elected officials; (7) implementation of the decision; and, subsequently, (8) evaluation of the effects of the decision implemented.

While such models have been thoroughly critiqued (e.g., Colebatch 2002; Bacchi 2009), they nonetheless provide a sense of how research may be more or less relevant at different points in the policy process. For example, research may identify a new problem, which then comes to the attention of the government. Or research may inform the government about the strengths and weaknesses of a particular potential solution, for example, its technical or economic strengths and weaknesses. Research might also inform the evaluation of a particular policy or program.

Finally, it is useful to mention the importance of recognizing the potential mismatch between the supply of, and demand for, science (Sarewitz and Pielke 2007). In this context, Sarewitz and Pielke (2007, p. 5) argue that "'better' science portfolios ... would be achieved if science policy decisions reflected knowledge about the supply of science, the demand for science, and the relationship between the two." To this end, they develop a matrix for reconciling supply and demand, which is based on responses to two questions: Is relevant information produced? Can users benefit from the research? Based on this matrix, science policy relations can be characterized as reflecting one of four scenarios: inappropriate research agenda; research agenda and user needs poorly matched (and potentially disenfranchised users); unsophisticated or marginalized users or other constraints that are obstacles to information use; and empowered users taking advantage of well-deployed research capabilities (Sarewitz and Pielke 2007, p. 12).

3.3 From Linear Knowledge Transfer to Dynamic and Interactive Ecologies of Knowledge Exchange

Having provided insights into the complexities of science and decision-making, we now turn to conceptualizing the relationship between science and decision-making by discussing three general models: the linear model, a cyclical process, and a dynamic multidirectional process (Ward et al. 2009).

3.3.1 The Linear and Cyclical Models

The linear model is effectively the default position for conceptualizing the relationship between science and decision-making and is so pervasive as to be taken for granted. It is also known by the following names: the deficit model (Rayner 2004), the traditional knowledge system (Roling and Jiggins 1998), science first (Kelsey 2003), and the loading dock approach (Cash et al. 2006). The core features of this model are its focus on the stepwise progression between identifiable beginning and end points (Ward et al. 2009) and an overwhelmingly linear and unidirectional orientation, whereby knowledge is viewed as generated by researchers and then transferred to others. Put simply, the model focuses on knowledge transfer and roughly assumes the following sequence: basic research > applied research > development > application (Weiss 1979). This model assumes that science is the major source of new ideas and technologies (Roling 1992).

For Kelsey (2003), the implications of this model are that it assumes a hierarchical relationship in which scientific knowledge, in particular the knowledge of the physical sciences, is elevated above other forms of knowledge. Further, under this assumption, the public is expected to respond to environmental problems, initially and accurately described by scientists, with solutions informed by science, negotiated by politicians, and enacted through various means of persuasion and regulation. Three variations of this model have been identified: (1) science is needed because the media and public oppose something and they do not have the appropriate scientific knowledge necessary to assess the benefits and risks involved (people need more facts); (2) there is a lack of public understanding of the processes of science (people need to understand the processes of science); and (3) there is a deficit of public trust (peoples' trust in science needs to be developed) (Rayner 2004).

Kelsey (2003, p. 2) sees problems with this model in privileging expert information as it "marginalises public knowledge" and "restricts the ability of the public to participate," which can "undermine the public's own belief in the value of their knowledge and participation." Instead, Kelsey advocates for "a willingness to adopt decision-making processes, timelines and organisational structures that reflect the different values on which alternative knowledge systems are based" (Kelsey 2003, p. 4). She also highlights the importance of recognizing that "knowledge is not transferred directly from one knower to another, but [instead] is actively built up by the learner" (Kelsey 2003, p. 9). Relatedly, for Roling (1992, p. 3), the key weaknesses of the linear model are that "it implies that there is a science-based fix for all societal problems: a promise that inhibits the search for other survival strategies" and that there are "strong incentives and political dynamics [which] keep it alive."

Ward et al. (2009, p. 6) also identify a cyclical model of knowledge transfer, which we consider a variation of the linear model in that "the individual components … are linked via a stepwise progression, but the process is depicted as interactive and ongoing." It differs from the linear model since there is no stopping point—evaluation leads to the identification of new problems, and so the process starts again. While clearly not as optimistic about science's capacity to provide *silver bullet* solutions, the model nonetheless emphasizes the centrality of science in decision-making. In broad terms, these models align with the policy–cycle model discussed in the previous section, as they involve linear stepwise approaches to problem solving.

3.3.2 The Dynamic, Interactive, and Multidirectional Model

The relationship between science and decision-making can also be conceptualized as "a dynamic, interactive and multidirectional process which involves many actors and activities" (Ward et al. 2009, p. 6). Support for such a view is evident in the argument by Raymond et al. (2010, p. 1766) that "to manage the scope, complexity and uncertainty of global environmental problems, it is important to take account of different types and sources of knowledge." This model is frequently labeled *knowledge systems* (Roling 1992; Cash et al. 2003; Coffey and O'Toole 2012) or *knowledge-action systems* (Cash and Buizer 2005; Weichselgartner and Kasperson 2010; Van Kerkoff and Lebel 2006), and represents a major reconceptualization of the relationship between science and decision-making. For Roling and Jiggins (1998), knowledge systems are mental constructs encompassing relatively stable networks of actors and coherent sets of cognitions, cosmologies, and practices, comprising seven elements—namely, an epistemology, ecology (belief about the way in which people interact with their biophysical environment), a set of practices (for managing agro-ecosystems), ways of learning (about agro-ecosystems), ways of facilitating and supporting such learning, supportive institutional frameworks and actor networks, and a conducive policy context. They argue that, taken together, these elements occur in unique, internally coherent combinations, which help to determine a particular type of knowledge system. Interest in this model is evident in diverse areas, including agricultural extension (Roling 1985; 1992), natural resource management (Campbell 2006; Ojha et al. 2008), sustainable development (Cash et al. 2003), biodiversity management (Kelsey 2003), public health (Van Kerkoff and Szlezak 2006), indigenous knowledge (Verran 1998; Mauro and Hardison 2000; King 2004; Houde 2007), business (Tsoukas and Mylonopolous 2004), innovation (Howells and Roberts 2000), knowledge management in firms (Lee and Van den Steen 2010), and information technology (Stefik 1995).

Coffey and O'Toole (2012) identify three ways in which knowledge systems have been investigated:

1. The nature and characteristics of particular knowledge systems are explored. For example, studies of traditional ecological knowledge (Kelsey 2003; Houde 2007).

2. Competing knowledge systems are compared and contrasted. For example, Roling and Jiggins (1998) contrast a traditional approach to agricultural extension (which is effectively the linear model discussed previously) with a "soft system" oriented approach, which they label "the ecological knowledge system."

3. Studies focus on the interactions between multiple knowledge systems (Erickson and Woodley 2005; Ojha et al. 2008). For example, Ojha and colleagues (2008, p. 3) identify "at least four different but overlapping systems of knowledge operating within the natural resource management sector in Nepal," which they consider have consolidated around techno-bureaucratic organizations, development agencies, politicians, and civil society. They argue that "in the processes of political interaction and deliberation over issues of natural resource governance, we see that these four systems of knowledge underpin the constitution of the four categories of social and political agents" (Ojha et al. 2008, p. 3).

Given that marine and coastal management involves multiple knowledge systems, this model is viewed as being particularly useful for conceptualizing the relationship between science and decision-making. The merits of such an approach are illustrated by Erickson and Woodley (2005), who outlined the benefits of using multiple knowledge systems (such as scientific, indigenous, traditional ecological, local, and practitioner knowledge) in the development of the Millennium Ecosystem Assessment (MEA). These benefits include the value of the insights provided from such knowledge systems, the value of using participation as a means for empowering local resource users, and the value of using multiple types of knowledge for improving the relevance, credibility, and legitimacy of the results generated. Put simply, recognizing multiple knowledge systems provides a richer understanding of the complex dynamics involved in marine and coastal management. However, the MEA was less helpful in suggesting practical mechanisms or processes for integrating multiple knowledge systems in ecosystem assessment.

Raymond et al. (2010) explicitly engage with the challenge of integrating different forms of knowledge. They discuss the ontological, epistemological, and applied challenges associated with integrating different types of knowledge and provide a framework to assist with considering and addressing these challenges. They carefully point out, though, that "there is no single optimum approach for integrating local and scientific knowledge and encourage a shift in science from the development of knowledge integration products to the development of knowledge integration processes" (Raymond et al. 2010, p. 1775). While acknowledging the value of the work of Raymond et al. (2010), Coffey and O'Toole (2012, p. 321) consider that the focus of Raymond et al. on projects, and their view of problems as "identifiable," constrains the focus and types of issues that can be explored. Instead, Coffey and O'Toole (2012, p. 321) argue that "achieving integration of different knowledge systems is likely to represent a considerable, if not insurmountable, challenge," such that "it remains necessary to focus on the interactions between different knowledge systems." This reflects a stronger focus on political mediation and negotiation, rather than any inherent resolution of the differences between knowledge systems. Further, it also situates science within the mix, rather than outside it.

As is evident from the preceding discussion, there is value in conceptualizing coastal knowledge relations in ways that are dynamic and multidirectional and recognizing the roles played by different stakeholders. In place of a linear and unidirectional transfer of knowledge from researchers to extension officers to clients, coastal knowledge relations need to be conceptualized as dynamic exchanges, whereby participants may be involved in knowledge generation, dissemination, and use, and different forms of knowledge are given due recognition. Thus it is useful to consider coastal management and knowledge relations as encompassing dynamic networks of multiple (intersecting) knowledge subsystems (each of which reflects diverse sets of values, worldviews, and practices) and which are advocated to varying degrees by different individuals and organizations. Clearly, such an approach is warranted given the challenges of implementing ICZM.

3.3.3 Myths and Models

When contrasting different conceptualizations of the relationship between science and decision-making, it is instructive to emphasize that the linear model informs people in fundamental ways, such that they may not be aware that they are embracing it. It is, therefore, useful to draw attention to some of the myths that inform science–policy relations, and also to highlight the different ways that people may respond to receiving scientific and other information.

As part of efforts to expand science policy debates in Australia, Harris and Meyer (2011, p. 9) outline various science policy myths that they see as retaining "disproportionate cultural power," despite having been "repeatedly discredited in recent decades." These myths and models of integrating science into policy are summarized in Table 3.4.

TABLE 3.4

Science Policy Myths

Name of Myth	Myth	Reality
Infinite benefit	More science and technology will necessarily lead to more public good. Any new knowledge is helpful.	Benefit is not a forgone conclusion. Science and technology may be useful, even harmful.
Serendipity	Because the benefits of science are unpredictable, we should not attempt to steer science in a particular direction.	Serendipity is an important part of research, but this does not prevent us from making well-reasoned choices about the kind of investments, institutions, and scientific practice likely to yield useful knowledge and technologies.
Authoritativeness	Scientific information provides an objective basis for resolving political disputes.	Science may inform policy and politics, but such disputes are based on values. Conflict based on values is unlikely to be resolved through science.
Accountability	Metrics of scientific quality (e.g., peer review, journal citations) are sufficient indication of worthwhile investments.	Policy makers (and scientists) justify research investment based on the promise of social benefits, thus taking on responsibility beyond scientific quality.
Linear model of science into society I	Basic research > applied research > development > social benefit	Benefit is not guaranteed. Interconnections and dynamics among basic, applied, and development are complex and nonlinear.
Linear model of science into society II	Science > reduced uncertainty > better policy	More science may increase uncertainty. Policy progress will not necessarily result from improved understanding, or additional data alone.

Source: Data from Harris, P. and R. Meyer. 2011. Science policy: Beyond budgets and breakthroughs. Discussion paper on enhancing Australian government science policy. Canberra: HC Coombs Policy Forum, Australian National University.

Policy actors also differ in how they respond to issues, as evidenced by the literature on cultural grid-group theory (Hoppe 2011; Thompson et al. 1990; Thompson 1997) and social functional psychology (Tetlock 2002). Not only do different policy actors view the world in different ways, but how they respond to the same stimulus varies. Put simply, people vary in how they respond to issues: some will seek to develop a collective response, others may adopt an individualistic response. Under cultural grid-group theory, four different cultural dispositions are inferred: fatalists, hierarchists, individualists, and egalitarians (Thompson 1997). A consequence of these different cultural orientations is that efforts to characterize a problem in a particular way may be self-defeating. The more effort that is taken to persuade particular stakeholders that a problem should be viewed in a particular way, the less likely it is that they will embrace that problem's characterization. This effect can be illustrated through the example of contrasting conversational styles, whereby "the behaviours of one social actor drive another into increasingly exaggerated expressions of incongruent behaviour in a mutually aggravating spiral" (Tannen 2005, p. 31). Relatedly, social functional psychology identifies five styles (Table 3.5) that people may draw on in responding to particular information (Alexander et al. 2012; Tetlock 2002). These styles illustrate that stakeholders think differently and may not respond to particular problem definitions in the way others would like.

TABLE 3.5

Social Functionalist Decision-Making Styles

Decision-Making Style	Explanation
Intuitive scientists	Driven by epistemic goals and the need to discover causal relationships in the pursuit of truth
Intuitive economists	Driven by goals of maximizing the benefits of resource use for themselves and/or the community, and hold a utilitarian ethic, where rational human decision-making is conceived as the result of comparing costs and benefits
Intuitive politicians	Attempt to cope with accountability demands from key constituencies in their lives, and need to establish, or preserve, a desired social identity and possess a reasonably reliable mental compass for navigating the self through role-rule structures
Intuitive prosecutors	Seek to enforce social norms, by directing accountability demands on those tempted to derive the benefits of collective interdependence without contributing their fair share or without respecting the role-rule regime
Intuitive theologians	Try to protect sacred values from secular encroachments, and have a need to believe that the prevailing accountability and social control regime is not arbitrary, but rather flows naturally from an authority that transcends accidents of history or whims of dominant groups

Source: Data from Alexander, K. et al. 2012. *Journal of Environmental Planning and Management* 55: 409–433; Tetlock, P., 2002. *Psychological Review* 109: 451–471.

3.4 Understanding Operating Environments and Knowledge–Governance Interfaces

So far in this chapter, we have highlighted the complexity of science and decision-making and argued that it is desirable to conceptualize the relationship between science and decision-making as dynamic and interactive. We now turn to exploring how these dynamic interactions play out in particular circumstances. We begin with the obvious point that coastal and marine environments are subject to multiple threats (Agardy and Alder 2005; Beeton et al. 2006; Commissioner for Environmental Sustainability 2008). These threats include, but are not limited to, flooding (McFadden et al. 2009), sea-level rise (Abel et al. 2011), estuary management (Hoare 2002), overharvesting of fish stocks and fisheries management (Hill et al. 2010; Ebbin 2011), and marine conservation and protected area planning (Osmond et al. 2010; Ritchie and Ellis 2010; Gray and Campbell 2008).

For Jentoft and Chuenpadgee (2009, p. 553), fisheries and coastal issues can be characterized as *wicked* which they argue affects their governability and leads them to conclude that there are limitations to how rational and effective fisheries and coastal governance can be. In characterizing fisheries and coastal issues in this way, they draw on Rittel and Webber's (1973) concept of wicked problems, which provides a powerful critique of rational, technocratic approaches to policy research and analysis. For Rittel and Webber (1973, pp. 161–166), ten properties distinguish wicked problems from *tame* problems (Box 3.1).

Rittel and Webber claim that "the search for scientific bases for confronting problems of social policy is bound to fail, because of the nature of these problems" (Rittel and Webber 1973, p. 155) and argue that

> The kinds of problems that planners deal with—societal problems—are inherently different from the problems that scientists and perhaps some classes of engineers deal with. Planning problems are inherently wicked. (Rittel and Webber 1973, p. 160)

and

> As distinguished from problems in the natural sciences, which are definable and separable and may have solutions that are findable, the problems of governmental planning—and especially those of social or policy planning—are ill-defined; and they rely on elusive political judgement for resolution. (Rittel and Webber 1973, p. 160)

Within this context, Stocker et al. (2012, p. 44) suggest that tackling coastal and marine oriented wicked problems requires

BOX 3.1   THE TEN PROPERTIES OF WICKED PROBLEMS

1. There is no definitive formulation of a wicked problem.
2. Wicked problems have no stopping rule.
3. Solutions to wicked problems are not true-or-false, but good-or-bad.
4. There is no immediate and no ultimate test of a solution to a wicked problem.
5. Every solution to a wicked problem is a "one-shot operation": because there is no opportunity to learn by trial and error; every attempt counts significantly.
6. Wicked problems do not have an enumerable (or exhaustively describable) set of potential solutions, nor is there a well-described set of permissible operations that may be incorporated into the plan.
7. Every wicked problem is essentially unique.
8. Every wicked problem can be considered to be a symptom of another problem.
9. The existence of a discrepancy representing a wicked problem can be explained in numerous ways. The choice of explanation determines the nature of the problem's resolution.
10. The planner has no right to be wrong.

Source: Data from Rittel, H. and M. Webber. 1973. *Policy Sciences* 4: 155–169.

- Accommodating multiple alternative perspectives rather than pre-scribing single solutions
- Functioning through group interaction and iteration rather than back office calculations
- Generating ownership of the problem formulation through stake-holder participation and transparency
- Facilitating a graphical (visual) representation of the problem space for the systematic group exploration of a solution space
- Focusing on the relationships between discrete alternatives rather than continuous variables
- Concentrating on possibility rather than probability

The significance of Rittel and Webber's contribution is demonstrated by Head's (2008, p. 101) assessment that they provided "the most challenging and wide-ranging critique of orthodox planning rationality ... evident at the time." However, Rittel and Webber's contrasting of wicked and tame problems means only two types of coastal and ocean problems occur. This provides a constrained view of the kinds of issues that exist. Head's (2008, p. 103) disentangling of wicked in terms of low, moderate, and high levels of complexity, uncertainty, and divergence provides a finer-grained reading of problems and offers greater insight into what kinds of interventions might be useful in particular circumstances.

The literature on problem structuring (Hisschemöller and Hoppe 1996; Hoppe 2011) is similarly useful, as it proposes four types of policy problem: structured, unstructured, moderately structured (goals), and moderately structured (means). Structured problems are, like Rittel and Webber's tame problems, those where "unanimity or near-consensus on the normative issues at stake" exists as well as considerable certainty about "the validity and applicability of claims to relevant knowledge" (Hoppe 2011, p. 72). Unstructured problems are like wicked problems. Moderately structured problems (ends) are defined as those where there may be "a great deal of agreement on the norms, principles, ends, and goals of defining a desirable future state, but simultaneously considerable levels of uncertainty about the reliability of knowledge claims about how to bring it about" (Hoppe 2011, p. 74). Finally, moderately structured problems (means) are characterized by high levels of agreement on the relevant and required knowledge, but ongoing dissent over the normative claims at stake (Hoppe 2011).

Problem structuring highlights that not all problem situations are the same, which means the types of interventions may need to vary, that is, there is no one-size-fits-all solution. Further, Hisschemöller et al. (2001, p. 465, cited in Wesselink and Hoppe 2011, p. 404) argue that "science use depends on the structure of the problem as constructed by dominant policy actors at a given moment." An insightful application of problem structuring in an ocean and coastal context is provided by Turnhout et al. (2008) who use it to explore science–policy relations in the Wadden Sea. Their analysis explores the different roles scientists can occupy, and shows how these may shift from accommodation to advocacy in different situations. Adding further complexity to the range of problem types, Levin et al. (2010, p. 3) identify "super wicked problems" as having four additional features that distinguish them from wicked problems—namely, urgency (time is running out), the lack of an adequate central authority to address the issue, that those who cause the problem also seek to create a solution, and the occurrence of hyperbolic discounting that pushes responses into the future when immediate actions are required to initiate longer-term policy solutions. Collectively, these approaches provide more nuanced insights into the nature of ocean and coastal problems and what might be done about them than does Rittel and Webber's simple framing of tame and wicked problems.

These approaches, however, assume that it is possible to objectively differentiate between different types of issues. Drawing on discourse theory (Hajer 1995; Bacchi 2009), we suggest that it is important to be mindful of how issues are represented. To this end it is possible to suggest that problem structuring can be considered in three ways: (1) there are different types of issues (the four problem structures); (2) different policy actors define issues differently, for example, a well-structured problem to a scientist may appear as unstructured to a member of parliament; and (3) it is possible for issues to move from one category to another, for example, from unstructured to well-structured or vice versa.

It is clear that great care, and considerable reflexivity, must be exercised in characterizing particular issues. Leith et al. (2014) explore this view in developing a diagnostic model for linking science, society, and policy for sustainability. Drawing on insights from a series of case studies, they argue that there are "a variety of ways in which interactions between science and decision-making are consistently structured by recurring characteristics" (Leith et al. 2014, p. 168). This leads them to focus on the analysis of operating environments as a means for informing the characterization of problems and the kinds of responses that might be proposed (Figure 3.1). They argue that

> Issues, stakes and boundary spanning can be considered to constitute the "operating environment" that affects whether and how sciences can have an impact on decision-making. An operating environment is an emergent property of elements as diverse as an advertising campaign, a well-networked policy entrepreneur, and a storm event that threatens coastal homes. It is neither deterministic, nor fully tractable to an analyst. Any analysis of an operating environment will be partial. Among stakeholders there will be diverse interpretations of operating environments, and much understanding will be tacit, vaguely articulated, or contested. (Leith et al. 2014, p. 169)

Through detailed and participatory explorations of specific operating environments, it is possible to characterize issues in ways that open up multiple avenues for boundary spanning (Table 3.6) and avoid proposing silver bullet solutions. Particular "operating environments" manifest in specific science–policy (Van Enst et al. 2014), science–policy–practice (Weichselgartner and Kasperson 2010), knowledge–governance (Clarke et al. 2013; Bremer and Glavovic 2013), and knowledge arrangement (Janssen et al. 2014; 2015) interfaces. Importantly, for Leith et al. (2014, p. 168), the effectiveness of such science–policy programs is not achieved through a single means but "through a combination of 'design elements'" which can be "mutually reinforcing."

A critical element in improving the impact of science on decision-making is that proposed information exchange activities are appropriate to the specific circumstances in which they are to be implemented: cookie cutter approaches to knowledge transfer and exchange should be avoided. Within

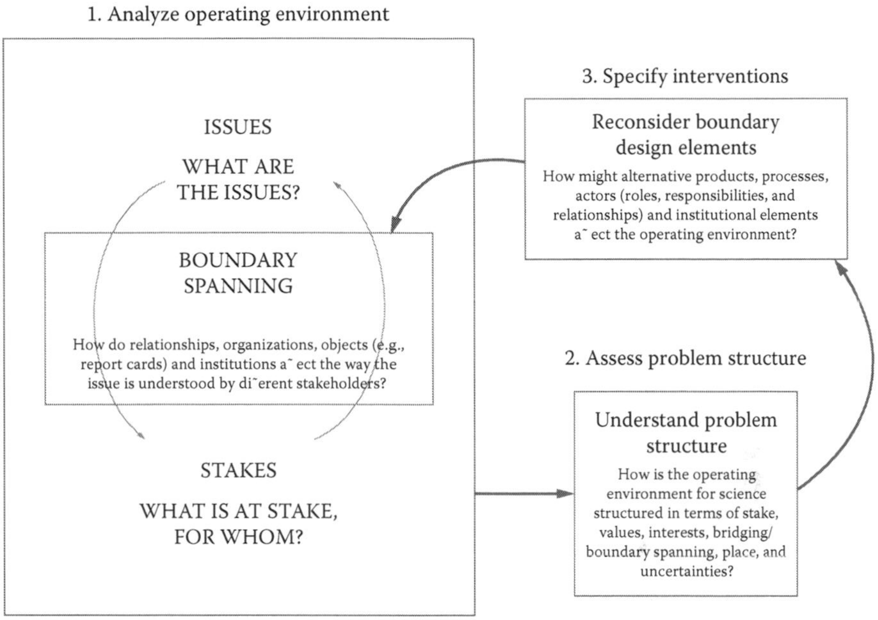

1. Analyze operating environment

ISSUES

WHAT ARE
THE ISSUES?

BOUNDARY
SPANNING

How do relationships, organizations, objects (e.g.,
report cards) and institutions a¯ect the way the
issue is understood by di¯erent stakeholders?

STAKES

WHAT IS AT STAKE,
FOR WHOM?

3. Specify interventions

Reconsider boundary
design elements

How might alternative products, processes,
actors (roles, responsibilities, and
relationships) and institutional elements
a¯ect the operating environment?

2. Assess problem structure

Understand problem
structure

How is the operating
environment for science
structured in terms of stake,
values, interests, bridging/
boundary spanning, place, and
uncertainties?

FIGURE 3.1
Schematic process for diagnosing and intervening in the operating environment for sciences.
(From Leith et al. 2014. *Environmental Science and Policy*, 39, 170.)

TABLE 3.6

Key Design Elements of Boundary Spanning

Element and Focus	Explanation
Science communication (product focus)	Development of boundary objects.
Informal linkages (relationship focus)	Where problems are poorly structured or unstructured, building informal linkages among key stakeholder groups can begin to create mutual understanding of stakes and values across groups, thereby allowing clearer definition of issues.
Brokering/intermediary (actor focus)	The building of capacity within organizations that manage problems in which science and community values are both important.
Temporary organization (structure/network focus—e.g., reference groups)	Temporary organizations or projects used to address complex issues and/or short-term imperatives.
Boundary organization (organization focus)	Long-lived, persistent wicked problems, managing complex conditions, often within multiple organizations.

Source: Leith, P. et al. 2014. *Environmental Science and Policy* 39: 162–171.

this context, O'Toole et al. (2013, pp. 208–209) highlight the importance of participatory stakeholder engagement and introduce the concept of *participatory logic*, which they summarize as follows:

- Institutionalize the processes that derive from stakeholder coproduction and comanagement (not just ways of bringing the public into technical decisions but significant deliberation over aspects of the design, which requires some convergence in terms of allocation of value).

- Enhance the capacity to make meaningful decisions about issues of importance in an ongoing way thus allowing for the update of science input as well as changing social knowledge.

- Include all stakeholders in the process, which is an issue of justice and equity.

- Ensure central policies enable participation by stakeholders at the local level, the outcomes of which are fed back to central policies and programs.

- Allow and develop pathways for the uptake of diverse knowledge systems, including building the capacity of all stakeholders to comprehend other forms of knowledge.

Informed by Stojanovic et al. (2009), O'Toole et al. (2013, p. 209) argue that institutionalizing interactive and participatory knowledge exchange is "not a matter of mere dissemination of knowledge, but rather the development of platforms where the range of stakeholder knowledge can be deliberated," where such platforms can be considered as an intersecting matrix of specific participatory mechanisms available within a particular situation.

3.5 Conclusion

While the role of science in coastal and ocean decision-making is clearly complex, it is possible to develop science policy programs that can enhance the contribution of science to decision-making, and that this can occur in ways that are respectful of other legitimate forms of knowledge. However, there is clearly significant room for improvement in the ways in which these mechanisms can be designed and implemented, as recognized by Cvitanovic et al. (2015), and Van Enst et al. (2014). For Cvitanovic et al. (2015, p. 32), such improvements could be gained from the generation of "quantitative empirical evidence" in order "to understand how the relationship between science and decision-making varies amongst locations and under different conditions." They also suggest that there is merit in developing methods to evaluate knowledge exchange activities and embedding them in research programs; determining the specific expertise and skills required by individuals to successfully engage in knowledge exchange; and understanding the potential of how new and evolving social media can be used to enhance knowledge exchange. We consider that while such suggestions may be useful in some circumstances (i.e., those where there is limited disagreement), they will be less useful in more complex and contested circumstances. By contrast, Van Enst et al. (2014, p. 20) identify issues for further research that might provide broader insights into marine and coastal knowledge dynamics—namely,

- What are the processes and strategies through which science–policy interactions take place and to what extent can they be influenced?
- How are science–policy interfaces enabled and constrained by social, economic, and political dynamics, and what other contextual factors influence the performance of science–policy interfaces (SPIs)?
- In what manner can design principles be formulated for science–policy interfaces in addressing a diverse set of problems in specific contexts, and in particular, to what extent can science–policy interfaces be complementary to each other?
- To what extent does an increased level of credibility, legitimacy, and salience in knowledge, established through the use of SPIs, lead to enriched decision-making on environmental issues?

We commend these suggestions as potentially fruitful avenues for investigation, particularly the focus on identifying what design elements might be of most value for science–policy programs in particular circumstances (Leith et al. 2014). We also see merit in learning from "successful" examples of science uptake (Keneley et al. 2013) and in broadening the scope of science–policy scholarship beyond its focus on the physical (natural) sciences, to consider the ways in which social science knowledge informs environmental governance in general, and ocean and coastal governance in particular. A stronger focus on understanding the contribution of social science knowledge would seem to be particularly useful given that effective ICZM is as much, if not more, a social challenge as it is a technical one.

More broadly, we consider that ethnographic and action research into the cultural and social dynamics operating within particular settings is likely to be insightful as it takes seriously the discursive dimensions of coastal knowledge exchange (Nursey-Bray et al. 2014). Such an approach shifts the focus from *knowledge exchange* to the *ecology of knowledge*, which broadens the range of "knowledges" considered and the range of actors recognized as having a stake in the debates, and focuses attention on the dynamics involved. It also highlights that knowledge dynamics always occur in particular circumstances, such that what works in one place may not work elsewhere. Consequently, we are mindful that such issues are always political. Perhaps the biggest gains in coastal and ocean management may be found in more democratic and transparent governance. Given this, it is pertinent to conclude with more wisdom from Latour (1998, p. 209):

> To the old slogan of science—the more disconnected a discipline from society, the better—now resonates a more realistic call for action: The more connected a scientific discipline, the better.

Acknowledgments

The research on which this chapter is based was undertaken as part of the Coastal Collaboration Cluster, which was supported by the Commonwealth Scientific and Industrial Research Organisation (CSIRO) Flagship Collaboration Fund. We thank Professor Marcus Haward and Dr. Peat Leith for their contributions in developing the model articulated in Leith et al. 2014.

References

Abel, N., R. Goddard, B. Harman, et al. 2011. Sea level rise, coastal development and planned retreat: Analytical framework, governance principles, and an Australian case study. *Environmental Science and Policy* 14 (3): 279–288.

Agardy, T. and J. Alder. 2005. Coastal systems. In *Ecosystems and Human Well-Being: Current State and Trends*, Volume 1, edited by R. Hassan, R. Scholes, and N. Ash, 513–549. Washington, DC: Island Press.

Alexander, K., A. Ryan, and T. Measham. 2012. Managed retreat of coastal communities: Understanding responses to projected sea level rise. *Journal of Environmental Planning and Management* 55 (4): 409–433.

Althaus, C., P. Bridgman, and G. Davis. 2012. *The Australian Policy Handbook*. 5th ed. Sydney: Allen and Unwin.

Bacchi, C. 2009. *Analysing Policy: What's the Problem Represented to Be?* Sydney: Pearson.

Beeton, R., K. Buckley, G. Jones, et al. 2006. Australia State of the Environment 2006, Independent Report to the Australian Government Minister for the Environment and Heritage. Canberra: Department of Environment and Heritage.

Berkes, F. and C. Folke. 1998. *Linking Social and Ecological Systems: Management Practices and Social Mechanisms for Building Resilience*. Cambridge: Cambridge University Press.

Bremer, S. and B. Glavovic. 2013. Mobilizing knowledge for coastal governance: Re-framing the science–policy interface for integrated coastal management. *Coastal Management* 41: 39–56.

Cahoon, L. and C. Dumas. 2011. Preface to "Making the connection: Translating science into effective coastal policy". *Coastal Management* 39 (3): 223–224.

Campbell, A. 2006. *The Australian Natural Resource Management Knowledge System.* Canberra: Land and Water Australia.

Cash, D., J. Borck, and A. Patt. 2006. Countering the loading dock approach to linking science and decision making: Comparative analysis of El Niño/Southern Oscillation (ENSO) forecasting systems. *Science, Technology and Human Values* 31 (4): 465–494.

Cash, D. and J. Buizer. 2005. Knowledge-action systems for seasonal to interannual climate forecasting: Summary of a workshop. Report to the Roundtable on Science and Technology for Sustainability Policy and Global Affairs. Washington, DC: National Academies.

Cash, D., W. Clark, F. Alcock, et al. 2003. Knowledge systems for sustainable development. *Proceedings of the National Academy of Sciences of the United States of America* 100 (14): 8086–8091.

Chalmers, A. 2010. *What Is This Thing Called Science?* Brisbane: University of Queensland.

Cicin-Sain, B. and R. Knecht. 1998. *Integrated Coastal and Ocean Management Concepts and Practices.* Washington, DC: Island Press.

Clarke, B., L. Stocker, B. Coffey, et al. 2013. Enhancing the knowledge–governance interface: Coasts, climate and collaboration. *Ocean & Coastal Management* 86: 88–99.

Coffey, B. and G. Marston. 2013. How neoliberalism and ecological modernization shaped environmental policy in Australia. *Journal of Environmental Policy and Planning* 15 (2): 179–199.

Coffey, B. and K. O'Toole. 2012. Understanding coastal knowledge dynamics: The potential of knowledge systems. *Conservation and Society* 10 (4): 318–329.

Colebatch, H. 2002. *Policy.* 2nd ed. Maidenhead: Open University Press.

Commissioner for Environmental Sustainability. 2008. *State of the Environment Victoria 2008 Summary.* Melbourne: Commissioner for Environmental Sustainability.

Cummins, V. and J. McKenna. 2010. The potential of sustainability science in coastal zone management. *Ocean & Coastal Management* 53: 796–804.

Cvitanovic, C., A. Hobday, L. Van Kerkoff, et al. 2015. Improving knowledge exchange amongst scientists and decision-makers to facilitate the adaptive governance of marine resources: A review of knowledge and research needs. *Ocean & Coastal Management* 112: 25–35.

Ebbin, S. 2011. The problem with problem definition: Mapping the discursive terrain of conservation in two pacific salmon management regimes. *Society and Natural Resources* 24 (2): 148–164.

Erickson, P. and E. Woodley. 2005. Using multiple knowledge systems: Benefits and challenges. In *Ecosystems and Human Wellbeing: Multiscale Assessments,* edited by D. Capistano, C. Samper, M. Lee, and C. Raudsepp-Hearne, 85–117. Washington, DC: Island Press.

Feyerabend, P. 1975. *Against Method: Outline of an Anarchistic Theory of Knowledge.* London: New Left Books.

Flyvberg, B. 2001. *Making Social Science Matter.* Cambridge: Cambridge University Press.

Funtowicz, S. and J. Ravetz. 1993. Science for the post-normal age. *Futures* 25 (7): 739–755.
Gray, N. and L. Campbell. 2008. Science, policy advocacy and marine protected areas. *Conservation Biology* 23 (2): 460–468.
Hajer, M. 1995. *The Politics of Environmental Discourse*. Oxford: Oxford University Press.
Harris, P. and R. Meyer. 2011. Science policy: Beyond budgets and breakthroughs. Discussion paper on enhancing Australian government science policy. Canberra: HC Coombs Policy Forum, Australian National University.
Harvey, D. 2005. *A Brief History of Neoliberalism*. Oxford: Oxford University Press.
Harvey, N. and B. Caton. 2003. *Coastal Management in Australia*. Melbourne: Cambridge University Press.
Head, B. 2008. Wicked problems in public policy. *Public Policy* 3 (2): 101–118.
Heynen, N. and P. Robbins. 2005. The neoliberalization of nature: Governance, privatization, enclosure and valuation. *Capitalism Nature Socialism* 16 (1): 5–8.
Hill, N., K. Michael, A. Frazer, et al. 2010. The utility and risk of local ecological knowledge in developing stakeholder driven fisheries management: The Foveaux Strait Dredge Oyster Fishery, New Zealand. *Ocean & Coastal Management* 53: 659–668.
Hisschemöller, M. and R. Hoppe. 1996. Coping with intractable controversies: The case for problems structuring in policy design and analysis. *Knowledge for Policy* 8 (4): 40–60.
Hisschemöller, M., R. Hoppe, P. Groenewegen, et al. 2001. Knowledge, power and political choice: A problem and restructuring perspective on real life experiments in extended peer review. In *Knowledge, Power, and Participation in Environmental Policy Analysis*, Policy Studies Review Annual 12, edited by M. Hisschemöller, R. Hoppe, W. Dunn, and J. Ravetz, 437–470. Piscataway, NJ: Transaction Publishers.
Hoare, A. 2002. Natural harmonies but divided loyalties: The evolution of estuary management as exemplified by the Severn Estuary. *Applied Geography* 22 (1): 1–25.
Hogwood, B. and L. Gunn. 1984. *Policy Analysis for the Real World*. Oxford: Oxford University Press.
Hoppe, R. 2011. *The Governance of Problems: Puzzling, Powering and Participation*. Bristol: Policy Press.
Houde, N. 2007. The six faces of traditional ecological knowledge: Challenges and opportunities for Canadian co-management arrangements. *Ecology and Society* 12 (2): 1–17.
Howells, J. and J. Roberts. 2000. From innovation systems to knowledge systems. *Prometheus* 18 (1): 17–31.
Howlett, M. and M. Ramesh. 2003. *Studying Public Policy: Policy Cycles and Policy Subsystems*. Oxford: Oxford University Press.
Hulme, M. 2001. Meet the humanities. *Nature Climate Change* 1 (July): 177–179.
Janssen, S. K. H., A. P. J. Mol, J. P. M. van Tatenhove, et al. 2014. The role of knowledge in greening flood protection: Lessons from the Dutch case study future Afsluitdijk. *Ocean & Coastal Management* 95: 219–232.
Janssen, S. K. H., J. P. M. van Tatenhove, H. S. Otter, et al. 2015. Greening flood protection—An interactive knowledge arrangement perspective. *Journal of Environmental Policy and Planning* 17 (3): 309–331.
Jentoft, S. and R. Chuenpagdee. 2009. Fisheries and coastal governance as a wicked problem. *Marine Policy* 33: 553–560.

Kay, R. and J. Alder. 2005. *Coastal Planning and Management*. London: Taylor and Francis.

Kelsey, E. 2003. Integrating multiple knowledge systems into environmental decision-making: Two case studies of participatory biodiversity initiatives in Canada and the implications for conceptions of education and public involvement. *Environmental Values* 12: 381–396.

Keneley, M., K. O'Toole, B. Coffey, et al. 2013. Stakeholder participation in estuary entrance management: The development of Victoria's Estuary Entrance Management Support System (EEMSS). *Australasian Journal of Environmental Management* 20 (1): 49–62.

King, L. 2004. Competing knowledge systems in the management of fish and forests in the Pacific Northwest. *International Environmental Agreements: Politics, Law and Economics* 4: 161–177.

Kooiman, J. 2008. Exploring the concept of governability. *Journal of Comparative Policy Analysis* 10 (2): 171–190.

Latour, B. 1998. From the world of science to the world of research. *Science* 280 (5361): 208–209.

Lee, D. and E. Van den Steen. 2010. Managing know-how. *Management Science* 56 (2): 270–285.

Leith, P., K. O'Toole, M. Haward, et al. 2014. Analysis of operating environments: A diagnostic model for linking science, society and policy for sustainability. *Environmental Science and Policy* 39 (May): 162–171.

Levin, K., B. Cashore, S. Bernstein, et al. 2010. Playing it forward: Path dependency, progressive incrementalism, and the "Super Wicked" problem of global climate change. Unpublished manuscript, 3 June 2010. http://citeseerx.ist.psu.edu/viewdoc/download?doi=10.1.1.464.5287&rep=rep1&type=pdf.

Mauro, F. and P. Hardison. 2000. Traditional knowledge of indigenous and local communities: International debate and policy initiatives. *Ecological Applications* 10 (5): 1263–1269.

McCarthy, J. and S. Prudham. 2004. Neoliberal nature and the nature of neoliberalism. *Geoforum* 35: 275–283.

McFadden, L. 2007. Governing coastal spaces: The case of disappearing science in integrated coastal zone management. *Coastal Management* 35 (4): 429–443.

McFadden, L., E. Penning-Rowsell, and S. Tapsell. 2009. Strategic coastal flood-risk management in practice: Actors' perspectives on the integration in flood risk management in London and the Thames Estuary. *Ocean & Coastal Management* 52 (12): 636–645.

Nursey-Bray, M., J. Vince, M. Scott, et al. 2014. Science into policy? Discourse, coastal management and knowledge. *Environmental Science and Policy* 38: 107–118.

Ojha, H., R. Chhetri, N. Timsina, et al. 2008. Knowledge systems and deliberative interface in natural resource governance: An overview. In *Knowledge Systems and Natural Resources: Management, Policy and Institutions in Nepal*, edited by H. Ojha, N. Timsina, R. Chhetri, and K. Paudel, 1–22. New Delhi: Cambridge University Press India; Ottawa: International Development Research Centre.

Osmond, M., S. Airame, M. Caldwell, et al. 2010. Lessons for marine conservation planning: A comparison of three marine protected areas planning processes. *Ocean & Coastal Management* 53: 41–51.

O'Toole K., M. Keneley, and B. Coffey. 2013. The participatory logic of coastal management under the project state: Insights from the case of the Estuary Entrance Management Support System (EEMSS) in Victoria, Australia. *Environmental Science and Policy* 27 (March): 206–214.

Prudham, S. 2004. Poisoning the well: Neoliberalism and the contamination of municipal water in Walkteron, Ontario. *Geoforum* 35: 343–359.

Raymond, C., I. Fazey, M. Reed, et al. 2010. Integrating local and scientific knowledge for environmental management. *Journal of Environmental Management* 91: 1766–1777.

Rayner, S. 2004. The novelty trap: Why does institutional learning about new technologies seem so difficult? *Industry and Higher Education* 18 (6): 349–355.

Ritchie, H. and G. Ellis. 2010. A system that works for the sea? Exploring stakeholder engagement in marine spatial planning. *Journal of Environmental Planning and Management* 53 (6): 701–723.

Rittel, H. and M. Webber. 1973. Dilemmas in a general theory of planning. *Policy Sciences* 4: 155–169.

Roling, N. 1985. Extension science: Increasingly preoccupied with knowledge systems. *Sociologia Ruralis* 25 (3/4): 269–290.

Roling, N. 1992. The emergence of knowledge systems thinking: A changing perception of relationships amongst innovation, knowledge process and configuration. *Knowledge and Policy* 5 (1): 42–56.

Roling, N. and J. Jiggins. 1998. The ecological knowledge system. In *Social Learning for Sustainable Agriculture*, edited by N. Roling and M. Wagemakers, 283–311. Cambridge: Cambridge University Press.

Sarewitz, D. and R. Pielke. 2007. The neglected heart of science policy: Reconciling supply and demand for science. *Environmental Science and Policy* 10: 5–16.

Smith, H. 2002. The role of social science in capacity building in ocean and coastal management. *Ocean & Coastal Management* 45: 575–582.

Sorenson, J. 1997. National and international efforts at integrated coastal zone management: Definitions, achievements, and lessons. *Coastal Management* 25: 3–41.

Stefik, M. 1995. *Introduction to Knowledge Systems*. San Francisco: Morgan Kaufman.

Stepanova, O. 2015. Conflict resolution in coastal resource management: Comparative analysis of case studies from four European countries. *Coastal Management* 103: 109–122.

Stocker, L., R. Kenchington, D. Kennedy, et al. 2012. Introduction to Australian coasts and human influences. In *Sustainable Coastal Management and Climate Adaptation*, edited by R. Kenchington, L. Stocker, and D. Wood, 1–27. Melbourne: CSIRO.

Stojanovic, T., I. Ball, R. Ballinger, et al. 2009. The role of research networks for science–policy collaboration in coastal areas. *Marine Policy* 33: 901–911.

Tannen, D. 2005. *Conversational Style: Analyzing Talk among Friends*. Oxford: Oxford University Press.

Tetlock, P. 2002. Social functionalist frameworks for judgement and choice: Intuitive politicians, theologians, and prosecutors. *Psychological Review* 109 (3): 451–471.

Thompson, M. 1997. Cultural theory and integrated assessment. *Environmental Modelling and Assessment* 2: 139–150.

Thompson, M., R. Ellis, and A. Wildavsky. 1990. *Cultural Theory*. Boulder: Westview.

Thompson, R. 2007. Cultural models and shoreline social conflict. *Coastal Management* 35 (2): 211–237.

Tsoukas, H. and N. Mylonopolous. 2004. *Organizations as Knowledge Systems: Knowledge, Learning and Dynamic Capabilities.* Houndmills: Palgrave MacMillan.

Turnhout, E., M. Hisschemöller, and H. Eijsackers. 2008. Science in Wadden Sea policy: From accommodation to advocacy. *Environmental Science and Policy* 11: 227–239.

Van Enst, W., P. Driessen, and H. Runhaar. 2014. Towards productive social-policy interfaces: A research agenda. *Journal of Environmental Assessment Policy and Management* 16 (1): doi: 10.1142/S1464333214500070.

Van Kerkoff, L. and L. Lebel. 2006. Linking knowledge with action for sustainable development. *Annual Review of Environment and Resources* 32: 1–33.

Van Kerkoff, L. and N. Szlezak. 2006. Linking local knowledge with global action: Examining the global fund to fight AIDS, tuberculosis and malaria through a knowledge system lens. *Bulletin of the World Health Organization* 84 (8): 629–635.

Verran, H. 1998. Re-imagining land ownership in Australia. *Postcolonial Studies* 1 (2): 237–254.

Ward, V., A. House, and S. Hamer. 2009. Developing a framework for transferring knowledge into action: A thematic analysis of the literature. *Journal of Health Services Research and Policy* 14 (3): 156–164.

Weichselgartner, J. and R. Kasperson. 2010. Barriers in the science-policy-practice interface: Toward a knowledge-action system in global environmental change research. *Global Environmental Change* 20: 266–277.

Weiss, C. 1979. The many meanings of research utilization. *Public Administration Review* 39 (5): 426–431.

Weiss, C. 1980. Knowledge creep and decision accretion. *Knowledge: Creation, Diffusion, Utilization* 1 (3): 381–404.

Wesselink, A. and R. Hoppe. 2011. If post-normal science is the solution, what is the problem? The politics of activist environmental science. *Science, Technology and Human Values* 36 (3): 389–412.

4

Science Information and Global Ocean Governance

Jake Rice

CONTENTS

4.1 Introduction

More than 60% of the world's ocean is beyond national jurisdiction. Governance of that portion of the ocean is necessarily global, as introduced in Chapter 2 and developed in more detail in this book. Moreover, for many reasons—some primarily political and some simply pragmatic—choices by governments for the policies they set and the management options that they choose are strongly influenced by governance decisions made at the global scale (Ridgeway 2014). Hence, the interactions of science processes and scientific information with policy-making at the global scale have implications not just at the global scale but also at regional, national, and subnational scales. In this chapter, I will explore how those interactions play out in the real world, primarily at the global scale, drawing from the scientific literature, reports of intergovernmental agencies, and personal experience.

Although a significant portion of my career has been spent as a research scientist in government and academia, for the past decade a primary duty has been to serve as science advisor to either Canadian delegations attending meetings of United Nations (UN) working groups and intergovernmental organizations (IGOs) or occasionally as a resource expert for the IGOs themselves. From that perspective, I have been able to experience how the science–policy interface actually operates in global governance. In this chapter, I will draw on literature and IGO documents when possible, as well as experiential knowledge, in my observations and conclusions.

4.2 Characteristics of Global Ocean Governance

At the global level, countries explicitly acknowledged the need for a sound science foundation for ocean policy in the World Summit on Sustainable Development (WSSD) Plan of Action, when it called on the UN to "establish by 2004 a regular process under the United Nations for global reporting and assessment of the state of the marine environment, including socio-economic aspects, both current and foreseeable, building on existing regional assessments" (UN 2002, para 36b). However, many previous agreements had unquestioned roots in sound science, as summarized in Garcia et al. (2014a). This does not mean that all available scientific information has been translated into appropriate ocean policy or that all ocean policy has firm scientific foundations. To understand where science–policy coherence is stronger and where it is weaker, it is necessary to consider the special characteristics of global ocean governance from the perspective of how it uses scientific information.

One of the characteristics of ocean governance with important implications for the role of scientific information in the science–policy interface is that governance and management are sectoral. This situation is increasingly criticized as leading to fragmentation of policy-making (Ban et al. 2014; Druell et al. 2012), but it has been and will continue to be the reality, even if mechanisms for greater high-level coordination are found. This means that there are many intergovernmental institutions, each with a mandate for policy and management of a particular sector such as fishing (e.g., the Food and Agriculture Organization of the United Nations [FAO], and regional fishery management organizations and arrangements [RFMO/As]), shipping (International Maritime Organization [IMO]), seabed mining (International Seabed Authority—ISA), dumping (London Convention & Protocol), and pollution (Regional Seas Conventions). Each organization requires science support for its policies and programs, and each has its own body to provide such support (Table 4.1).

There may have been a time when it was reasonable for each organization to get science support from a custom-designed source. Their separate mandates meant that they were developing policies and programs to address different issues. Thus, each organization required a different mix of expertise and could restrict the "relevant data" to the core information needed for its restricted mandate. The ecosystem approach has rendered that view less tenable. In a check of the websites of 14 major IGOs with at least partial

TABLE 4.1

Illustrative List of IGOs That Require "Official" Science Advice in Order to Discharge Some Parts of Their Mandates, Including the Type of Group from Which They Routinely Receive Advice, and the General Operating Procedures of the Advisory Group

Organization	Area of Activity	Type of Group	Operations
FAO	Fisheries	Expert consultation	Invited scientific experts matched to specific meeting terms of reference. May be supported by contracted experts. Topics prioritized by Committee on Fisheries (COFI); experts invited by the Secretariat.
CBD	Biodiversity conservation	Expert workshops and ad hoc technical expert groups	Invited scientific experts matched to specific meeting terms of reference. May be supported by contracted experts. Topics for meetings referred by the biannual conference of parties. Experts nominated by parties and final selection by the Secretariat.
ISA	Seabed mining	Legal and Technical Commission (LTC) and working groups	LTC members elected by member states for fixed terms; members advise on scientific and legal issues referred to them by the Council. May be supported by expert working groups or special panels invited to address specific issues, with experts nominated by states and final selection by the Secretariat
IMO	Shipping and transport	Maritime Environmental Protection Committee	Membership appointed by states that are members of IMO. Responds to requests for advice from the Council.
IOC	Oceanography and ocean science	Mixture of thematic and regional science programs and subcommittees	Work and science products usually coordinated by permanent staff supported by experts and working groups with members nominated by states and selected by IOC. Program of work approved by the Council.
RFMOs	Fisheries	Generally have standing scientific committee or subcommittee; or use scientific body such as ICES	Generally scientific committee made up of members appointed by member states. Requests for advice come from the Policy or Management Council of the RFMOs. May be supported by ad hoc working groups created by Council with memberships appointed by member states.

(Continued)

TABLE 4.1 (CONTINUED)

Illustrative List of IGOs That Require "Official" Science Advice in Order to Discharge Some Parts of Their Mandates, Including the Type of Group from Which They Routinely Receive Advice, and the General Operating Procedures of the Advisory Group

Organization	Area of Activity	Type of Group	Operations
Regional Seas organizations	Marine environmental quality	Generally have standing scientific committee or subcommittee; or use standing scientific body such as ICES	Generally scientific committee made up of members appointed by member states. Requests for advice come from the Policy or Management Council of the Regional Seas organizations. May be supported by ad hoc working groups created by the Council with memberships appointed by member states.
UN General Assembly	Regular Process for World Ocean Assessments	Group of experts and teams of authors	Group of experts nominated by states and selected by UN groups of countries. Members of teams of authors nominated by states, selected by the group of experts with approval by the UN Ad Hoc Working Group of the Whole Bureau.
GESAMP	All marine science	Scientific experts in many disciplines	Steering body of 16 experts who act in an independent and individual capacity. Studies and assessments are usually carried out by dedicated working groups, with members nominated by IGOs or states.

Note: For RFMOs and Regional Seas organizations, the corresponding row entries are typical practices. In all cases, there are exceptions to these general procedures, but the generic scientific culture is described.

ocean mandates, every one of them stated they had adopted the ecosystem approach as a part of their overarching policy framework. The details of how the ecosystem approach was defined differed among organizations, but in all cases it included both taking the state of environmental forces into account and accepting responsibility for the larger footprint of their sector on the ocean (Rice et al. 2014b). Now there is substantial overlap in both the expertise and information needed by agencies with differing mandates. Inconsistencies in how the same information is interpreted by separate science advisory agencies to different organizations undermines the credibility of the advisory processes of both agencies, as the scientific information may be viewed as politicized and no longer objective from either perspective.

A direct consequence of IGOs adopting an ecosystem approach is that there is now a necessary overlap in the science support needed by the various organizations. One type of overlap arises because the basic physical oceanographic drivers of each sector are likely to be similar—for example, expressions of temperature, salinity, stratification, and currents—even if the details relevant to the individual sectors are not the same. Another more subtle but important type of overlap arises from the footprint aspects of the ecosystem approach, as the footprint of one sector—habitat impacts, pollution, changes in biotic community composition—may be an important driver of the inputs to some other sectors.

Superficially, the move to an ecosystem approach might be taken as positive for the science–policy interface, since it presents obvious opportunities for coherence of science input to sectoral management. However, the history of separate science advisory processes for the various sectors casts a long shadow. Separate scientific subdisciplines have arisen to support different sectors. Different sectors often prefer different data streams, use different technical terms, give greatest credence to different analytical methods, and manage risk from different perspectives. These differences have been well documented for areas as similar as assessing trends in exploited fish stocks and in species potentially at risk of extinction (Mace et al. 2014), as have some of the implications that the differences have for policy-making.

In addition to the methodological and technical language barriers to greater commonality of science support for the ocean sectors, there is another legacy of the long history of each governance body having its own science advisory process—the legacy of trust. This topic will be explored in greater depth as part of the illustrative case below. However, its influence on the science–policy interface for global ocean governance cannot be understated. At a fisheries-biodiversity workshop sponsored by the Convention on Biological Diversity (CBD) in cooperation with FAO, a list of possible ways to improve the consideration of biodiversity in fisheries assessments was developed. Some participants whose entire careers had been aligned with the fisheries sector opposed including in the list of options the establishment of a practice of inviting CBD to provide marine biodiversity experts from its network to participate in fish stock assessment meetings of RFMOs (CBD 2012). Similar perspectives have been expressed at biodiversity workshop steering committees on which I have sat, where committee members may seek academic fisheries expertise but not if it comes from experts involved in the actual management of fisheries. The rationale is the same in both cases: there is an acknowledgment of the need for broad input but a distrust of the reliability of experts who are closely aligned to the "other" sector.

A second feature of global ocean governance that has important implications for the use of scientific information is that the more global an organization is, the more likely it operates as a consensus body. For regional sectoral groups such as RFMOs, there may be formal objection procedures (e.g., Article XIV, para 2, of the Northwest Atlantic Fisheries Organization (NAFO) Convention, NAFO 1978), and even for global bodies, countries that cannot subscribe to a near-consensus can enter a reservation in meeting decisions (e.g., table 2 of the Annex to CBD Conference of Parties (COP) Decision XII/22, CBD 2014). However, such footnotes or objections are usually registering some point of substantial national interest that may modify how a conclusion is applied, but do not change the conclusion fundamentally.

Because global governance bodies generally work by consensus, they will not move faster than the most skeptical states or parties participating in the meeting. This often tries the patience of science advisors who are asked to input scientific and technical information to inform the negotiations. From the perspective of the experts' discipline, the information may appear clear and relevant to the debate, and they are confident the methods used to collect, analyze, and interpret the information were scientifically sound. The implications of the scientific input for the parts of the governance conclusions that should be consistent with sound science should, therefore, be simple and straightforward, yet consensus may still require long debates and substantial compromise. As is perhaps best documented in the debates at the Intergovernmental Panel on Climate Change (IPCC), in the end the scientists may not recognize the outcome as built on their inputs (cf. Stavins, 2014) and certainly do not feel full use was made of the science advice.

Having often been in this situation personally, I can highlight three aspects of the consensus process that contribute to this perceived underutilization of the input of scientific information. Firstly, the science community tends to bound the problem to be addressed in policy with the boundaries of their scientific discipline. Although this is hard to document from the meeting floors where negotiations are occurring, journal articles where science is presented in policy contexts may invoke *externalities* when developing their policy conclusions as if the externalities were actually outside the scope of the problem, and not just outside the scope of the information they use to find a solution. The bounding can be justified in the world of research, but the ocean policy agenda is not neatly partitioned into scientific disciplines, even if management is largely sectoral. The same fisheries policy issue may be about food security and livelihoods to some countries and economic optimality in others (HLPE 2014). Therefore, different perspectives among countries on what factors bound the policy question make finding consensus a slow and often painful process, leaving many science advisors wondering what value is really given to their inputs.

Secondly, consensus is not some final outcome but a standard that has to be reached at many steps on the pathway to a final agreed text. The first consensus is primarily on the soundness of the basic science; it is rarely reached explicitly at the negotiation sessions, but nevertheless is a consideration in play during the negotiations. Dueling experts at a policy session are not unknown, but as the earlier referenced debates about species-at-risk listing for commercially exploited marine species has highlighted, they are unhelpful. Experienced science advisors endeavor to find ways to resolve such issues before formal policy negotiations commence.

The second step in the consensus process is often not acknowledged, despite its importance. Some formulations of decision theory (e.g., Montgomery 2005) highlighted that when errors do occur, they can be either errors of omission or of commission; that is, misses or false alarms. In the world of global ocean policy, these notions of misses and false alarms have close parallels. Misses are policies that fail to effectively address issues where practice has to change if uses are to be sustainable. False alarms are policies that are overly intrusive, restricting human activities beyond the degree needed to achieve sustainability. The costs of misses and false alarms are distributed very differently, with the costs of misses largely borne (in the short term) by the environment, and the cost of false alarms largely borne by the economic and social interests (Connor and Cooper 2014, Rice and Legacè 2007). The relevance of this decision framework is that different participants in global governance are very likely to have different risk tolerance profiles for misses versus false alarms. This difference could be the case whether the policy interactions are between bodies with primarily conservation mandates and those with primarily industry-regulation mandates, or between parties within a governance body where some countries are willing to defer generating wealth to reduce environmental impacts whereas others are willing to accept the possibility of greater environmental impact in order to increase economic or livelihood returns.

Often the scientists producing the input information for a policy discussion assume that if the data and analyses are accepted as sound, a science-based consensus is within sight. This assumption does not allow for legitimate differences among agencies or parties in risk tolerance for misses and false alarms. Even if there is consensus on the information and the relative risks of misses and false alarms, there is no reason to assume there will be consensus on what balance of misses and false alarms is acceptable (Mace et al. 2014; Rice and Legacè 2007). The negotiations can be long and tension filled, as the participants seek consensus on policy reflecting an acceptable balance of those risks. No amount of scientific effort to increase clarity about the magnitude or nature of the risks can be expected to resolve, on some objective basis, those underlying differences in the tolerances for misses and false alarms. Scientists raised in the "knowledge is power" mind-set may have trouble accepting this limitation on that "power."

The third aspect of consensus-based decision-making in most global governance bodies is that the relationship among scientists, formal science advisory processes, and governments is not a global constant. Related to that, the boundary between what is scoped within legitimate science and science advisory activity, and what is scoped as within the domain of policy makers and outside the legitimate business of science advisors varies much in parallel. This may be well illustrated by the reports of the UN Ad Hoc Working Group of the Whole (AHWGW) on the Regular Process for Global Reporting and Assessment of the State of the Marine Environment, including Socio-Economic Aspects (Regular Process). The groundwork for the

Regular Process was laid in the Assessment of Assessments (UNEP and IOC-UNESCO 2009). In this report, the global landscape of marine assessments of ecological, economic, and social aspects of the ocean was reviewed, and best practices for a fully integrated assessment were presented and justified.

Based on the foundation described above, the Group of Experts appointed by the UN to conduct the first World Ocean Assessment (WOA) expected timely approval of both an outline of the assessments and the modalities under which it would be prepared (UN 2015a). Such was not the case; all states strongly supported a science-based WOA. However, ideas about what a science-based assessment would and would not address were far from universal. The workshop reports highlight the protracted discussions among states about the parameters within which the WOA would be conducted. Germane to this chapter, one point of contention among states was the extent to which the WOA would provide policy advice. At the crucial point in this discussion a series of progressively more restrictive options were laid out. All began with the equivalent of "the WOA will assess the status and trends of the ocean environment and economic and social benefits from uses of the ocean at global and regional scales." The options for what else would be done on these scales can be paraphrased as

1. The WOA will tabulate policies in place, relate the policies to the trends, and recommend policies that have resulted in improved ocean status and improved economic and social benefits.
2. The WOA will tabulate policies in place, relate the policies to the trends, and report their findings.
3. The WOA will tabulate policies in place and merely report the types of policies currently in place, but not link them to the status and trends.
4. The WOA will report status and trends (as per above) and "factors" that may be causing the trends, but will neither selectively tabulate policies nor explain trends as due in part or in whole to the presence or absence of specific policies.

Each of the options had support from some states, with regional differences prominent in the discussion. A portion of the states active in the AHWGW argued that only Option 4 was appropriate for a science-based assessment, and anything more would be straying into the domain of policy-making. As a consensus process, that option was the final guidance given to the Group of Experts by the AHWGW as the only option that *all* states agreed was within the legitimate scope of operation of scientific and technical experts. Any of the options for more attention to policy tabulation and analysis were beyond the tolerance of some states, and therefore outside the final consensus.

Some of the differences among countries in the location of the science–policy boundary might be attributed to differences in science capacity to both inform policy development and conduct policy analysis (Anon 2012). However, some portion of the differences around the global are cultural and institutional, and as experts work in truly international settings, they find the boundaries between science and policy that characterize Western Europe, the United States, Canada, and a few other countries are located differently in other regions. These differences have important implications for both how and how much scientific information can support policy development. Easy resolutions to these cultural and governance differences are not obvious.

A final feature of global ocean policy-making also influences the impact of scientific and technical information substantially. Global agendas are becoming increasingly complex and integrative, which has both positive and negative aspects. On the positive side, the increasing drive to integrate can contribute to the desired coherence of policies, whether across sectors of human activities or among the social, economic, and environmental aspects of policies and decisions. The greater coherence, in turn, is expected to increase the likelihood that the policies and programs will actually deliver expected outcomes, or at least move them in the desired direction (Charles et al. 2014).

On the negative side, it is becoming increasingly difficult to evaluate policies and plan future pathways based on the scientific and technical information that is provided. This difficulty is not solely an issue of social sciences needing to be integrated with the biological, physical, and chemical sciences, although such integration is certainly required (Rice et al. 2014b; UNEP and IOC-UNESCO 2009). Agendas of human rights, equity of distribution of benefits, indigenous peoples, and traditional/experiential knowledge are now inseparable parts of global governance discussions. At the UN meetings of the Ad Hoc Open-Ended Informal Working Group to study issues relating to the conservation and sustainable use of marine biological diversity beyond areas of national jurisdiction (BBNJ Working Group), access to marine genetic resources and sharing the benefits from their commercial development are dominating much of the discussion (UN 2015b). Scientific information on the current commercial products derived from marine genetic sources, their value in trade, and projections of the potential for development have informed debate at the BBNJ Working Group meetings and discussions during the negotiation of the Nagoya Protocol for the CBD (CBD 2010). The policy debate is certainly informed by the scientific reports. However, the scientific information is not able to resolve the question of whether these resources are available according to the freedom of the seas or are the common heritage of (hu)mankind, yet the policy discussion blends these issues to a significant extent.

This blending of scientific and technical issues with issues of human rights, gender, equity, and so forth, occurs at the sectoral level as well as the UN working group level. The FAO work on small-scale fisheries ended up producing guidelines that characterized not just such fisheries, their challenges and their benefits, but placed the guidance in the context of food security and poverty eradication, with environmental considerations of small-scale fisheries receiving little attention (FAO 2014).

It is certainly possible to disentangle the natural sciences, social sciences, and humanitarian/human values dimensions of these global issues. The 2014 report of the FAO High-Level Panel on meeting food security needs that addressed the role of marine and freshwater food products very effectively partitioned the various aspects of the complex issue of food security and the seas (HLPE 2014). The relevance to the topic of scientific information and global governance is that the report could not be prepared without dealing with the full spectrum of issues associated with food security. It is a comparatively straightforward scientific challenge to document the nutritional value of fish to coastal fishing communities. It is more complex when the policy decision includes, as part of the problem, the rights implied for those coastal communities, given the nutritional information. The time when natural science experts could partition off the parts of complex policy issues informed by natural science, address them, and leave the social science and human rights issues for other experts in other fora is fading fast. There is every reason to expect that these issues will become even more intertwined in the future, with further implications for how scientific and technical information will be used in the science–policy interface.

4.3 Case History: Ecologically and Biologically Significant Areas and Vulnerable Marine Ecosystems

Interest in spatially based approaches to conservation and management of the world's marine areas has been growing for at least the past two decades (Ehler and Douvere 2009). Correspondingly, both nongovernmental organizations (NGOs) with marine conservation interests and IGOs with marine regulatory or conservation mandates have recognized the value of prioritizing areas of particular importance for conservation and protection efforts. The names associated with the priority areas vary with the organization. However, in all cases the role of scientific information in the prioritization process is explicitly acknowledged, usually through the identification of criteria to be applied in meetings of scientific experts. Aspects of these processes and their outcomes have been described in a number of publications (Clark et al. 2014; Dunn et al. 2014), but the ways that scientific information has been used have not been reviewed.

For several reasons, the FAO processes and criteria for identifying vulnerable marine ecosystems (VMEs) and the CBD processes and criteria for describing areas that meet the criteria for ecologically and biologically significant areas (EBSAs) are a particularly useful pair of case histories that illustrate the effects of many of the contextual issues presented in the preceding part of this chapter. First, both processes have similar goals—to identify areas of the ocean where management needs to be more risk averse than whatever standard is being applied in the larger areas in which the VMEs or EBSAs are being identified. Second, the same basic scientific and technical information was used to develop criteria for both types of areas, and the same information is needed in the regional-specific applications. Third, there is even overlap in the experts who have participated in both developing the criteria and in their application. With these strong similarities, one would expect substantial similarity in how VMEs and EBSAs inform management. Such expectations are only partially realized. To understand the causes of the differences in practice, it is necessary to review the origins of the two initiatives. Their histories are presented in detail in Rice et al. (2014a), but the highlights relative to this theme can be summarized briefly.

The VME initiative arose from a multiyear debate at the UN level on a proposal to ban bottom trawling on the high seas. The rationale was that such fishing gears may cause substantial alteration to the seafloor, and recovery of seabed habitats and benthic communities from such alterations could take many decades to centuries, if possible at all; some interested groups referred to bottom trawling as categorically a "destructive fishing practice" (FAO 2010). More generally, it was also argued that the species exploited in high seas fisheries were easily depleted and slow to recover (Hutchings and Kuparinen 2014; Wright 2014), although this was not a reason to selectively ban one specific fishing method. Counterarguments were that almost every method of fishing *might* have detrimental impacts, but the impacts could be managed and made sustainable if the place, time, and method of bottom trawling were appropriately controlled. The debate culminated in UN Resolution 61/105, which required states and RFMOs to only allow fishing with bottom-contacting gears if the locations of VMEs within the general area being fished were known and those areas were either avoided or the gear was used in ways that did not cause serious adverse impacts to such ecosystems. This resolution resulted in the FAO developing Voluntary Guidelines for the Management of Deep-Sea Fishing on the High Seas (FAO 2009), including criteria for determining what constituted a VME.

The CBD initiative arose from the momentum to identify and designate marine protected areas (MPAs). The general interest in MPAs received global policy endorsement at the WSSD, which called for states to "[d]evelop and facilitate the use of diverse approaches and tools, including ... the establishment of marine protected areas ... [and] representative networks by 2012 and time/area closures for the protection of nursery grounds and periods" (UN 2002, para 32c). Although many conservation interests picked up that call, the first UNEP-CBD Ad Hoc Open-Ended Working Group on Protected Areas produced the request for criteria to inform the selection of areas to be prioritized for inclusion in MPAs and networks of MPAs (UNEP 2005).

Both the FAO and CBD followed the same general science practices of expert workshops, followed by a technical meeting of parties, each with a mandate to conduct quality assurance of the scientific workshop products and organize the results for policy action, and the adoption at a policy forum for parties (Table 4.2a). Not surprisingly for science-based processes, the criteria for describing areas that may be EBSAs or VMEs have initial similarities (Table 4.2b). However, the contextual considerations had already

TABLE 4.2

Scientific Information Pathways and Criteria for VMEs and EBSAs

a: Pathways of Scientific Information to Policy Outcomes in FAO and CBD Processes		
Action	*CBD*	*FAO*
Charge to undertake a project	Conference of Parties	Committee on Fisheries
Assembly of input information	Notice of project and call for input to parties, IGOs, and NGOs	Notice of project and call for input to parties, IGOs, and NGOs
Summarize and interpret science	Expert workshops by invitation from national nominations	Expert by invitation; nominations in only some cases
Interpretation of science products for policy consideration	Secretariat	Secretariat
Quality assurance and adaptation of products for policy consideration	Subsidiary body for scientific, technical, and technological information with participation by all interested parties and observers	Technical consultation with participation by all interested states and observers
Policy negotiations and adoption	Conference of Parties	Committee on Fisheries

b: Criteria for VMEs and EBSAs	
FAO—VME	*CBD—EBSA*
Uniqueness or rarity	Uniqueness or rarity
Functional significance of the habitat	Special importance for life-history stages of species
Life-history traits of component species that make recovery difficult[a]	
	Importance for threatened, endangered, or declining species and/or habitats[a]
Fragility	Vulnerability, fragility, sensitivity, or slow recovery
Structural complexity	Biological productivity
	Biological diversity
	Naturalness

[a] The VME criterion applies to a specific type of species but not their habitats, whereas the EBSA criterion applies to a class of habitats for species that have been selected based on other criteria.

influenced the products and processes in several ways that had important consequences.

The differences in the wording of the criteria reflect directly the tolerances of the organizations for misses versus false alarms in describing areas in need of more risk adverse management. There are fewer VME criteria than EBSA criteria, and the VME criteria are more generic. There is as little overlap as possible among their coverage, although features such as structural complexity and functional significance will often coincide. The EBSA criteria have much more overlap, with productivity and diversity crossing nearly completely with the other criteria when applied. At the stages of the scientific workshops, great similarity was present in the types of areas identified as needing to be covered by criteria. At each subsequent stage, however, the FAO and CBD applied their respective risk tolerance profiles for misses and false alarms. The dialogue at the FAO technical consultation focused on avoiding duplication, so ideally a given area would have to meet a single criterion very well, and areas would not be called VMEs just for being somewhat above average in several features. For the CBD final expert group and the Subsidiary Body on Scientific, Technical and Technological Advice (SBSTTA), there was concern that no areas be missed, so the criterion with explicit mention of threatened or endangered species was added just in case some area known to support populations of such species would be an EBSA, even if no other criteria were met. Consequently, the same scientific information and the same general goal were met in two different ways, with one focused on avoiding the requirement to unnecessarily regulate an activity, and the other focused on as high an inclusiveness as possible.

The differences between the VME and the EBSA processes become even greater when the outcomes of their application are reviewed. Both CBD and RFMOs have used expert workshops as their fora for application of their criteria, and in several cases they include some of the same experts, as well as others with traditional working affiliations to the respective hosts. A review in 2010 found that every RFMO with competence for bottom fisheries had commenced application of the VME criteria, and most had identified some areas as VMEs (FAO 2011). However, in every case, the RFMOs had also noted that application of the criteria was hampered by inadequate guidance on metrics and threshold values, incomplete databases, and lack of capacity. With few exceptions, when the RFMO processes encountered those challenges, the VME identification process was suspended. The areas of high uncertainty were usually not identified as VMEs, although often states noted they were not presently fishing the specific areas, so there was time to improve the information before a decision on VME status was needed.

In the case of CBD workshops, incomplete databases and less than full guidance on application of the EBSA criteria were also encountered. However, the CBD criteria were applied in regional workshops organized by the CBD Marine and Coastal Secretariat, together with regional hosts. Workshops in regions weak in scientific and technical capacity were preceded by capacity building workshops, and each application workshop opened with presentations on the criteria and the processes for their application. As experience was gained, those presentations became increasingly detailed, with each workshop empowered to go as far as possible in applying the criteria to the available information. In fact, over time, the single concept of an EBSA was partitioned into several different types of EBSAs to deal with biological features whose position varied seasonally or interannually, and to accommodate different degrees of uncertainty about the exact location of a feature. Boundaries were interpreted generously, which meant that the discussion of more risk aversion in management would be triggered whenever there was some evidence that EBSA criteria could be met. From my perspective as cochair of three regional workshops, it seemed that areas that were suggested as EBSAs were not accepted by the working group only in cases of nearly complete absence of data. The difference in outcomes can be illustrated with Figure 4.1, which shows the VMEs identified in the northwest Atlantic by NAFO (thick lines) and the EBSAs in nearly the same area identified by the respective CBD regional workshop (dashed lines).

This contrast of processes and outcomes is not intended to present either set as the "right way," but to illustrate how these global considerations affect the actual scientific and technical use of information (and uncertainty) as well as the ultimate decisions. In the case of identifying an area as a VME, there was an immediate and obligatory management consequence, as a result of UN Resolution 61/105. Bottom fishing had to be carefully managed or else the area would be closed to such fishing. In the case of identifying an area as meeting EBSA criteria, the only immediate policy or management consequences were that the UN General Assembly, other IGOs with marine mandates, and parties were notified of where areas meeting EBSA criteria were located and what criteria the areas had met. What the General Assembly, IGOs, and parties did with that information was up to the organization or party. Each group could then place the EBSAs within their own histories and have the experts with whom they had greatest comfort decide on appropriate follow-up. Even so, the consensus process of global governance affected the EBSA description, as several specific areas were removed from the Annex tabled at CBD COP XII because single parties questioned their identification on various grounds (CBD 2014).

In addition, the differences among parties in their perception of the boundary between science and policy played a major role in negotiating the text on follow-up activities for areas that had been accepted as meeting one or more EBSA criteria. The initial text of the paragraph outlining follow-up activities that parties might consider included the possibilities of assessing "status of," "trends in," and "pressures on" biodiversity in the areas meeting the criteria, and of preparing inventories of human activities in those areas that might place pressure on biodiversity. All those terms were within the comfort zone of some participants in the discussions. However, full consensus could only be reached on assessing that status of biodiversity in the areas, and that is the

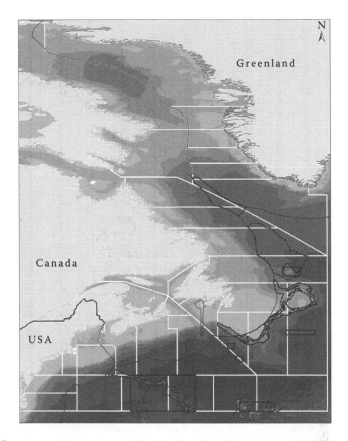

FIGURE 4.1

Map of the Northwest Atlantic Ocean showing the VMEs that have been identified by NAFO (thick solid lines) and the areas that have been accepted as meeting EBSA criteria by CBD COP XII (dashed lines). Note that (with one exception) neither the VME identification process nor the CBD EBSA description process considered areas within the 200 mile exclusive economic zone of the bordering countries (thin solid lines). The gray areas represent the depth contour shading.

text of the final decision (CBD 2014). It should be noted that parties are free to conduct other biodiversity assessments as they feel appropriate, taking account of their jurisdictions and competencies of organizations with whom they partner. However, the consensus call for action is quite restricted, and even that call is modified by some "as appropriate" and "if they choose."

The purpose of this case history is not to highlight any particular strengths or weaknesses in either the FAO or CBD processes. Beauty will remain in the eye of each beholder. Rather, the intent has been to draw out how differently organizations can travel paths from a common corpus of scientific and technical information to outputs that synthesize that information and interpret it in the context of spatially-based conservation and management. Both journeys tried at each step to practice sound science and to produce products that meet the needs of their parties. However, differences in all the contextual considerations—mandates, histories, risk aversion, perceived boundaries between science and policy, the need to achieve consensus, and perhaps above all, culture and trust—influenced each step on the journey and shaped quite different outcomes from similar staring points. The case histories illustrate that a common information base and a dedication to sound science by both organizations does not ensure identical outcomes.

4.4 Implications for the Future

How will these science–policy nuances play out in the future? One can only speculate, but if the differences are as entrenched in practice as I argue above, there are implications for most global ocean policy initiatives. I will speculate on those implications for two of the major ocean governance issues currently on the global agenda: the possibility of negotiating an "implementing agreement" to support the United Nations Convention on the Law of the Sea, and the development of the next round of sustainable development goals.

The sectoral approach to managing human activities in the open ocean has had critics for decades. Criticism has escalated since at least WSSD that such sectoral approaches make efforts to conserve marine biodiversity at least inefficient, and potentially not possible at all (e.g., Gjerde et al. 2008; Tladi 2011). To explore this issue, the UN called for the Ad Hoc Open-ended Informal Working Group, which first met in 2006 (UN 2006), "to study issues relating to the conservation and sustainable use of marine biological diversity beyond areas of national jurisdiction." The BBNJ Working Group held wide-ranging discussions on issues of marine biodiversity and conservation, with the chairs' report informing provisions of the annual UN resolution on oceans and law of the sea.

From the outset, some states and NGO interveners argued that existing governance structures were inadequate for protection of biodiversity and that a third "implementing agreement" was needed to function in parallel with, and at the same level as, the UN Fish Stocks Agreement (UN 1995) and the International Seabed Authority Agreement (UN 1994). Regardless of the position individual states took on the adequacy of existing governance instruments, each used science, when convenient, to support their position. Those arguing that the challenges are primarily insufficient implementation of existing agreements could use science-based arguments to show how those existing instruments can use scientific information on status of and threats to marine biodiversity to manage those threats. Those arguing that a new agreement is needed could use science-based arguments to show how many important biodiversity features are exposed to multiple threats, and managing any one of these is not sufficient to protect marine biodiversity.

The dueling experts can become strident over issues such as high seas marine protected areas (Caveen et al. 2014; Costello 2014). Under UNCLOS, there is no single agency aside from the UN General Assembly empowered to create a marine protected area in areas beyond national jurisdiction (ABNJ), but each human use that could threaten biodiversity in a particular area does have a body empowered to regulate its impacts, including effects on biodiversity. The choice of options is thus not a science issue of whether a particular instrument does or does not work, but a governance issue of whether regulators for different sectors choose to apply their instruments in coherent manners (Garcia et al. 2014b). There may be social scientific information to inform choices of whether that willingness can better be stimulated from the top down by a new implementing agreement, or motivated from the bottom up by the increasing accountability of each sector that comes from adopting an ecosystem approach. However, such information has been not brought to the debate so far.

The BBNJ Working Group meetings evolved during the 2000s to discuss a set of three issues that collectively have become "the package" (UN 2009; 2010): high seas marine protected areas, a common standard (or possibly forum) for environmental impact assessment, and access and benefits sharing to marine genetic resources. The debate has converged on a view that they cannot be resolved individually, but agreement must nonetheless be found on all issues in the package. This packaging of issues has added some complexity to following the threads of scientific information through this global ocean governance dialogue. Individually, the dialogue at BBNJ Working Group meetings acknowledged that each issue has a foundation in scientific and technical information. Correspondingly, it arranged a series of expert workshops to bring relevant information forward on each separate issue (UN 2013). Table 4.3 illustrates the diversity of scientific information and perspectives that were presented. These presentations were augmented by nine additional written submissions to the workshops from one state, one IGO, and three NGOs or institutes.

The workshops were well received by BBNJ Working Group delegates, but the reception provides insight into how scientific information is being used in this initiative. In particular, the scientific information is not being used to resolve any of the items on the agenda. Rather, the scientific information is being used to scope the issues under consideration. In discussions since the workshops, some experts have expressed disappointment that their presentations were not taken up as sources of solutions, but the restricted use is consistent with the features of global governance summarized above. Using scientific information to support scoping decisions is consistent with not conceding institutional or state mandate or jurisdiction. The diversity of experts invited to speak at each workshop ensured that the workshops touched the comfort zone of the range of participants. From that base, it was left to the negotiation stage to sort out any contrasting views among experts on whether to include or exclude items on the inventory of issues raised by the workshop as a whole. Likewise, the consensus process has reached only the simpler consensus that the range of perspectives are relevant to

TABLE 4.3

Abbreviated Titles of Presentations Made to the Scientific and Technical Workshops for the BBNJ Working Group Meeting, with the Country of the Presenter

Short Title	Country	Institution
Basics of marine genetic resources (MGR)	China	G
MGR in ABNJ—Clarifying terminology and constraining expectations	Canada	A
Marine microbiological research and possible applications	Japan	G
Why should marine genetic resources be conserved?	Portugal	A
Requirements and approaches for managing the future MGR	*	N
Marine genetic resources: technical challenges values	France	G
Environmental aspects of marine genetic resources	*	I
Access to MGR: collecting organisms and facilitating samples and data	Norway	G
Exploring different benefits and benefit-sharing approaches	*	N
Marine genetic resources: benefit sharing and obstacles	USA	A
Global regimes on GR [Genetic Resources]: the CBD and the Nagoya Protocol	*	I
Global regimes on GR: the food and agriculture, and health sectors	France	G
Regional regimes on genetic resources, experiences and best practices	Belgium	A
Scientific data about plankton ecosystems—governance implications	EU	A
Exchange of information on marine biodiversity research	Barbados	G
Scientific cooperation and research projects on the Tara expeditions	EU	A
Addressing collective marine biotech and bioprospecting challenges	EU	G
Relevant activities of the International Seabed Authority	*	I
Key ecosystem functions and processes in ABNJ	Chile	A
Impacts and challenges of high-seas fisheries to marine biodiversity ABNJ	Nigeria	A
Human impacts on fisheries productivity in ABNJ	UK	A
Impacts on, and challenges to, marine biodiversity ABNJ	*	I
Overview of new and emerging uses of the ocean ABNJ	*	I
Trends of new and emerging uses of biodiversity in ABNJ	*	A
Area-based management tools	Netherlands	A
Fisheries and spatial management measures in ABNJ	*	I
EIAs [Environmental Impact Assessments], SEAs [Strategic Environmental Assessment], and Biodiversity in ABNJ—Current arrangements	Canada	G
Gaps and options in EIA on marine biodiversity	Philippines	A
Social and environmental considerations for management in ABNJ	Barbados	G

TABLE 4.3 (CONTINUED)

Abbreviated Titles of Presentations Made to the Scientific and Technical Workshops for the BBNJ Working Group Meeting, with the Country of the Presenter

Short Title	Country	Institution
Scientific expertise and infrastructure for marine biodiversity management	Norway	G
Existing regimes, experiences, and best practices	*	N
Ecosystem services and area-based management	Japan	A
Trends in cooperation for research, management, and capacity building	Australia	A
OBIS [Ocean Biogeographic Information System] and capacity-building needs for marine biodiversity data	USA	A

Note: Country provided if listed as affiliated with a national or European Union (EU) site; whether the presenter's institution was academic or an international research center (A), a government department or institute (G), an intergovernmental organization (I), or an NGO or corporation (N), (*) means that the author(s) did not list a national affiliation.

the discussion, but no consensus on where the balance point among them might lie.

Even if this is a limited impact of scientific information on the global issue of an implementing agreement, it is not a trivial one. Many experienced negotiators conceded that after five years of policy debate, the presentations finally helped the BBNJ negotiators commence developing a roadmap for the issues to be resolved; this is a major step forward. The unrealistic expectation of some experts is that the scientific information itself will be the substance of the resolutions to the debates. As the IPCC debates have clearly illustrated, although the Fifth Assessment Report (IPCC 2014) represents a strong scientific consensus on trends, drivers, and options for climate change adaptation and mitigation, parties still negotiate policies taking a range of other issues into account. Solutions are sought that are within the scope of the scientific information, but until the uncertainty in scientific information is very low and diversity of interpretations of the information narrow, it is unrealistic to think scientific information *could* actually determine the outcomes of negotiations, let alone *would* determine them.

With regard to sustainable development goals, the Millennium Development Goals were adopted in 2000. Over a decade later, the Rio+20 Summit reviewed progress toward their achievement and concluded that although there were widespread examples of the alleviation of poverty and hunger, the goals had not been achieved (Rio+20 2012). In the follow-up document for implementation, the UN explicitly requested that states "must invest in the unfinished work of the Millennium Development Goals, and use them as a springboard into the future we want, a future free from poverty and built on human rights, equality and sustainability" (UN 2014b, para 18).

Moreover, we see in paragraph 37c of the Synthesis Report of the Secretary General (UN 2014b) that one of the main lessons learned in that unfinished work was that "the academics and scientists convened through the Sustainable Development Solutions Network recommended the adoption of a science-based and action-oriented agenda, integrating four interdependent dimensions of sustainable development (economic, social, environmental and governance)"; so here again scientific information should be at the heart of the path forward.

The 17 Sustainable Development Goals (SDGs) approved for consideration by the Open Working Group of the General Assembly on SDGs are about outcomes that are aspirational and qualitative (UN 2014a). As such, it would be inappropriate to judge the degree to which they are truly science-based. Nevertheless, at least 11 of the 17 goals require strong natural science foundations to identify effective pathways for progress, including SDG 14 on oceans, and the natural sciences have clear contributions to make to at least half the others (Table 4.4). The social sciences, of course, are essential to progress on all of them.

The SGDs are at an early stage of development, with respective targets and metrics, and a comprehensive implementation plan for their achievement is still a work in progress. However, based on the arguments made about the role of scientific information in global ocean governance, it is possible to sketch out some of that likely role. Science will have a strong role in outlining possible pathways forward. For goals such as 9 (resilient infrastructure), 12 (sustainable production and consumption), and 13–15 on achieving healthy oceans and terrestrial ecosystems and combating climate change, science may to some extent help set a pace for progress. Even there, however, the lack of information may become an excuse for deferral or very cautious action, due to the costs of misses and false alarms being distributed differently among countries viewed primarily as donors and those viewed primarily as recipients of development aid, and their risk profiles for tolerating those risks will be at least as different.

The presence of information may serve to guide consensus that some possible pathways toward one or more goals are not viable, but only if multiple lines of research agree that a particular method is likely to fail to result in progress. Based on the debates among scientists about the degree to which marine protected areas benefit commercial fisheries (Caveen et al. 2014; Hilborn et al. 2004; Rice et al. 2012), and what fishing practices are "destructive" (FAO 2010), even that scale of contribution may be modest at best. All the issues of different perspectives and institutions wanting to receive their science support from their familiar sources will be in play. It is perhaps the technical information on how particular technologies work under varying circumstances that may make the most unquestioned contributions to implementation of the SGDs. Here though, the contributions are not really to policy, but to guiding the *how-to* after the policy questions have been resolved.

TABLE 4.4

The 17 Sustainable Development Goals Approved for Further Refinement, with a Judgement of Their Dependence on Scientific Information and Advice

Goal	Outcome	Science
1	End poverty in all its forms everywhere	2
2	End hunger, achieve food security and improved nutrition, and promote sustainable agriculture	3
3	Ensure healthy lives and promote well-being for all at all ages	3
4	Ensure inclusive and equitable quality education and promote life-long learning opportunities for all	2
5	Achieve gender equality and empower all women and girls	1
6	Ensure availability and sustainable management of water and sanitation for all	3
7	Ensure access to affordable, reliable, sustainable, and modern energy for all	3
8	Promote sustained, inclusive and sustainable economic growth, full and productive employment, and decent work for all	3
9	Build resilient infrastructure, promote inclusive and sustainable industrialization, and foster innovation	3
10	Reduce inequality within and among countries	2
11	Make cities and human settlements inclusive, safe, resilient, and sustainable	3
12	Ensure sustainable consumption and production patterns	3
13	Take urgent action to combat climate change and its impacts	3
14	Conserve and sustainably use the oceans, seas, and marine resources for sustainable development	3
15	Protect, restore, and promote sustainable use of terrestrial ecosystems, sustainably manage forests, combat desertification, and halt and reverse land degradation, and halt biodiversity loss	3
16	Promote peaceful and inclusive societies for sustainable development, provide access to justice for all, and build effective, accountable, and inclusive institutions at all levels	1
17	Strengthen the means of implementation and revitalize the global partnership for sustainable development	1

Source: UN (United Nations). Report of the open working group of the general assembly on sustainable development goals. A/68/970. 2014.

Note: All judgments are for the natural sciences, as political and other social sciences are relevant to all goals. (1) Scientific information may be useful but is not central to the goal; (2) scientific information can inform development of options, but progress is possible with limited science input; (3) little progress can be made without a strong science underpinning.

For the many goals where equity is featured, even science efforts to establish metrics of equity may have limited influence. Scientific information may again be useful to justify *not* taking particular actions because outcomes are uncertain. However, from my observations of debates about equity at a global scale, there is little appetite for serious discussions of what pathways may lead *toward* equity if there is not already agreement between those perceived as disproportionately favored and those perceived as disadvantaged on what a truly equitable outcome will be. Again, the issues of risk tolerance, jurisdiction, and scope of what is inside and outside "the solution" are likely to supersede issues of the scientific evidence regarding how effective it may be to follow any particular pathway.

4.5 Conclusion

At the global level, scientific information supports policy development across the full spectrum of issues in both conservation of ecosystems and their components, and in management of human activities that use or impact those ecosystems. However, although scientific information can inform policy-making, it does not exclusively determine policy outcomes. In fact, even the degree to which global policy outcomes are constrained by scientific information is limited by several characteristics of policy-making at the global level. These include the increasingly overlapping mandates and jurisdictions of the many organizations and agencies with roles in global governance; the diversity of legitimate science perspectives on many ocean issues, even some which may initially appear simple; differences in distributions of risks of overly intrusive and overly permissive policy instruments and regulatory measures among industry sectors and ocean ecosystems; differences in tolerances for those same risks among the various participants in global policy-making; and simply familiarity with and trust in science advisors associated with the various governance streams.

All these factors play out in the global science–policy interface. They can lead to policy outcomes that frustrate or disappoint the expert advisors who may have had expectations of the science advice they input having greater impact than was realized in practice. This should not be viewed as failures of the science–policy interface at the global level, however. Rather, it is a reminder that experts in specific science disciplines should always be aware that policies routinely have to address diverse agendas simultaneously, even when, from a disciplinary perspective, it may appear possible to disentangle scientific and technical aspects of an issue from other aspects. The boundaries on what constitutes a "problem" in need of policy attention will always be context specific, and at global scales the players in policy-making rarely all bring the same context to the table. Therefore, science inputs may be most effective when they focus on providing a sound but neutral (to the greatest extent possible) starting point for policy development, and describe what pathways are available, where they lead, and what impediments may be encountered along the way. As science inputs focus on trying to influence the determination of policy outcomes, the effectiveness of these inputs may be increasingly overshadowed by these other considerations.

The dynamics discussed here may also play out at regional, national, and subnational scales. However, at more local scales, the range of perspectives of those who use the scientific information is unlikely to increase, and may become narrower. Thus, the inputs of scientific information to policy-making at these less global scales may have greater influence at the later stages of policy-making than they do at the global scale. Nevertheless, there are important global governance issues on the near horizon, including the possibility of negotiating a new implementing agreement to support UNCLOS and the development of the next series of sustainable development goals. A good understanding of the roles that can be played effectively by those who provide scientific information will help ensure that the scientific information gets used to the fullest extent possible in these important initiatives.

Acknowledgments

I would like to thank Ellen Kenchington and Camille Lirette of the Bedford Institute of Oceanography, Dartmouth, Nova Scotia, Canada, for preparing Figure 4.1. I also acknowledge very helpful input from the editors and reviewers in improving the descriptions of many of the science–policy interactions that are discussed.

References

Anon. 2012. Research policy: How to build science capacity. *Nature* 490: 331–334.

Ban N. C., N. J. Bax, K. M. Gjerde, et al. 2014. Systematic conservation planning: A better recipe for managing the high seas for biodiversity conservation and sustainable use. *Conservation Letters* 7: 41–54.

Caveen, A., N. Polunin, T. Gray, et al. 2014. *The Controversy over Marine Protected Areas: Science Meets Policy*. Dordrecht: Springer.

CBD (Convention on Biological Diversity). 2010. Nagoya protocol on access to genetic resources and the fair and equitable sharing of benefits arising from their utilization to the convention on biological diversity. Accessed 15 January 2015. http://www.cbd.int/abs/text/.

CBD (Convention on Biological Diversity). 2012. Report of joint expert meeting on addressing biodiversity concerns in sustainable fisheries. UNEP/CBD/SBSTTA/16/INF/13, CBD, Montreal.

CBD (Convention on Biological Diversity). 2014. Marine and coastal biodiversity: Ecologically or biologically significant marine areas (EBSAs) decision XII/22. UNEP/CBD/COP/DEC/XII/22.

Charles, A. T., S. M. Garcia, and J. C. Rice. 2014. A tale of two streams: Synthesizing governance of marine fisheries and biodiversity conservation. In *Governance of Marine Fisheries and Biodiversity Conservation*, edited by S. M. Garcia, J. C. Rice, and A. T. Charles, 413–428. Oxford: Wiley Blackwell.

Clark, M. R., A. A. Rowden, T. A. Schlacher, et al. 2014. Identifying ecologically or biologically significant areas (EBSA): A systematic method and its application to seamounts in the South Pacific Ocean. *Ocean & Coastal Management* 91: 65–79.

Connor, B. M. and A. B. Cooper. 2014. Determining decision thresholds and evaluating indicators when conservation status is measured as a continuum. *Conservation Biology* 28: 1626–1635.

Costello, M. J. 2014. Long live marine reserves: A review of experiences and benefits. *Biological Conservation* 176: 289–296.

Druell, E., J. Rochette, R. Billé, et al. 2012. A long and winding road. International discussions on the governance of marine biodiversity in areas beyond national jurisdiction. IDDRI Studies N°07/13. Paris: Institute for Sustainable Development and International Relations.

Dunn, D. C., J. A. Ardron, N. C. Bax, et al. 2014. The convention on biological diversity's ecologically or biologically significant areas: Origins, development, and current status. *Marine Policy* 49: 137–145.

Ehler, C. and F. Douvere. 2009. Marine spatial planning: A step-by-step approach toward ecosystem-based management. Intergovernmental Oceanographic Commission and Man and the Biosphere Programme, IOC Manual and Guides No. 53, IOCAM Dossier No. 6. Paris: UNESCO.

FAO (Food and Agricultural Organization of the United Nations). 2009. International guidelines for the management of deep-sea fisheries on the high seas. Accessed 15 January 2015. http://www.fao.org/fishery/topic/166308/en.

FAO (Food and Agricultural Organization of the United Nations). 2010. FAO/UNEP expert meeting on impacts of destructive fishing practices, unsustainable fishing, and illegal, unreported and unregulated (IUU) fishing on marine biodiversity and habitats. FAO Fisheries and Aquaculture Report No. 932. Rome: FAO.

FAO (Food and Agricultural Organization of the United Nations). 2011. Report of the FAO workshop on the implementation of the FAO international guidelines for the management of deep-sea fisheries in the high seas—Challenges and ways forward. FAO Fisheries and Aquaculture Report No. 948. Rome: FAO.

FAO (Food and Agricultural Organization of the United Nations). 2014. Voluntary guidelines for securing sustainable small-scale fisheries in the context of food security and poverty eradication, Chairperson's Report of the Technical Consultation on International Guidelines for Securing Sustainable Small-scale Fisheries, Appendix E. COFI/2014/Inf.10. Accessed 15 January 2015. http://www.fao.org/cofi/23885-09a60857a289b96d28c31433643996c84.pdf.

Garcia, S. M., J. C. Rice, and A. T. Charles. 2014a. Governance of marine fisheries and biodiversity conservation: A history; and annex 1. In *Governance of Marine Fisheries and Biodiversity Conservation*, edited by S. M. Garcia, J. C. Rice, and A. T. Charles, 3–17 and 429–469. Oxford: Wiley Blackwell.

Garcia, S. M., J. C. Rice and A. T. Charles. 2014b. Governance of marine fisheries and biodiversity conservation: The integration challenge. In *Governance of Marine Fisheries and Biodiversity Conservation*, edited by S. M. Garcia, J. C. Rice, and A. T. Charles, 37–51. Oxford: Wiley Blackwell.

Gjerde, K. M., H. Dotinga, S. Hart, et al. 2008. *Regulatory and Governance Gaps in the International Regime for the Conservation and Sustainable Use of Marine Biodiversity in Areas beyond National Jurisdiction*. Gland: IUCN.

Hilborn, R., K. Stokes, J.-J. Maguire, et al. 2004. When can marine reserves improve fisheries management? *Ocean & Coastal Management* 47: 197–205.

HLPE (High Level Panel of Experts on Food Security and Nutrition). 2014. Sustainable fisheries and aquaculture for food security and nutrition. A report by the high level panel of experts on food security and nutrition of the committee on world food security. HLPE Report No. 7. Rome: FAO.

Hutchings, J. A. and A. Kuparinen. 2014. Ghosts of fisheries-induced depletions: Do they haunt us still? *ICES Journal of Marine Science* 71: 1467–1473.

Intergovernmental Panel on Climate Change (IPCC). 2014. Fifth assessment report. Accessed 15 September 015. http://www.ipcc.ch/.

Mace, P. M., C. O'Criodain, J. C. Rice, et al. 2014. Conservation and risk of extinction of marine species. In *Governance of Marine Fisheries and Biodiversity Conservation*, edited by S. M. Garcia, J. C. Rice, and A. T. Charles, 181–194. Oxford: Wiley Blackwell.

Montgomery, D. C. 2005. *Introduction to Statistical Quality Control*. New York: Wiley.

NAFO (Convention on Future Multilateral Cooperation in the Northwest Atlantic Fisheries). 24 October 1978. 1135 UNTS 369.

Rice, J. C. and È. Legacè. 2007. When control rules collide: A comparison of fisheries management reference points and IUCN criteria for assessing risk of extinction. *ICES Journal of Marine Science* 64: 718–722.

Rice, J. C., E. Moksness, C. Attwood, et al. 2012. The role of MPAs in reconciling fisheries management with conservation of biological diversity. *Ocean & Coastal Management* 69: 217–230.

Rice, J. C., J.-H. Lee, and M. T. Tandstad. 2014a. Parallel initiatives: The CBD's ecologically or biologically significant areas (EBSAs) and FAO's vulnerable marine ecosystems (VMEs) criteria and processes. In *Governance of Marine Fisheries and Biodiversity Conservation*, edited by S. M. Garcia, J. C. Rice, and A. T. Charles, 195–208. Oxford: Wiley Blackwell.

Rice, J. C., S. L. Jennings, and A. T. Charles. 2014b. Scientific foundation: Towards integration. In *Governance of Marine Fisheries and Biodiversity Conservation*, edited by S. M. Garcia, J. C. Rice, and A. T. Charles, 124–136. Oxford: Wiley Blackwell.

Ridgeway, L. A. 2014. Global level institutions and processes: Assessment of critical roles, foundations of cooperation and integration and their contribution to integrated marine governance. In *Governance of Marine Fisheries and Biodiversity Conservation*, edited by S. M. Garcia, J. C. Rice, and A. T. Charles, 148–164. Oxford: Wiley Blackwell.

Rio+20 (United Nations Conference on Sustainable Development). 2012. Outcome document: The future we want. A/CONF.216/L.1.

Stavins, R. 2014. Is the IPCC government approval process broken? *Huffington Post*. 27 April. Accessed 15 September 2015. http://www.huffingtonpost.com/robert-stavins/is-the-ipcc-government-ap_b_5223421.html.

Tladi, D. 2011. Ocean governance: A fragmented regulatory framework. In *Oceans: The New Frontier—A Planet for Life*, edited by P. Jacquet, R. Pachauri, and I. Tubiana, 99–111. New Delhi: Teri Press.

UN (United Nations). 1994. Agreement relating to the implementation of Part XI of the United Nations Convention on the Law of the Sea of 10 December 1982. Accessed 15 September 2015. http://www.un.org/depts/los/convention_agreements/texts/unclos/closindxAgree.htm.

UN (United Nations). 1995. The United Nations agreement for the implementation of the provisions of the United Nations Convention on the Law of the Sea of 10 December 1982 relating to the conservation and management of straddling fish stocks and highly migratory fish stocks (In force as from 11 December 2011) overview. Accessed 15 September 2015. http://www.un.org/depts/los/convention_agreements/convention_overview_fish_stocks.htm.

UN (United Nations). 2002. World summit on sustainable development plan of implementation. Johannesburg, South Africa, 26 August–4 September 2002. Accessed 15 September 2015.

UN (United Nations). 2006. Report of the ad hoc open-ended informal working group to study issues relating to the conservation and sustainable use of marine biological diversity beyond areas of national jurisdiction. A/61/65. Accessed 15 September 2015.

UN (United Nations). 2009. Oceans and the law of the sea. Report of the secretary-general. Addendum. Accessed 15 September 2015.

UN (United Nations). 2010. Letter dated 16 March 2010 from the chairperson of the Ad Hoc open-ended informal working group to study issues relating to the conservation and sustainable use of marine biological diversity beyond areas of national jurisdiction. A/65/68.

UN (United Nations). 2013. Letter dated 23 September 2013 from the co-chairs of the ad hoc open-ended informal working group to the president of the general assembly. A/68/399.

UN (United Nations) 2014a. Report of the open working group of the general assembly on sustainable development goals. A/68/970.

UN (United Nations). 2014b. Synthesis report by the secretary general on the post-2015 sustainable development agenda. A/69/700.

UN (United Nations). 2015a. A regular process for global reporting and assessment of the state of the marine environment, including socio-economic aspects (regular process). Accessed 14 September 2015. http://www.un.org/depts/los/global_reporting/global_reporting.htm.

UN (United Nations). 2015b. Ad hoc open-ended informal working group to study issues relating to the conservation and sustainable use of marine biological diversity beyond areas of national jurisdiction. Accessed January 2015. http://www.un.org/depts/los/biodiversityworkinggroup/biodiversityworkinggroup.htm.

UNEP (United Nations Environment Programme). 2005. Report of the ad hoc open-ended working group on protected areas, Montecatini, Italy, 13–17 June 2005. Nairobi: UNEP.

UNEP (United Nations Environment Programme) and IOC-UNESCO (Intergovernmental Oceanographic Commission of the United Nations Educational, Scientific and Cultural Organization). 2009. An assessment of assessments: Findings of the group of experts. Start-up phase of a regular process for global reporting and assessment of the state of the marine environment including socio-economic aspects. Paris: IOC-UNESCO.

Wright, P. J. 2014. Are there useful life history indicators of stock recovery rate in gadoids? *ICES Journal of Marine Science* 71: 1393–1406.

5

Risk Refined at the Science–Policy Interface: The International Risk Governance Framework Applied to Different Classes of Coastal Zone Risks

Kevin Quigley and Kate Porter

CONTENTS

5.1 Introduction

Risk is understood to be a function of probability and consequence. Up until the 1980s, the study of risk was dominated in the West by scientists, engineers, economists, and decision analysts. Their views were overwhelmingly influenced by a rational actor paradigm (RAP) (Jaeger et al. 2001, pp. 19–22), in which risk is an objective condition that can be understood from a rational and individual perspective. From this standpoint, determining risk means determining the probability of an event and multiplying it by the consequence, usually measured in dollars or operational deficiency. There is an optimism that the data are obtainable and uncontroversial. These calculations loaned themselves to risk prioritization: those events with the highest risk score (probability × consequence) could be identified as the greatest risks and therefore first in need of attention. From this traditional view, risk is largely understood to be a negative concept; specialists seek to identify, segment, and eliminate it.

Challenges to this traditional view have emerged from the fields of psychology, sociology, and anthropology. For psychologists, risk is understood through an individual's lens, and it is assumed that risk is a subjective (that is, personal, intimate) construction; different people will have different understandings and views about the likelihood and consequence of events, irrespective of empirical data. Using the availability heuristic, for example, people will believe that an event is more likely to occur if they are able to imagine or recall it (Slovic et al. 1979). Slovic et al. (1979) note, for instance, that fear of shark attacks increased dramatically after the release of the movie *Jaws*, despite the fact that there was no empirical evidence to suggest that shark attacks had become more probable. For sociologists, institutional and community arrangements significantly influence our understanding of risk. The complexity of modern society and its networks cannot be reduced and ordered according to simple risk calculations. We have to think carefully about social processes. Socioeconomic status, for instance, makes a difference in someone's ability to understand and respond to risk. The better-off, for example, are more likely to have various forms of insurance, and are generally healthier, better educated, and live longer than the less well-off. For anthropologists, risk is socially constructed through an institutional setting. According to Mary Douglas (2001, p. xix), "certainty is only possible because doubt is blocked institutionally. Most individual decisions about risk are taken under pressure from institutions." For Douglas, these views of risk can be fundamentally incompatible. A regulator working in a bureaucracy, for example, will see an unregulated market as a risk; a business person working in that unregulated market, in contrast, will see the regulator's desire to regulate as a risk.

Multidisciplinary analysis of risk demonstrates that there are many ways to interpret and understand the concept of risk. Each approach makes different assumptions about the nature of risk, which impacts the tools and mechanisms required to manage risk. For rational actors, like engineers and actuaries, managing risk is about statistics and formal models. For psychologists, managing risk is about understanding and managing perceptions. For sociologists and anthropologists, risk is about social context and culture. While taking multiple views into account almost certainly enriches our understanding of risk, it also introduces potentially incompatible notions that have to be managed and trade-offs that have to be weighed. While taking one view is potentially narrow, taking all views is potentially unwieldy.

The International Risk Governance Council's (IRGC) framework is a normative and process-oriented approach to managing risk (Renn 2006). For the IRGC, risk governance can be defined as the totality of actors, rules, conventions, processes, and mechanisms concerned with how relevant risk information is collected, analyzed, and communicated, and how management decisions are taken. To a degree, the IRGC's framework accommodates contributions from different disciplines when examining and responding to risk; this framework combines technical risk analyses with issues of perception and process. The framework divides the components of risk governance into two broad categories: *risk assessment* and *risk management*. Assessment focuses on generating knowledge concerning risk, and management concerns making decisions and implementing those decisions. The framework further subdivides this process into four phases: *preassessment, risk appraisal, tolerability and acceptability judgment*, and *risk management*. Each phase includes concepts that should be applied to understand and manage risk.

These four stages are similar to other holistic approaches to risk. In addition to these four stages, however, the IRGC framework further divides risks into four classes: *simple, complex, uncertain*, and *ambiguous*. The classification of risk is "not related to the intrinsic characteristics of hazards or risks themselves but to the state and quality of knowledge available about both hazards and risks" (Renn and Walker 2008, p. 18). As such, efforts to classify risk refine our understanding of risk. These four categories help to improve the risk governance process and, in particular, how we communicate about risk and where we place our emphasis in the risk governance process.

The purpose of this chapter is to (1) briefly describe and discuss each of the four stages of the IRGC framework, (2) further develop the four types of risk—namely, simple, complex, uncertain, and ambiguous, and (3) show how these categorizations can help us to manage different types of coastal zone risks more effectively. We use the examples of the Nova Scotia shrimp fishery (simple), offshore oil and gas exploration (complex), port security (uncertain), and aboriginal fishing rights (ambiguous) to illustrate fundamental differences in risk problems and to link directly to the theme of this book—namely, how information operates at the science–policy interface in coastal and ocean management.

There are several challenges to conducting risk analyses in the modern state: problems of quantification, plurality of knowledge claims, stochastic events, counterintuitive implications, and inadequacy of trial and error learning (Renn 2015). Public and stakeholder engagement has increasingly become an important method by which to understand various perspectives and in so doing develop more reliable solutions; many policy makers struggle, however, in understanding how, when, with whom, and why to engage. The IRGC framework distinguishes between types of risk and, in so doing, assists policy makers in understanding the engagement process. The framework underscores how our state of knowledge of a particular risk can influence how to proceed. While it usefully describes an iterative process that includes a capacity for negotiation and learning, particularly with ambiguous risks, it also signals that we should limit engagements with simple and—to a lesser extent—complex risks. These distinctions provide a better chance of more stable and efficient risk management solutions.

5.2 International Risk Governance Council Framework

Figure 5.1 illustrates the basic interconnections of the phases within the International Risk Governance Council (IRGC) framework and will act as a graphical reference for this section. While the framework is divided into separate phases and presented in a sequential manner, risk governance does not necessarily occur in a tidy, sequential manner. At times, each of the phases can occur simultaneously, and communication flows back and forth, further

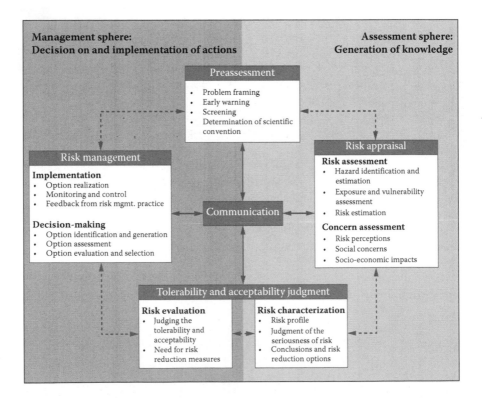

FIGURE 5.1
The International Risk Governance Council framework. (Adapted from Renn, 2006. Risk governance: Towards an integrative approach. White Paper No. 1. Geneva: International Risk Governance Council.)

informing or shaping the overall risk governance response, which is why the arrows indicate a movement in both directions.

5.2.1 Phases of Risk Governance

5.2.1.1 Preassessment Phase

The preassessment phase captures the variety of issues that stakeholders might associate with a certain risk, as well as existing indicators, routines, and conventions that may prematurely narrow—or act as a filter for—what will be considered a risk (Renn 2006). This phase examines, in particular, the manner in which data are (or are not) collected and shared, including early warning signs, research methods, screening practices, and scientific conventions. At this stage, the framework tries to capture the risks from the point of view of all parties affected by the threats: the official agencies (government), the risk and opportunity producers (private enterprises), those affected by the risks and opportunities (producer employees, spin-off businesses, importers, and those who live nearby) and interested bystanders (media, and environmental groups). It is concerned with the "systematic search for detecting hazards and threats, in particular new emerging risk events" and the capacity to share information with other interested parties (Renn and Aven 2010, p. 70).

5.2.1.2 Risk Appraisal

The purpose of the risk appraisal phase is to determine whether the endeavor that creates the risk is worth pursuing, and, if so, what steps can be taken to mitigate or contain the risk (Renn 2006). The risk appraisal phase consists of two major components: a scientific assessment of the risk, or risk assessment, and an assessment around societal concerns about the risk, or concern assessment.

The risk assessment component of risk appraisal aims to identify potential hazards, assess the level of exposure and vulnerability, and estimate the end risk using the best scientific models available (Renn 2006). It accomplishes these aims through scientific modeling of risks and the traditional determination of probabilities. Risk assessment generally uses existing data along with scenario modeling to determine various scenarios for risk to estimate the probability of a given occurrence. Concern assessment aims to gain an understanding of some of the issues that underlie the risk, such as how the risk affects different socioeconomic groups and how the risk is perceived by members of society, including an analysis of any cognitive biases that may exist around the risk (Renn 2006). By combining risk and concern assessment, risk managers are able to gain insight into the values and evidence required to assess the public's tolerance for risk exposure.

5.2.1.3 Tolerability and Acceptability Judgment

The third phase of the IRGC framework, the tolerance and acceptability judgment phase, is about determining the appetite for risk given the likelihood and the consequence of its occurrence. Tolerance with regard to risk looks at whether the endeavor that creates the risk is worth pursuing, given the potential consequences of disaster. Acceptability refers to the level of residual risk allowable after measures are put in place to mitigate or minimize exposure (Renn 2006). Something that is intolerable should be avoided, something that is tolerable requires risk reduction measures until it becomes acceptable, and something that is acceptable requires no action. This phase is often viewed as the most difficult; the lines between intolerable, tolerable, and acceptable are rarely clear because making a decision about tolerability and acceptability requires one to weigh values and evidence. For this reason, the phase is divided into two categories: *risk characterization* and *risk evaluation*.

Risk characterization is the collection and summarization of "all relevant evidence necessary for making an informed choice on tolerability or acceptability of the risk in question and suggesting potential options for dealing with the risk from a scientific perspective" (Renn 2006, p. 41). This component is generally completed by experts in the field. The risk evaluation component filters the risks through societal values and norms to make a judgment on the tolerability and acceptability of the risks and, subsequently, to judge the need for further risk reduction. It is when a risk is deemed to be tolerable and in need of methods to reduce exposure to the consequences of the risk, that risk governance enters the risk management phase.

5.2.1.4 Risk Management

The risk management phase of the IRGC framework takes the information obtained from the other phases and uses it to make decisions about the actions required to manage the risk (Renn 2006). Ultimately, the goal should be to make the tolerable risk acceptable over time. The categories within this phase endeavor to create learning organizations among both the risk producers and the government organizations in charge of regulation. These categories do this by instituting a risk management regime with feedback loops in which risk managers identify, assess, evaluate, select, and implement options for moving toward risk *acceptability*. Finally, the chosen options should be monitored and measured for "intended and unintended consequences" (Renn 2006, p. 20). The information taken from monitoring should then be fed back into the beginning process of identifying policy options, and the cycle should repeat.

Many acknowledge that this framework is a step forward in addressing a gap in the approach to risk, given the increased levels of complexity and uncertainty in modern society (North 2005; Rosa 2008). It is considered a useful tool that helps those unfamiliar with recent academic research to understand different approaches to risk management and acts as a guide to identify the types of issues that risk managers should be taking into consideration (North 2005; Rosa 2008). Despite this recognition, some critics note the lack of a clear definition of risk (North 2005; Boholm et al. 2012). Some argue that the framework is too complex or too simple to be useful as a practical tool (Boholm et al. 2012), if the framework is applied rigidly (North 2005; De Vries et al. 2011). Others note a bias toward systemic risks and that its emphasis on stakeholder engagement opens the door for lobbying (Boholm et al. 2012); it also seems to put more emphasis on the system as a whole and, as a result, may lose sight of the importance of micro-end decision-making (Boholm et al. 2012).

The framework relies on taxonomies that force risks into categories; due to the systemic nature of risks, it can be challenging to interpret them through a formal model (De Vries et al. 2011). Also, in attempting to look at the risk from as many aspects as possible, it does not consider the context of how the risk has been managed and the process and definitions of risk that shaped the current situation (Boholm et al. 2012).

5.3 Four Types of Risk: Examples from the Coastal Zone

Other risk frameworks are available to guide us through an assessment and management process (see O et al. 2015; ISO 2009; Treasury Board of Canada 2010). However, in addition to the IRGC framework, Renn (2006) has usefully divided risks into four classes: simple, complex, uncertain, and ambiguous. The classification of risk is "not related to the intrinsic characteristics of hazards or risks themselves but to the state and quality of knowledge available about both hazards and risks" (Renn and Walker 2008, p. 18). While some critics have noted the lack of clarity concerning the definition of risk, as mentioned earlier, Renn's effort to refine our understanding of risk into four categories creates considerable opportunity to improve the risk governance process, in particular how we communicate about risk and where we place our emphasis in the risk process.

5.3.1 Simple Risks at the Science–Policy Interface: Shrimp

Simple risks are risks for which "the number of predicted events are frequent and the causal chain obvious" (Renn and Walker 2008). The shrimp fishery in Nova Scotia is a good example of simple risk. The fishery is largely stable and predictable. Shrimp are very abundant in the North Atlantic. In 2013, 141,291 tonnes of shrimp were landed in Atlantic Canada, down from a record high of 185,974 tonnes in 2007, but up from 1990s levels (Fisheries and

Oceans Canada 2015). Shrimp ecology and its relationship to the fishery are also well understood. Shrimp stocks may be impacted by high surface water temperatures (Appollonio et al. 1986), predation by groundfish (Worm and Myers 2003), and the fishery itself. The Department of Fisheries and Oceans (DFO) attributes the current abundance of shrimp to a combination of cold surface water, unrecovered groundfish stocks, and sound fishery management practices (Fisheries and Oceans Canada 2014).

Risk management is generally more important than risk assessment within this fishery because much is already known about the risks involved, for example, overfishing, groundfish predation, and bycatch. DFO takes an integrated approach to managing Nova Scotia's shrimp fishery. Commercial trawlers and DFO carry out shrimp surveys each June. Based on a combination of these survey results and commercial data, that is, fishing logs and catch samples, each year a total allowable catch (TAC) is set. Risks associated with overfishing, for example, are tolerable, so while data collection needs to continue to avoid future problems, monitoring is simple because proven scientific methods exist and are already in use.

In addition to TAC, a variety of other conservation measures are enforced. These measures include individual quotas, limited licensing, minimum mesh sizes on nets, Nordmore separator grates, and monitoring (Fisheries and Oceans Canada 2015). Taken altogether, these conservation measures minimize bycatch and ensure the long-term sustainability of Nova Scotia's shrimp fishery. Problem framing is relatively simple, as stakeholders (namely, DFO and industry) agree on what the probability and consequences of the threats are, and agree they are tolerable in light of the conservation methods and the economic benefits of the fishery. Compared to other local fisheries, for example, the lobster fishery, the shrimp fishery is uncontroversial. Data suggest that the crustaceans can be extracted at the present levels with little risk to the environment or their population for the foreseeable future (Fisheries and Oceans Canada 2014).

In sum, the management of simple risks is relatively straightforward and can often be left to the market, albeit with some regulations. The dialogue that characterizes risk governance of a simple risk is instrumental. It involves agency staff, external experts, and external stakeholders; the analysis is largely statistical. The data are not contested or particularly controversial; the standards are largely accepted, albeit with some small variation from year to year.

5.3.2 Complex Risks at the Science–Policy Interface: Offshore Oil and Gas Exploration

Complex risks are those where there is difficulty "identifying and quantifying causal links between a multitude of potential causal agents and specific observed effects" (Renn and Walker 2008, p. 19). This difficulty may arise from interactive effects among potential causal agents, long delay periods between cause and effect, interindividual variation, that is, greater differences from case to case, and intervening variables.

Offshore oil and gas exploration is a salient example of complex risk. Every stage of offshore oil and gas exploration involves diverse risks. The complex interplay of variables makes it difficult to predict the probability of risk outcomes. Risk models are used extensively (e.g., Foreman 2005), however, these models are generally less reliable than in simple risk contexts. With complex risks, too many variables are at play, and while we may have experience in managing these risks, the interactions of the variables are not necessarily well understood.

The Deepwater Horizon oil well blowout and spill demonstrates how a series of technical failures may intersect in complex—and unanticipated—ways, resulting in tragedy. On 20 April 2010, the Deepwater Horizon, an offshore drilling platform, exploded, killing 11 people (Smithsonian Institution 2015). Two days later the rig sank. It took 87 days to cap the well, which in that time leaked an estimated 4.9 million barrels of oil into the Gulf of Mexico (Spier et al. 2013). No single event led to the Deepwater Horizon blowout, explosion, and resulting spill. British Petroleum, the company that was leasing the Deepwater Horizon platform at the time of the spill, identified eight causal agents—namely, dodgy cement, valve failure, misinterpreted pressure tests, failure to spot the leak, another valve failure, an overwhelmed separator, failure of the gas alarm, and failure of the blowout prevention system (Mullins 2010). No single failure caused the spill; rather, the complex interplay of these eight causal agents led to the disaster.

Some impacts of the spill were immediately self-evident. Other impacts are still unfolding. Ongoing research aims to determine how the Gulf of Mexico, was—and continues to be—impacted by this particular oil spill (e.g., Incardona et al. 2014). There can be a long delay between the occurrence of a spill and its effects. It took several years for the Alaskan shoreline to recover from the *Exxon Valdez* oil spill (Shigenaka 2014). Furthermore, a dearth of baseline ecological data at the location of future accidents may make it impossible to gauge impacts, and therefore recovery, from an oil spill. In the wake of the Deepwater Horizon spill, the President of the United States created a national commission to determine the causes of the spill, lay out a long-term recovery plan, and come up with strategies to mitigate and respond to future spills related to offshore drilling. Quite tellingly, in the foreword to its report to the president, the commission quotes the board that investigated the *Columbia* space shuttle disaster: "Complex systems fail in complex ways" (National Commission on the BP Deepwater Horizon Oil Spill and Offshore Drilling 2011, p. viii).

Renn suggests that the risk appraisal stage is crucial for analyzing complex risks because of the potential for technical analyses to improve our knowledge of risk. Risk modeling is a commonly used method for analyzing simple risks and complex risks. Whereas simple risks are associated with phenomena that are relatively frequent with fairly well understood causal links, extending these rational quantitative methods can become increasingly unreliable as the risk situation becomes more complex. The fundamental process of subdividing a problem into constituent parts is often inadequate to capture interactive effects between system elements. Furthermore, each cause and effect relationship in the complex system is typically inferred assuming prompt linear reactions, yet many systems are characterized by nonlinear interactions and delayed feedback. This latter aspect has been shown to confound attempts to grasp the full extent of consequences of a hazard (Hobbs et al. 2002). Marine ecosystems are inherently complex and respond in various ways to perturbations such as oil spills, as shown by many studies (GESAMP 1993; Committee on Oil in the Sea 2003).

To compensate for deficiencies in historical data which preclude developing statistically valid cause and effect inferences, modelers turn to probability theory to estimate likelihoods based on limited data and/or expert opinion. The expected value, or expected utility, underpinning a rational risk assessment model must be viewed judiciously given these limitations in the data and relationships. Comparisons with other apparently similar scenarios are often made to help define the system scope and cause and effect relationships; however, complex systems are rarely mirrored very well in other contexts.

In sum, while the model can help inform policy discussion by prioritizing problems and examining vulnerabilities and sensitivity, the actual evolution of an incident may be quite different from anticipated model outputs. Nevertheless, modeling to gain insight into potential outcomes from the failure of a complex system can be instrumental for building in redundancy to reduce the likelihood of failure propagation and adding buffers to mitigate the impacts. The caution here is the human tendency to overestimate our ability to understand, model, and control the complexities of a large system. Modelers require humility; they need also to improve their capacity to communicate their findings to a lay audience and in so doing help to inform the concern assessment, which is also part of the risk appraisal stage. As risk psychologists will note, how people feel about the risk is an important consideration for policy makers, irrespective of whether or not those feelings align with the formal predictions of the model and modelers.

5.3.3 Uncertain Risks at the Science–Policy Interface: Port Security

Uncertain risks are those where there is "a lack of clear scientific or technical basis for decision-making," which "often results from an incomplete or inadequate reduction of complexity in modelling cause-effect chains" (Renn and Walker 2008, pp. 18–19). Furthermore, uncertain risks often go unreported by business and industry. This diminishes the confidence level of traditional objective measures of risk estimation and risk analysts become more reliant on *fuzzy* or subjective measures of risk estimation (Renn and Walker 2008, pp. 18–19). According to Renn and Walker (2008), uncertainty can be subdivided into the following categories: epistemic, which is the result of imperfect knowledge, and can include target variability and systematic and random error in modeling, and aleatory, which includes indeterminacy or genuine stochastic events, system boundaries, and ignorance or nonknowledge.

A salient example of an uncertain risk is seaport security. Seaports are critical hubs in the global supply chain; 70% of the world's imports are moved by sea (Burns 2013). Ports compete against one another for business and, therefore, have to keep goods moving as efficiently as possible. At the same time, they are exposed to considerable threats. Security threats range from those that capture the public's attention, such as terrorism, drug smuggling, people trafficking, and people smuggling, to those that have perhaps more serious business implications, such as piracy, cargo theft, and cybercrimes, to the more mundane and probable, such as trespassing and petty crime. Many risks relate to broader questions of the underground economy, economic and political stability in parts of the developing world, and access to key trade routes in international markets (Quigley and Mills 2014). The somewhat open and accessible nature in which seaports operate also creates security threats.

The data we have for these types of risks are only partial, and unreliable. Terrorist attacks at a seaports, such as the attack on the USS *Cole*, are extremely rare. On 12 October 2000, the *Cole* was attacked in the port of Aden, Yemen, by suicide bombers in a small, explosive-laden boat; 17 U.S. sailors were killed and 39 were injured (CNN 2014). This was the deadliest attack on a U.S. naval ship since the Iran–Iraq War (National September 11 Memorial and Museum 2015).

Terrorism and criminal activity in particular create unique challenges to risk analysts because they are contending with adaptive adversaries; in other words, unlike a natural disaster, a terrorist will adapt his or her behavior in light of the risk management strategy that the port staff adopt. Moreover, managing low-probability risks in a robust manner can rarely be justified at the firm level (Seidenstat 2002); security is usually seen as a negative and usually unnecessary expense. Market-sensitive organizations shipping goods to and from ports will often not take pronounced steps to protect against low-probability/high-consequence events (Jaeger et al. 2001), which thereby enhances the vulnerabilities.

With uncertain risks, formal, rational models are unlikely to capture the full scope of the challenge. Uncertain risks can frequently generate surprises or realizations that are not anticipated or explained explicitly within a risk modeling framework. There are simply not enough data to understand the likelihood and consequences of the risk.

The absence of data that can help officials to be more specific about the magnitude of the risk requires that governments employ a precautionary approach, particularly when the harm is potentially catastrophic or irreversible (Sunstein 2009). Uncertain risks also require that governments avoid high vulnerability as well as they can.

In sum, it is unrealistic to think that a plan would dictate that sufficient human resources would constantly be available to respond to worst-case scenarios generated by uncertain risks. Solutions will require risk modeling coupled with a reflective discourse by policy makers, experts, industry, and affected stakeholders that attempt to strike the balance between over- and under-managing the response to the event. Scenario planning exercises can help, provided they infuse an element of the unpredictable into the scenarios, and are not merely test scenarios for which everyone is prepared, among friendly and convenient partners who are prepared to join the exercise.

5.3.4 Ambiguous Risks at the Science–Policy Interface: Aboriginal Fishing Rights

Ambiguous risks are a result of divergent or contested perspectives on the justification, severity or wider "meanings" associated with a given threat (Renn and Walker 2008). Categories of ambiguity include interpretative (i.e., different interpretations of the same results) and normative (i.e., different concepts of what can be considered tolerable) (Renn 2006).

The case of aboriginal fishing rights in Canada is an example of ambiguous risk. Under section 35 of the Constitution Act, 1982, "[t]he existing aboriginal and treaty rights of the aboriginal peoples of Canada are hereby recognized and affirmed." These rights include the right to fish. There is considerable disagreement among stakeholders, however, over what those existing rights are and how to interpret treaty rights in modern contexts.

There are numerous stakeholders in the Canadian fishing industry. These stakeholders include the federal government (DFO), Aboriginal peoples, non-Aboriginal peoples, environmental groups, media, the general public, and the fishing industry itself. Each of these parties may be affected in different ways by risks and opportunities within the industry. For example, section 35 of the Constitution Act, 1982 affected Aboriginal and non-Aboriginal fishers in different ways. To protect Aboriginal peoples' right to fish, DFO reallocated fishing licenses belonging to non-Aboriginal fishers (Fisheries and Oceans Canada 2012). Since the early 1990s, roughly 900 commercial licenses have been reallocated, and 1,300 seasonal jobs have been created, for Aboriginal fishers (Fisheries and Oceans Canada 2012). Many non-Aboriginal fishers had already lost their licenses in the collapse of the northern cod (and other groundfish) fisheries (Parliament of Canada 1999). While both Aboriginal and non-Aboriginal fishers recognized the need to regulate fisheries to ensure their recovery, a great deal of tension existed around who should be in charge of fisheries management (Parliament of Canada 1999). Some First Nations communities were reluctant to recognize the authority of DFO to manage their communal fisheries (e.g., Burnt Church, New Brunswick; after all, the fisheries collapsed under DFO's management). This rejection of DFO's authority was widely perceived by non-Aboriginal fishers as a threat to the fisheries (Nixon 2001).

The same piece of legislation that provided opportunities for Aboriginal peoples represented a threat to non-Aboriginal fishers. Different stakeholders may perceive threats differently; an opportunity for one stakeholder group may be seen as a threat by another. To deal with differences in perceived risk, Renn (2006) recommends that managers identify emerging threats and communicate these threats to stakeholders. In this particular case, the Government of Canada needed to communicate more clearly to non-Aboriginal fishers how this new legislation would impact them.

In numerous instances disagreements over Aboriginal fishing rights have been decided by the Supreme Court of Canada (for examples, see Allain 1996). The Donald Marshall decision and the ensuing strife in the community of Burnt Church, New Brunswick, exemplify the conflict and ambiguity involved in these disagreements. Donald Marshall Jr. was convicted of eel fishing out of season, without a license, with nonregulation equipment, and selling the eels he caught (Parliament of Canada 2001). In the past, the Supreme Court had usually upheld these kinds of convictions against First Nations fishers, albeit with some disagreement among judges (Allain 1996). In this instance, Marshall was acquitted by the Supreme Court.

Judging by DFO's lack of contingency plans (Parliament of Canada 2001), the Department appeared unprepared for the Supreme Court ruling in favor of Donald Marshall and its broader implications for Aboriginal fishing rights. In the wake of the Court's decision, many First Nations communities in eastern Canada, including Burnt Church, began to exercise their fishing rights. This led to conflict between these coastal communities, non-Aboriginal fishers, and DFO. The conflict centered around two main issues: (1) access to resources and (2) conservation management.

Each stakeholder group had a different interpretation of their rights and activities. Misunderstandings among stakeholders contributed to a lack of clarity about fisheries regulation and poor communication between non-Aboriginal and First Nations fishers. DFO was unprepared to manage threats to the fishing industry and rights of the various parties involved. Although scientific models existed for managing the lobster fishery, these models did not allow for the emerging role of First Nations fishers and traditional ecological knowledge in this industry. So DFO was left scrambling to assess the risks posed by people fishing without a license, outside the established fishing season. Also, they were not prepared to address the concerns of non-Aboriginal fishers and the general public.

Eventually, DFO managed to reach agreements with 30 of the 34 affected First Nations communities (Parliament of Canada 2001). Negotiations broke down between DFO and Burnt Church when the community rejected all government regulation in favor of pursuing its own management plan (Parliament of Canada 2001); the conflict between First Nations and non-Aboriginal fishers escalated to the point where shots were fired, although no one was harmed (CBC News 2001). Although DFO launched the Aboriginal Fisheries Strategy in 1992 to reach management agreements with Aboriginal communities (Fisheries and Oceans Canada 2012), insufficient funds were set aside to reallocate licenses to Aboriginal peoples. The conflict at Burnt Church may have been avoided if the DFO had framed the issue more broadly; it was not strictly an issue of licensing and fishing seasons, but rather one with legitimate competing and arguably incompatible views among stakeholders.

In the case of ambiguous risks, there is little disagreement on the data; there is disagreement, however, on what the data mean. How the risk is framed is a key consideration when responding to an ambiguous risk. For this type of risk, broad consultation is important and solutions can sometimes be only provisional until more reliable data become available.

In sum, when modeling the risks, how the risk is framed (or characterized) is important, as is the process stakeholders establish to resolve conflicts and arrive at a stable solution. Such risks can frequently pit one group against another and can include extreme reactions by ideologically driven groups. Risk modeling alone will not solve this problem. Solutions rely on modeling coupled with political bargaining between stakeholders and trade-offs between different risks. The process should involve agency staff, industry,

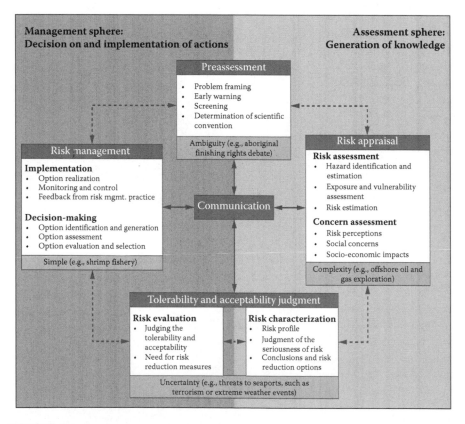

FIGURE 5.2
The International Risk Governance Council framework and classes of risk. (Adapted from Renn. 2006. Risk governance: Towards an integrative approach. White Paper No. 1. Geneva: International Risk Governance Council.)

and stakeholders, and awareness, if not explicit support, from the general public. If there is broad-based consensus that competing groups have legitimate claims, then risk governance processes normally proceed with caution and continue to gather information until a resolution can be achieved. Provisional solutions are put in place. Like uncertain risks, ambiguous risks can easily default into the precautionary principle. This approach is not without controversy, as will be discussed further in the discussion section. Figure 5.2 summarizes the stages at which policy makers should spend more time when examining a specific type of risk.

5.4 Discussion and Conclusion

In complex situations and environments (such as the coastal zone), it is important to understand as much as we can about the risks involved with human activities and natural events, and to improve our level of knowledge. Such efforts can help minimize social and economic disruption and increase opportunity. The IRGC framework and its classifications are useful. The simple, complex, uncertain, and ambiguous classifications help us to be more efficient in our analysis, process, and words (see Table 5.1).

Simple risks generate reliable data that help to inform our view about the risk; we can be more confident about the extent to which the threat will materialize and the consequences of that threat. As a result, when considering a simple risk, the discussion between policy makers and scientists is largely instrumental; market failure logic and limiting government intervention (to that which is optimal in market terms) can be a helpful way to develop a regulatory approach. Alas, governments rarely find themselves dealing with simple risks, or if they do, it is rarely the stuff of headlines, debate, or political consequence. Rather, governments find themselves drawn much more often into debates about complex risks, and indeed, the even more contentious uncertain and ambiguous risks.

Complex risks depend largely on expert opinion and formal modeling. Formal models help to explain in rational terms the interactions between many variables; they can help concentrate the best minds in a particular field on technical challenges that can bring about significant achievements; technical risks associated with space travel, the power grid, and the Internet can be described as complex risks. Expert processes can also allow us to focus on the existing data, however imperfect, and in so doing, increase transparency and remove the politics and sometimes petty negotiations. Complex risk problems are the domain of scientists, or medics; these professions are trusted more than most (Canadian Pharmacists Association 2009) and, therefore, the solutions they generate offer a better opportunity for acceptance from the community at large.

TABLE 5.1

Risk Management and Stakeholder Involvement

Risk Problem	Simple	Complex	Uncertain	Ambiguous
Example	Shrimp fishery	Offshore oil and gas exploration	Port security	Aboriginal fishing rights
Type of discourse	Instrumental	Epistemological	Reflective	Participative

Actors	• Agency staff
	• External experts
		• Agency staff
		• External experts
			• Agency staff
			• External experts
			• Stakeholders
			  • Industry
			  • Directly affected groups
				• Agency staff
				• External experts
				• Stakeholders
				  • Industry
				  • Directly affected groups
				  • General public

Type of conflict		Cognitive	Cognitive Evaluative	Cognitive Evaluative Normative
Remedy	Statistical risk analysis	Probabilistic risk modeling	Risk balancing necessary • Probabilistic risk modeling	Risk trade-off analysis and deliberation necessary • Risk balancing • Probabilistic risk modeling

Source: Adapted from Renn, O. 2006. Risk governance: Towards an integrative approach. White Paper No. 1. Geneva: International Risk Governance Council.

Formal models, the tools of the experts, have important limitations. From a normative standpoint, formal models embed key assumptions. To start, we assume complex technological and ecological systems are accessible to detailed human comprehension and that a reductionist approach is the best way to understand these systems (Jaeger et al. 2001, p. 91). Formal models can sometimes completely overlook important social, and even moral, considerations. We saw the failure to take broader social concerns into account in the Deepwater Horizon event.

While social concern is part of risk appraisal, complex risks tend to overemphasize the perspectives of the experts. People often have blind faith in numerical analysis and computer models; these processes, however, are subject to bias and can manipulate information through the manner in which data are presented (Jaeger et al. 2001, pp. 81–82). Experts need to show humility, recognize the limitations to their models, the contexts in which decisions are made, the privileged position they hold, and the consequences of bad advice. The risk management plan must also have a degree of robustness, lest the system fail.

Finally, while formal models offer the hope of transparency, rigorous analysis, and optimal outcomes, the models fail to include the more subtle dynamics in decision-making, such as strategic reasoning, power plays, interests, and institutional responses (Jaeger et al. 2001, p. 82). In this sense, models offer important insights but do not provide a full account of decision-making. Dietz and Stern (1995) note, for example, that relatively complex mathematics does not correspond with what we know about human behavior with respect to decision-making. People are good at pattern recognition, classification, and applying rules of thumb; this undermines the usefulness of the model altogether and frustrates the experts who developed the model with the intention of reducing the influence of seemingly irrational human behaviors. This gap between the scientists and the lay people, including policy makers, aggravates everyone and threatens to undermine the legitimacy of each group from the other's point of view.

As we move along the continuum to uncertain and ambiguous risks, the data become even more unreliable and contested. The IRGC framework recommends a precautionary approach. The precautionary principle should be limited to catastrophic and irreversible events and even then should be applied with care (Sunstein 2009). There are risks on both sides of any equation. There are risks if one acts, just as there are risks if one does not. Too often, advocates of the precautionary principle narrow their examination too quickly and neglect to consider the trade-offs and opportunity costs that must be considered in any risk management plan. Kheifets et al. (2001) found considerable variation in the manner in which the precautionary principle is used and the concept is deployed in practice. For example, the strength of evidence required to justify action under the precautionary principle can vary. The principle may be adopted (1) when there is "sufficient evidence" that an action or substance is harmful; (2) when there is no conclusive scientific proof one way or the other; or (3) when the substance or action has been suggested as a possible cause of harm (Kheifets et al. 2001). The necessary action can also vary. Definitions of the precautionary principle (see also Chapters 2 and 19 in this volume) imply a wide range of actions that should be taken, once the strength of evidence requirement has been satisfied. These actions range from (1) prevention or elimination of exposure to (2) adoption of cost-effective action, or (3) mere consideration of action (Kheifets et al. 2001). Another important variable is who bears the burden of proof: (1) the opponents of a possibly harmful action, or (2) the proponents of a possibly harmful action (Kheifets et al. 2001).

When we consider uncertain risks, we need to have a focused discussion about risk trade-offs and how much protection is too much. The likelihood of a terrorist attack at a port is low, but the consequences socially and economically can be significant. Private industry—the owners and operators of most critical infrastructure—is also unlikely to take the possibility of such an attack seriously; it does not make sense from a business standpoint. In such a case, a degree of resilience—a capacity to bounce back—must be built into the risk management process. Public engagement is also necessary; if data are unreliable, it is important for policies to be generally supported by the community.

Indeed, as we consider uncertain and ambiguous risks, engagement with those external to the agency becomes an important part of the risk governance process. Commentators have raised questions about the extent to which the IRGC framework, in its effort to make the process more accessible, makes the process vulnerable to lobbying (Tait 2008). This is a concern with ambiguous risks, in particular, where there are competing and legitimate views. Fairness becomes an important consideration.

Despite the controversy over lobbying, political scientists have largely been silent on the question of risk governance. This is surprising. Political scientists can help to sharpen our focus on interest group dynamics. Wilson (1980), for example, describes different degrees of power and influence in lobbying. Client politics is typically the preferred position for interest groups: the benefits are concentrated among specific groups and the costs are shared by many, thereby making the cost for any specific group small. As a result, the opposition is largely indifferent and ineffective. Client politics normally describes the dynamics at play with large, powerful, organized, and well-funded interests with privileged access to policy makers. The offshore oil-drilling industry exemplifies client politics, as became apparent following the Deepwater Horizon disaster. Industry frequently received privileged access to decision makers, and drilling usually trumped environmental concerns (Davenport 2015; Kusnetz 2014). In the case of interest group politics, in contrast, costs and benefits are concentrated among different groups; in other words, one group is trying to secure the benefits at the other's expense. This is obviously a dynamic that leads to more conflict. The Aboriginal fishing rights example noted above exemplifies this dynamic. Aboriginal and non-Aboriginal fishers each sought benefits at the other's expense.

While the process articulated in the IRGC framework can be useful, the interactions between lobbyists and policy makers must be appropriate. The process must seek to establish trust among the stakeholders and the community at large. Peters et al. (1997) identified three dimensions that generate trust: (1) knowledge and expertise, (2) care and concern, and (3) openness and honesty. When we are dealing with client politics, for example, trust in the process can be enhanced by increasing transparency in the decision-making process, creating sufficient distance between industry and policy makers, ensuring adequate expertise within government offices to challenge industry, empowering third party oversight, and providing access to legitimate groups that are concerned about the consequences of these events (such as environmental NGOs; see Chapter 18 in this volume). In interest group politics, trust can be increased in largely the same way, however, those responsible for the risk governance process must ensure that all parties have adequate support to represent and defend themselves. There may also be a need for a short- and medium-term transition plan as resources are shifted from one group to another.

Over the past two decades, numerous risk governance frameworks have taken a more holistic approach to managing risk by incorporating contributions from several disciplines. These risk governance frameworks go well beyond simple risk calculations by signaling the importance of perception, process, and social context. Many of the assumptions about the nature of knowledge and human nature within these academic disciplines are fundamentally at odds. As a result, there will always be a tension at the heart of these models. They are roadmaps with signposts, not perfect solutions. As this chapter has shown, the risk models can be very helpful as decision makers deal with information and possibly competing issues in ocean and coastal management.

References

Allain, J. M. 1996. Aboriginal fishing rights: Supreme court decisions. BP-428E. Ottawa: Parliamentary Information and Research Service. Library of Parliament. Accessed 18 August 2015. http://www.parl.gc.ca/Content/LOP/ResearchPublications/bp428-e.htm.

Appollonio, S., D. K. Stevenson, and E. E. Dunton, Jr. 1986. Effects of temperature on the biology of the northern shrimp, *Pandalus borealis*, in the Gulf of Maine. NOAA Technical Report NMFS No. 42. U.S. Department of Commerce, NOAA/National Marine Fisheries Service. Accessed 24 August 2015. http://aquatic-commons.org/2780/.

Boholm, A., H. Corvellec, and M. Karlsson. 2012. The practice of risk governance: Lessons from the field. *Journal of Risk Research* 15 (1): 1–20.

Burns, C. 2012. Implicit and explicit risk perception. Paper presented at the European Academy of Occupational Health Psychology, Zurich.

Canadian Pharmacists Association. 2009. Canadians trust doctors and pharmacists most: Poll finds two health professions rate highest on honesty and ethics. Last Modified 9 September 2011. http://www.nanosresearch.com/library/polls/POLNAT-S09-T388.pdf.

CBC News. 2001. Shots fired in Burnt Church fishing dispute. Last updated 16 September 2001. Accessed 29 August 2015. http://www.cbc.ca/news/canada/shots-fired-in-burnt-church-fishing-dispute-1.299558.

CNN. 2014. USS *Cole* bombing fast facts. Last modified 8 October 2014. Accessed 19 September 2015. http://www.cnn.com/2013/09/18/world/meast/uss-cole-bombing-fast-facts/.

Committee on Oil in the Sea: Inputs, Fates, and Effects; Ocean Studies Board; and Marine Board. 2003. *Oil in the Sea III. Input, Fate and Effects.* Washington, DC: National Academies Press.

Constitution Act, 1982, being Schedule B to the Canada Act 1982 (UK), 1982, c. 11. Accessed 20 August 2015. http://laws.justice.gc.ca/eng/Const/page-15.html.

Davenport, C. 2015. New sea drilling rule planned, 5 years after BP oil spill. *New York Times*, 10 April. Accessed 15 September 2015. http://www.nytimes.com/2015/04/11/us/new-sea-drilling-rule-planned-5-years-after-bp-oil-spill.html.

De Vries, G., I. Verhoeven, and M. Boeckhout. 2011. Taming uncertainty: The WRR approach to risk governance. *Journal of Risk Research* 14 (4): 485–499.

Dietz, T. and P. C. Stern. 1995. Toward a theory of choice: Socially embedded preference construction. *Journal of Socio-Economics* 24 (2): 261–279.

Douglas, M. 2001. Dealing with uncertainty. Paper presented at the Multatuli Lecture, Leuven, Belgium. *Ethical Perspectives* 8 (3): 145–155. http://www.ethical-perspectives.be/viewpic.php?LAN=E&TABLE=EP&ID=118.

Fisheries and Oceans Canada. 2012. Aboriginal fisheries strategy. Last modified 24 September 2012. http://www.dfo-mpo.gc.ca/fm-gp/aboriginal-autochtones/afs-srapa-eng.htm.

Fisheries and Oceans Canada. 2014. Shrimp (*Pandalus borealis*)—Scotian Shelf—As of 2013. Accessed 10 June 2015. http://www.dfo-mpo.gc.ca/fm-gp/peches-fisheries/ifmp-gmp/shrimp-crevette/shrimp-crevette-2013-eng.htm.

Fisheries and Oceans Canada. 2015. Shrimp. Accessed 18 August 2015. http://www.dfo-mpo.gc.ca/fm-gp/sustainable-durable/fisheries-peches/shrimp-crevette-eng.htm.

Foreman, M. G. G. 2005. A review of models in support of oil and gas exploration off the North Coast of British Columbia. Canadian Technical Report of Fisheries and Aquatic Sciences No. 2612. Sydney, BC: Institute of Ocean Sciences, Fisheries and Oceans Canada.

GESAMP (Joint Group of Experts on the Scientific Aspects of Marine Environmental Protection). 1993. Impact of oil and related chemicals and wastes on the marine environment. GESAMP Reports and Studies No. 50. London: International Maritime Organization.

Hobbs, J. E., A. Fearne, and J. Spriggs. 2002. Incentive structures for food safety and quality assurance: An international comparison. *Food Control* 13 (2): 77–81.

Incardona, J. P., L. D. Gardner, T. L. Linbo, et al. 2014. Deepwater Horizon crude oil impacts the developing hearts of large predatory pelagic fish. *Proceedings of the National Academy of Sciences* 111 (15): E1510–E1518.

ISO (International Organization for Standardization). 2009. *Risk Management—Principles and Guidelines: BS ISO 31000:2009 = Management Du Risque—Principes Et Lignes Directrices*. Geneva: ISO.

Jaeger, C. C., T. Webler, E. A. Rosa, et al. 2001. *Risk, Uncertainty and Rational Action*. London: Earthscan.

Kheifets, L. I., G. L. Hester, and G. L. Banerjee. 2001. The precautionary principle and EMF: Implementation and evaluation. *Journal of Risk Research* 4 (2): 113–125.

Kusnetz, N. 2014. Governors and oil industry work hand in hand in offshore drilling group. The Center for Public Integrity, 24 November. Accessed 11 September 2015. http://www.publicintegrity.org/2014/11/24/16312/governors-and-oil-industry-work-hand-hand-offshore-drilling-group.

Mullins, J. 2010. The eight failures that caused the Gulf oil spill. *New Scientist*, 8 September. https://www.newscientist.com/article/dn19425-the-eight-failures-that-caused-the-gulf-oil-spill/.

National Commission on the BP Deepwater Horizon Oil Spill and Offshore Drilling. 2011. *Deep Water: The Gulf Oil Disaster and the Future of Offshore Drilling.* Report to the President. Accessed 3 September 2015. http://www.gpo.gov/fdsys/pkg/ GPO-OILCOMMISSION.

National September 11 Memorial & Museum. 2015. USS *Cole* bombing. Accessed 11 September 2015. http://www.911memorial.org/uss-cole-bombing.

Nixon, A. 2001. The Marshall decision and the Atlantic fishery. TIPS-63E. Ottawa: Parliamentary Information and Research Service, Library of Parliament. Accessed 18 August 2015. http://www.parl.gc.ca/Content/LOP/ ResearchPublications/tips/tip63-e.htm.

North, D. W. 2005. Comments on the IRGC framework for risk governance. In *Global Risk Governance: Concept and Practice Using the IRGC Framework,* edited by O. Renn and K. D. Walker, 93–99. Dordrecht: Springer.

O, M., R. Martone, L. Hannah, et al. 2015. An ecological risk assessment framework (ERAF) for ecosystem-based oceans management in the Pacific region. Canadian Science Advisory Secretariat Research Document 2014/072. Ottawa: Fisheries and Oceans Canada. http://www.dfo-mpo.gc.ca/csas-sccs/publications/resdocs-docrech/2014/2014_072-eng.pdf.

Parliament of Canada. 1999. The Marshall decision and beyond: Implications for management of the Atlantic fisheries. Second Report of the Standing Committee on Fisheries and Oceans. Ottawa: Parliament of Canada. Accessed 11 September 2015. http://www.parl.gc.ca/HousePublications/Publication.aspx?DocId=1031 629&Mode=1&Parl=36&Ses=2&Language=E&File=2.

Parliament of Canada. 2001. The marshall decision and the atlantic fishery. Accessed August 18, 2015. http://www.parl.gc.ca/Content/LOP/ResearchPublications/ tips/tip63-e.htm.

Peters, R., V. Covello, and D. McCallum. 1997. The determinants of trust and credibility in environmental risk communication: An empirical study. *Risk Analysis* 17 (1): 43–54.

Quigley, K. and B. Mills. 2014. An analysis of transportation security risk regulation regimes: Canadian airports, seaports, rail, trucking and bridges. Halifax: Dalhousie University. http://cip.management.dal.ca/wp-content/ uploads/2014/02/680692-QuigleyMills-Kanishka-Transportation.pdf.

Renn, O. 2006. Risk governance: Towards an integrative approach. White Paper No. 1. Geneva: International Risk Governance Council.

Renn, O. 2015. Risk and innovation in a time of rapid transformations. Plenary address presented at the World Congress on Risk, Singapore.

Renn, O. and T. Aven. 2010. *Risk Management and Governance. Concepts, Guidelines and Applications.* Berlin: Springer-Verlag.

Renn, O. and K. D. Walker. 2008. *Global Risk Governance: Concept and Practice Using the IRGC Framework.* Dordrecht: Springer.

Rosa, E. A. 2008. White, black and gray: Critical dialogue with the International Risk Governance Council's framework for risk governance. In *Global Risk Governance: Concept and Practice Using the IRGC Framework,* edited by O. Renn and K. D. Walker, 101–118. Dordrecht: Springer.

Seidenstat, P. 2002. Terrorism, airport security, and the private sector. *Public Administration Review* 2 (1): 275–291.

Shigenaka, G. 2014. *Twenty-Five Years after the Exxon Valdez Oil Spill: NOAA's Scientific Support, Monitoring, and Research.* Seattle: NOAA Office of Response and Restoration.

Slovic, P., B. Fischhoff, and S. Lichtenstein. 1979. Rating the risks. *Environment: Science and Policy for Sustainable Development* 21 (3): 14–39.

Smithsonian Institution. Museum of Natural History. 2015. Gulf oil spill. Accessed 3 September 2015. http://ocean.si.edu/gulf-oil-spill.

Spier, C., W. T. Stringfellow, T. C. Hazen, et al. 2013. Distribution of hydrocarbons released during the 2010 MC252 oil spill in deep offshore waters. *Environmental Pollution* 173: 224–230.

Sunstein, C. R. 2009. *Worst-Case Scenarios*. Cambridge, MA: Harvard University Press.

Tait, J. 2008. Risk governance of genetically modified crops—European and American perspectives. In *Global Risk Governance: Concept and Practice Using the IRGG Framework*, edited by G. Renn and K. Walker, 133–153. Dordrecht: Springer.

Treasury Board of Canada. 2010. Framework for the management of risk. Last modified 10 August 2010. http://www.tbs-sct.gc.ca/pol/doc-eng.aspx?id=19422.

Wilson, J. Q. 1980. *The Politics of Regulation*. New York: Basic Books.

Worm, B. and R. A. Myers. 2003. Meta-analysis of cod–shrimp interactions reveals top-down control in oceanic food webs. *Ecology* 84: 162–173.

6

Governing the Marine Environment through Information: Fisheries, Shipping, and Tourism

Hilde M. Toonen and Arthur P. J. Mol

CONTENTS

6.1 Introduction

Fisheries as well as long distance and overseas transport of goods and people have been key economic sectors for coastal nations for several centuries, even millennia. More recently, economic activities at sea have expanded exponentially, not only in these two sectors but in a broad range of economic domains, including mining of sand and gravel; exploitation of oil, gas, and other natural resources; energy production through offshore wind arrays and other forms of so-called ocean energy; tourism; and coastal or sea-based aquaculture (mariculture) (e.g., Young et al. 2007). It is widely acknowledged that the seas and oceans are currently at environmental risk through the (cumulative) effects of these—and numerous land-based—economic expansions (Crowder and Norse 2008; Halpern et al. 2008; Worm et al. 2006). With these competing claims and increasing pressure on the marine ecosystem, environmental issues and nature protection moved up the public and political agendas of nation-states, international organizations, civil society, and lately even economic actors (e.g., World Ocean Council 2014a,b).

Although the three-dimensional nature of sea protection and the fact that resources, activities, and pollution are often a moving target in such a liquid environment and are complicating characteristics, these can to a certain extent be found on land, too, for instance with respect to governing air pollution, freshwater deterioration, or car mobility flows (Mol and Spaargaren 2005). Whether at sea or on land, the governance of environmental flows fits in well with Ulrich Beck's (1986, 2009) ideas about a *risk society*. The risk society thesis relates to ways of dealing with risks in today's world, and implies that we face risks that are inherently part of and produced by human activities, rather than being exposed to problems induced by external forces. The question of who decides about risks, and about measures for risk minimization or adaptation, are often difficult to answer and therefore a topic of continuous debate (see Chapter 5 in this volume). However, the governance of risks and environmental problems relating to oceans and seas is markedly different from the environmental protection of land because of the distinctive sovereignty challenge: nation-states have decreasing authority the farther one moves offshore (Burn et al. 2015; Suárez-de Vivero 2013). Also, the sea is special compared to land in that much of the environmental deterioration happens out of direct experience and sight. To define (and redefine) risk management in the marine environment, we depend on observations and experiences intermediated by technology and experts, rather than relying on our own senses. Hence, Beck's notion of "expropriation of the senses" in a risk society has an additional dimension at sea (Beck 2009, p.116).

Information is arguably important in governing marine environmental protection, even more than compared to terrestrial environmental protection, for two main reasons. First, due to the limited state authority regarding major parts of the oceans, we witness a stronger reliance on non-state actors in environmental protection activities. While governance, defined as processes of decision-making and steering, is traditionally associated with governmental authority, this so-called *governance-by-government* is now best seen as one, albeit important, way of steering (Kooiman 2003). Well-known examples of marine governance by non-state actors are self-governance initiatives through voluntary sector-wide agreements from fisheries (Gray 2005), shipping (DeSombre 2006), and oil and gas production industries (Van Leeuwen 2010). Moreover, nongovernmental organizations (NGOs) also take up governance tasks by pushing sectors to act more sustainably. It is now commonly understood that these non-state forms of environmental governance (often labeled private governance, e.g., Pattberg 2007; Tysiachniouk 2012) rely on information resources considerably, especially at sea. Whether or not information is used only to spin a sustainable image (so-called green washing), both market parties and NGOs widely use information in order to substantiate, accept, and communicate the ecological rationality of proposed private governance measures, as well as to ensure and check compliance.

The second reason for the importance of information in marine governance is that detecting and monitoring environmental deterioration in the oceans requires different information systems and practices when compared to terrestrial environmental protection. Deterioration of the marine environment is not easily sensed by citizens and communities, as they do not experience or easily consider the oceans as their *heimat* or backyard (Toonen and Lindeboom 2015). Even though sea-faring users, fishermen in particular, are known and renowned for their rich knowledge of the marine environment, their perceptions can result in overlooking or misunderstanding environmental conditions in their noticing/assessing fish stock depletion, discharge of ballast water, oil leakage, and biodiversity decrease (Verweij et al. 2010). Technologically advanced monitoring and information systems are needed even more if we want to trace back to the economic actors who caused the deterioration.

For two decades, scholars have investigated environmental governance through information, often labeled informational regulation or informational governance. In informational governance of the environment, the disclosure and use of environmental information (through labels, product information systems, certification, bench marking, company environmental reporting, pollutant registries, sustainability rankings, etc.) are considered the main mechanism for (re)directing behavioral changes, rather than direct state regulation (e.g., environmental laws, licenses, state enforcement) or market incentives (e.g., subsidies, levies, payment for environmental services) (Mol 2006, 2008). Because governance through information is not exclusively linked to governments, but well suited to steering by non-state actors, one could expect that forms of informational governance will also—or perhaps even especially—prevail at sea.

This chapter explores the current practices, experiences, and future outlook of informational governance of marine resources. We start with theorizing informational governance (Section 6.2). Subsequently, we analyze the function and prevalence of informational governance in three marine sectors: fishing, shipping, and tourism (Sections 6.3, 6.4, and 6.5). After comparing the three sectors on prevalence, practices, and challenges of informational governance (Section 6.6), we reflect on the contribution of these sector-based informational governance systems to integrated ecosystem-based approaches in marine governance.

6.2 Environmental Governance through Information

6.2.1 Informational Governance

It is only since the mid-1990s that information disclosure, right-to-know, company environmental reporting, pollution release and transfer registers, sustainability rankings, and informational labels and certifications have been interpreted as making a major contribution to new forms of environmental governance (Karkkainen 2001; Mol 2006; Stewart 2001). In the legal, economic, and sociological international literature, the influence of the wider production and availability of environmental information on environmental regulatory and governance processes has been brought together under the notion of informational regulation or informational governance (Case 2001; Cohen 2000; Kleindorfer and Orts 1999; Konar and Cohen 1997; Mol 2008; Tietenberg 1998). The concept of informational governance, as we draw on in this chapter, refers to the idea that information (and informational processes, systems, technologies, institutions, and resources linked to it) is fundamentally changing processes, institutions, and practices of governance, thus making them essentially different from conventional modes of governance (Mol 2006, 2008). Conventional regulation and governance rely heavily on authoritative resources, belief in information control, and state power. In informational governance, information is becoming a crucial (re)source with transformative powers for a variety of actors, although nobody is in full control of the collection, verification, and use of information.

Information has been part of environmental governance since its establishment in the 1960s. In conventional regulatory systems, state authorities rely on state-run, expert-led, and (natural) science-based monitoring systems to see whether, where, and when state regulation is effective, enforcement needs to be intensified, and policies have to be adapted. However, in informational governance, new information systems and mechanisms as well as mandatory or voluntary information disclosure stimulate new governance and enforcement practices and dynamics. What makes information governance arrangements different from conventional regulatory governance is that information itself starts to become a constituting and transformative force in environmental governance instead of just being an enabling condition for formulating, implementing, and monitoring state policies. Moreover, while scientific information (central in conventional state governance) continues to play an important role in this governance through information, it is often blended with information from practitioners, local experts, citizen-consumers, and the media. Marine governance witnesses many examples of mixing different kinds of informational sources, for instance, the joint fact-finding and inclusion of scientific and stakeholder knowledge in fisheries management (e.g., Holm 2002; Toonen and Mol 2013; Verweij et al. 2010; Wilson 2009).

Information disclosure to competitors, chain actors, customers, the state, consumers, and the public at large is then to be understood as an act of governance and an enforcement mechanism supports and complements (and in specific cases even replaces) conventional governance and enforcement via the state. Polluters feel motivated and/or forced to clean up in order to safeguard reputation and markets. But informational governance moves beyond disclosure of polluters. State agencies, international organizations,

companies, utilities, NGOs, retailers, consumers, and the like govern—
and are being governed—through the production, use, release, framing,
accessibility, demand, and verification/certification of information. These
forms of informational governance prevail especially in contexts where
conventional nation-state governance is not considered adequate, effective, or desirable (e.g., where governance responsibilities are shifted to
self-governing communities, or governance transcends the nation-state
and becomes international/global, as in the case of marine environmental protection). According to Graham (2002), mandatory and voluntary
disclosure strategies differ in three ways from conventional governmental environmental policies. First, these strategies influence environmental risks not through legislative or regulatory processes by the state, but
through non-state (thus, societal and market) pressure. Information then
becomes the main tool to affect others, such as the public in their capacity of citizens and/or consumers, or chain partners, in order to achieve
behavioral change of polluters. Second, the regulators are not only governments; numerous non-state actors who are empowered by knowledge
and information also attempt to influence the purchasing of products and
services, credits and investments, handing out insurances, voting, collective actions, and so on. Third, these systems extend beyond the reach of
the government and beyond national boundaries, and thus have competitive governance strengths beyond sovereign territories and an international outreach.

6.2.2 Challenges of Informational Governance

Informational governance is articulated as a relatively new phenomenon
of how actors aim to govern the environment under conditions of globalization and the Information Age (the centralization of digitalization/com-
puterization in all aspects of societal development, Castells 1996/1997; Mol
2008). Informational governance emerged relatively recently, but its presence is not equally distributed over all environmental issues and places/
countries. Moreover, informational governance should not be understood
as a better form of governance compared to conventional or other forms
of environmental governance. Informational governance has no overall
normative preference. The emergence of informational governance raises
a number of questions and challenges, of which the most important ones
are (Esty 2004; Fung et al. 2007; Gupta and Mason 2014; Howes 2002; Mol
2006, 2009)

1. Governing through information works only under certain conditions: when information becomes freely available and widely accessible, and when actors, practices, and markets are responsive to
   disclosed environmental information. This is not always and every-
   where the case.

2. The collection, distribution, and access to data and information for informational governance are related to power. Hence, some actors, countries, and regions with poor information production and access capacities are disadvantaged, resulting in mechanisms of inclusion and exclusion. Hence, we partly witness new forms of inequalities, and partly the strengthening of existing forms of inequality in environmental governance.

3. Science is no longer the only information provider (see Chapter 10 in this volume). With multiple information producers and distributers of different kinds, it is essential to safeguard the quality and credibility of information and data for the quality of informational governance. How to ensure high quality and credibility of information, how to make high-quality data and information more influential and ensure delegitimizing of low-quality data, and how to prevent information quality requirements that block data disclosure are important questions. With the sharp increase of information availability on the Internet, reliability of and trust in information comes increasingly with reliability of and trust in information providers, whereby reputation and reputational capital are becoming crucial resources.

4. Governance through information runs against the danger of informational overflow: there is a risk in producing and distributing too much (contrasting) information so that information loses its power as it becomes impossible to distinguish true from false information. This "drowning in information" reduces the usefulness and steering capacity of information disclosure. Informational overflow is not only a negative side effect but can be—and is—also constituted strategically and purposefully, to ensure informational governance failure.

Hence, what we see is that conflicts about protecting the (marine) environment are partly relocated to the information scape, where different actors try to obtain favorable information power positions. Eco-labelling, certification, green standards and classifications, benchmarking, sustainability rankings, and other informational instruments are not so much the result of scientific calculations presented in a neutral way, but are powerful acts of governance that come with major controversies.

6.2.3 Assessing Informational Governance in Marine Management

Informational governance works primarily through the disclosure of information on specific actors and practices. The most well-known and well-studied terrestrial forms of informational governance are related to major industrial polluters, financial investors, energy and other utility companies, agricultural food producers and food chains, and the like (e.g., Gupta and Mason 2014; Mol 2010; Zhang et al. 2016). In order to gain a better insight in the prevalence, strengths and weaknesses, and future of informational governance in the oceans, we have selected three marine sectors to be analyzed on their actual and potential informational governance practices at sea: fishing, shipping, and tourism. To what extent do we see informational governance arrangements regarding these sectors, and what form do these arrangements take? Do actors and activities involved in environmental disruptions at sea change their behavior/practices when they are confronted with such informational governance arrangements? Why, how, and to what extent do they do that, and does that also have possible negative or controversial side effects?

Our assessment on informational governance initiatives and arrangements is based on earlier marine research by us and our colleagues at the Environmental Policy Group (Wageningen University), as well as a further (primary and secondary) literature review. We mainly focus on a qualitative assessment of informational governance, as quantitative data are hardly available and difficult to compare (for illustrative purposes, we present some numbers in Tables 6.1 through 6.3; however, this is based on information provided by the organizations themselves, because independent sources about performance are lacking). Our emphasis in this chapter is on so-called third-party verification. Third-party verification of labeling, certifications, benchmarks, sustainability rankings, emission disclosures, wallet cards, and the like is carried out by an (often accredited) auditing organization independent from the value chain. Modes of informational governance can also include first-party assessments (hence, self-assessments or self-disclosure of environmental information) and second-party verification (actors within the production chain are auditing and certifying their chain partners). However, discussions about credibility of information providers and users (a main theme in this edited volume) are more self-evident in first- and second-party assessments than in third-party verification.

6.3 Certifications and Seafood Guides in Sustainable Fisheries

In line with the growing realization that many environmental issues are global challenges, attention to the well-being of the world's oceans mounted in the 1980s. Overfishing in particular became a top priority on the sustainability agenda because fishery efforts and effectiveness increased substantially, due to improved gear and technologies (Dayton et al. 1995; Pauly et al. 2005; Worm et al. 2006). Upscaling in fisheries also led to more damage to, and even destruction of, habitats and benthic communities (Dayton et al. 1995). Moreover, Lewison et al. (2004) showed a significant increase of (incidental)

TABLE 6.1

Outreach Informational Governance in Fisheries: Five Examples

Scheme[a]	Earth Island Institute's Dolphin-Safe Tuna[b]	Marine Stewardship Council[c]	Monterey Bay Aquarium's Seafood Watch[d]	Friend of the Sea[e]	Greenpeace Carting Away the Oceans[f]
Type	Certification	Certification	Recommendation list	Certification	Ranking
Launched in	1990	1997	1999	2006	2008
Typology leading actor	NGO	NGO	NGO	NGO	NGO
Scope	Single issue/species	Broad	Broad	Broad (including social concerns)	Broad
Geographical reach	Global	Global	National (United States; other examples: Aquarium of the Pacific, United States; Vancouver Aquarium, Canada)	Global (over 50% of certification is small-scale/artisanal fisheries)	National (United States; also Canada[g] among others)
Outreach: chain actors *Note: scheme's own claim[b,c,d,e,f]*	To date, more than 515 companies in 71 countries	To date, 265 fisheries certified, 108 fisheries in assessment	—	To date, 320 companies worldwide	In 2014, 26 of the largest supermarkets across the United States profiled
Outreach: consumers *Note: scheme's own claim[b,c,d,e,f]*	—	To date, more than 27,000 certified products sold	To date, 45 million guides distributed; app downloaded over one million times	To date, 10 million tonnes of certified seafood sold; 600 certified products in 26 countries	—

Sources: [a] Facts and outreach data based on organization's own information, available online, see [b,c,d,e,f]. Assessment of *typology leading actor, scope, and geographical reach* is based on Parkes, G., J. A. Young, S. F. Walmsley, R. Abel, J. Harman, P. Horvat, A. Lem, et al. 2010. Behind the signs—A global review of fish sustainability information schemes. *Reviews in Fisheries Science* 18 (4): 344–356 and on view/judgment by the authors of this chapter.

[b] Earth Island Institute. 2015. Dolphine Safe Fishing. Accessed 7 July 2015. http://www.earthisland.org/.

[c] Marine Stewardship Council. 2015. The MSC in numbers. Accessed 7 July 2015. http://www.msc.org/.

[d] Monterey Bay Aquarium's Seafood Watch program. 2015. What is Seafood Watch? Accessed 7 July 2015. http://www.seafoodwatch.org/.

[e] Friend of the Sea. 2015. About us. Accessed 9 July 2015. http://www.friendofthesea.org.

[f] Mitchell, J. 2014. How sustainable is your supermarkets seafood? Accessed 9 July 2015. http://greenpeaceblogs.org/2014/05/13/sustainable-supermarkets-seafood/.

[g] Greenpeace 2014.

TABLE 6.2

Outreach Informational Governance in Shipping: Five Examples

Scheme[a]	The Blue Angel (*Der Blauer Engel*)[b]	Qualship 21[c]	Shipping Efficiency—A to G GHG Emission Rating[d]	Clean Shipping Index[e]	Green Marine Environmental Program[f]
Type	Certification	Certification	Ranking list	Ranking	Ranking
Launched in	1978	2001	2001	2006	2008
Typology leading actor	Mix	State (US Coast Guard)	NGO	Mix	Mix
Scope	Broad	Broad	Single issue	Broad	Broad
Geographical reach	Global	National (United States)	Global	Global	Global
Outreach: chain actors Note: *scheme's own* claim[b,c,d,e,f]	To date, 1,500 companies have been awarded	In 2015, 22 flag administrations fully qualified for entrance into the program	To date, information on over 70,000 vessels in the system	Data from 2011: 11 of the 14 largest shipping companies presented information to the database	In 2014, 94 evaluation reports submitted
Outreach: consumers Note: *scheme's own* claim[b,c,d,e,f]	To date, around 12,000 products and services awarded	–	–	–	–

Sources: [a] Facts and outreach data based on organization's own information, available online, see [b,c,d,e,f]. Assessment of typology leading actor, scope, and geographical reach is based on Svensson, E. and K. Andersson. 2012. Inventory and evaluation of environmental performance indices for shipping. Department of Shipping and Marine Technology Gothenburg: Chalmers University of Technology. Report No. R11:132 and on view/judgment by the authors of this chapter.

[b] Der Blauer Engel. 2015. Our label for the environment. Accessed 15 July 2015. http://www.blauer-engel.de/en.

[c] QUALSHIP 21 Initiative. Accessed 15 July 2015. http://www.uscg.mil/hq/cgcvc/cvc2/safety/qualship.

[d] Carbon War Room. 2015. Background. Accessed 15 July 2015. http://www.shippingefficiency.org.

[e] Clean Shipping Index. 2015. Clean Shipping Index. Accessed 15 July 2015. http://www.cleanshippingindex.com.

[f] Green Marine. Benchmarking environmental performance. Accessed 15 July 2015. http://www.green-marine.org.

TABLE 6.3

Outreach Informational Governance in Coastal and Marine Tourism: Five Examples

Scheme[a]	Blue Flag[b]	Marine Conservation Society's Good Beach Guide[c]	EarthCheck Initiative[d]	Green Key[e]	Green Coast Award[f]
Type	Certification	Recommendation list	Certification	Certification	Certification
Launched in	1985	1987	1987	1994	1996
Typology leading actor	NGO	NGO	Market	NGO	Mix
Scope	Broad	Broad	Broad	Broad	Broad
Geographical reach	Global (carried out on national level)	National (United Kingdom)	Global	Global	Regional (Wales, United Kingdom)
Outreach: chain actors Note: *scheme's own claim*[b,c,d,e,f]	—	—	To date, have worked with partners in over 70 countries	To date, 2,400 hotels and other sites in 49 countries	—
Outreach: consumers Note: *scheme's own claim*[b,c,d,e,f]	To date, 4,000 beaches and marinas in 49 countries	To date, website received over 700,000 visits per year; 1,145 beaches in the online directory	To date, have influenced the decisions of over 6 million conscious travellers every week	—	47 beaches in 2011

Sources: [a] Facts and outreach data based on organization's own information, available online, see b,c,d,e,f. Assessment of *typology leading actor, scope,* and *geographical reach* is based on view/judgment by the authors of this chapter.
[b] Foundation for Environmental Education. 2015. Blue Flag beaches/marinas. Accessed 15 July 2015. http://www.blueflag.org.
[c] Marine Conservation Society. 2015. Good beach guide. Accessed 15 July 2015. http://www.goodbeachguide.co.uk/.
[d] EarthCheck. 2015. About. Accessed 9 July 2015. http://www.earthcheck.org/.
[e] Foundation for Environmental Education. 2015. Awarded sites. Accessed 9 July 2015. http://www.green-key.org/.
[f] Keep Wales Tidy. Welsh Beaches. Accessed 9 July 2015. http://www.keepwalestidy.org.

takes of nontarget species such as whales, dolphins, sharks, turtles, and sea-birds since the mid-1980s. In the 2014 edition of *The State of World Fisheries and Aquaculture*, the Food and Agriculture Organization of the United Nations (FAO) pointed out that the number of overfished stocks have continued to increase since the late 1970s, even though the pace slowed after 1990 (FAO 2014). FAO's estimations from 2011 indicate that 28.8% of the total number of stocks assessed were overfished (in FAO terms "fished at a biologically unsustainable level"), and about 61.3% were fully fished (FAO 2014). Also, there is still an ongoing threat to marine biodiversity due to bycatch in fisheries (see, e.g., Wallace et al. 2013; Worm et al. 2013).

In the late 1980s, NGOs were no longer convinced that conventional state-focused strategies would help spur marine conservation (Ward and Philips 2008). Drawing on a market approach, they started to develop new consumer-oriented lines of action, especially in the United States (Iles 2004; Konefal 2013), where the wider public had a specific concern about dolphin bycatch during tuna fishing and, following a large consumer boycott of canned tuna, "dolphin-safe" or "dolphin-friendly" labels appeared on canned tuna. In 1988, the Flipper Seal of Approval was launched by the Hawaii-based NGO EarthTrust, and the California-based NGO Earth Island Institute established a dolphin-safe label in 1990. Also, the U.S. Department of Commerce moved in this new direction and launched a dolphin-safe label in 1990 (Konefal 2013; Teisl et al. 2002).

In the following decade, a proliferation of seafood labels occurred in other parts of the world, most notably in Europe, but also in Japan (Parkes et al. 2010). This led to a huge variety in NGO-led, state-led, industry-driven, and joint partnership programs. Leading actors of the most established schemes, however, are mainly not-for-profit organizations—together dubbed the sustainable seafood movement (Konefal 2013). This movement includes general environmental NGOs, specialist NGOs that focus on specific issues in marine conservation or food safety/quality, and certifying organizations, which are not necessarily not-for-profit organizations (see Chapter 18 in this volume). Although the initial focus was on one single (bycatch) species, the scope soon broadened and many certification schemes now include broader sustainability requirements, such as the need to address stock overexploitation of a target species and minimizing harmful impacts of fisheries on the marine ecosystem. A well-known example of a "broad" standard is the global certification programs of the Marine Stewardship Council (MSC), which was set up in 1996 by a partnership of the World Wide Fund for Nature (WWF) and the multinational corporation Unilever, and was replaced in 1998 by an independent certifying body (Gulbrandsen 2009; Ward and Philips 2008). Because it draws on third-party assessments with independently accredited auditors, MSC is regarded as one of the most credible certification schemes (Gulbrandsen 2009). However, MSC also receives criticism for being, among other things, lenient and unable to show real impact (Christian et al. 2013; Cressey 2013; Gulbrandsen 2009; Ward and Philips 2008).

Another informational mode in fishery governance that emerged at the start of the new millennium is the so-called *seafood guides* (Iles 2004; Ward and Philips 2008; Roheim 2009). These guides are mostly nation- or region-specific, linking up to consumers in terms of language and diet wishes. Based on a traffic-light system, the guides help consumers in buying "best choice" fish (marked green), a "good alternative" (seafood with some sustainability concerns, marked orange), or in avoiding fish products not derived from sustainable fisheries (marked red). Some seafood guides include a fourth category, adding whether a fishery has been certified, for example by the MSC or the Earth Island Institute's dolphin-safe label (de Vos and Bush 2011; Roheim 2009). Seafood guides were first only available as printed wallet cards, but now they are also accessible via mobile apps. Seaman (2009) estimated that up to 200 guides are in use, mostly driven by not-for-profit organizations. WWF, for example, launched national seafood guides in more than 10 countries. Aquariums are another prominent player strengthening the sustainable seafood movement in this respect. The most famous example is the Seafood Watch of Monterey Bay Aquarium. Its outreach is immense, both in terms of numbers of wallet cards handed out and in terms of other seafood guides following the aquarium's design and/or recommendations (Seaman 2009; Roheim 2009).

Performance indices or league tables are a different kind of market-based tool used to show the relative sustainability performance of retailers and how (large) processing companies perform. These indices are designed to influence a company's reputational capital and to push consumer preferences toward sustainable brands. Compared to standards and guides, rankings seem not very widespread in fishery governance, although Greenpeace has several league tables, such as a sustainability ranking on tuna fish and various nation-specific supermarket rankings on seafood (Greenpeace 2013, 2014).

Information challenges faced in fisheries management are equally evident in governing through certification, guides, and rankings, as in the science–policy interface in traditional (state-led) decision-making (see Chapters 4 and 8 in this volume). Discussions about the extent to which fisheries alone are responsible for low stocks and ecological damage remain persistent, also in informational governance arrangements. Also, uncertainties related to unknown or missing data are complicating decision-making processes (Ruckelshaus et al. 2008; Garibaldi 2012; Wilson 2009). More than in governmental steering, in informational governance arrangements the authority of leading actors is explicitly linked to information quality. *Informational authority*, that is, the decision-making power on what/whose information is needed and used, is reliant on whether the public at large considers the information provider trustworthy (Iles 2004; Auld and Gulbrandsen 2010; Mol 2008; Ward and Philips 2008). Because Seafood is publicly visible commodities, problems with traceability and mislabeling are major threats for informational governance programs (Jacquet and Pauly 2007; Miller and Mariani 2010).

Most well-established programs are led by not-for-profit organizations, and fall in the category of third-party verification. Authority can be safeguarded in two ways: by ensuring independence of the auditors and by making use of science-based information a key principle (e.g., through scientific review panels, and hiring independent scientists for assessments) (Gulbrandsen 2009; Toonen and Mol 2013). But these strategies have consequences for inclusion and exclusion of fisheries, for instance, if data are unavailable or financial burdens become too high. This is especially challenging with regard to the participation of small-scale fisheries from the global South (Bush et al. 2013).

Another pending question is whether information-based tools bring about sustainable change in global fisheries (Jacquet and Pauly 2007, 2008; Jacquet et al. 2010). In a 2008 FAO review, it was concluded that uptake of seafood eco-labels remains modest and limited to specific countries or regions, such as northwest Europe and the United States, in case of MSC (Washington 2008). Table 6.1 shows key characteristics, including recent numbers about outreach, of five well-known examples of informational governance arrangements using third-party verification. Although these figures show impact, it remains difficult to measure success in addressing issues of overfishing, declining marine biodiversity, and habitat destruction, both by individual schemes, and in general (as even numbers about uptake by chain partners and consumers of most individual schemes are not easily accessible). Moreover, given the mushrooming of initiatives, the wide variety of labels for consumers and the degree of competition between programs remain key challenges (Roheim 2009; Jacquet et al. 2010). According to Parkes et al. (2010), overall consumer confidence in labels is undermined due to inconsistent approaches and conflicting advice of the many programs and schemes. Significant differences exist, for example, in the way performance is assessed, with voluntary certification schemes providing in-depth assessments on individual stocks, while seafood guides give general information at the species level. There is a growing call for more harmonization of the various schemes (e.g., by Parkes et al. 2010).

6.4 Governing through Voluntary Shipping Standards

About 80% of the globally traded volumes are transported over sea. This cargo shipping has various problematic impacts on the marine (and wider atmospheric and terrestrial) environment related to resource depletion, especially energy, but also end-of-life-cycle materials, accidental or deliberate pollution of water (waste and waste water discharge at sea, oil leakage, introduction of invasive species through ballast water) and air (greenhouse gases, sulfur dioxide, nitrogen oxides), and disturbance of natural habitats related to, among others, marine protected areas and coastal protected areas (Lai et al. 2011; Yang et al. 2013) (see also Chapter 13 in this volume).

There are a limited but growing number of governmental and international regulations that aim to reduce environmental impacts from shipping. The most influential regulations are the more than 20 international conventions governed by the International Maritime Organization (IMO) (such as the International Convention for the Prevention of Pollution from Ships (MARPOL 1973/1978) and the Convention on the Prevention of Marine Pollution by Dumping of Wastes and Other Matter (1972/1996), as amended). However, these national and international regulations fall short due to their time-consuming process of international decision-making, lack of (incentives for) compliance for shipping companies, and lack of effective enforcement mechanisms (DeSombre 2006; Wuisan et al. 2012).

Over the last 20 years (starting with the 1994 Green Award of the Port of Rotterdam), but especially over the last 10 years, a variety of private and public–private initiatives can be identified that aim to govern the greening of shipping through non-state systems of informational governance. These performance indices, labeling systems, certification systems, and management systems are (largely) voluntary initiatives taken by different stakeholders in the value chain of shipping, which target the environmental and sustainability performance of cargo shipping through second- and third-party verification. Private classification societies (especially those that are members of the International Association of Classification Societies—IACS) class ships so they can benefit from preferential access to, for example, insurance, port entry, and registry, also include environmental criteria related to ship construction (DeSombre 2006, pp. 181ff). Most systems are based on (a set of) environmental standards and requirements related to installed equipment (e.g., double hull, air pollution abatement technology), operational measures, management systems, legal compliance, and/or environmental performance. EMSA (2007) and Svensson and Andersson (2012) listed 47 and 38 different voluntary environmental performance indices/systems, respectively, ranging from single issue systems on, for instance, energy performance (such as the Energy Efficiency Operational Indicator of the IMO) to systems that include a variety of environmental variables into one index/label (such as the Clean Shipping Index or Lloyd's Environmental Protection).

Informational governance arrangements for sustainable shipping vary in geographical reach (national versus international systems), and in target groups and intended users (port owners, shipowners, cargo owners such as those working together in the Clean Cargo Working Group, governmental authorities, etc.). There is a huge variety in application, as well as in the actors initiating, designing, and operating the system. Many of the more popular and advanced systems have third-party verification, enhancing quality control of the data used for performance assessment, and increasing credibility. Various studies have also identified or proposed criteria to assess the quality and usefulness of these environmental shipping indices or performance

standards (e.g., EMSA 2007; Jivén and Jivén 1998). These criteria are comparable to, and reflect, more general requirements for voluntary environmental
performance indicators/indices or labels.

These voluntary systems on greening shipping are used for governing the
environmental performance of shipping in various ways. Some systems are
linked to financial incentives such as port dues, registration fees, or tonnage
tax (e.g., Maritime Singapore Green initiative, World Ports Climate Initiative,
Green Award Foundation). Other systems are related to the possibility to
obtain a specific insurance or insurance premiums (such as those of the
American Bureau of Shipping or Nippon Kaiji Kyokai), or even to port access
(such as classifications of ships by IASC members). Systems are also developed and used by cargo owners as a prerequisite for being able to ship their
cargo; for example, indices developed by the Clean Shipping Project (Wuisan
et al. 2012) or the Clean Cargo Working Group (Svensson and Andersson
2012). Walmart and IKEA are examples of cargo owners that apply these tools
in selecting shipping companies (Lai et al. 2011). Larger associations of cargo
owners also use some of these voluntary informational systems, such as the
Oil Companies International Marine Forum and the Chemical Distribution
Institute, representing oil companies and chemical companies, respectively
(DeSombre 2006). Indices are also used by ship owners and their associations (such as INTERTANKO and INTERCARGO) and transport buyers to
regulate and differentiate themselves from other transporting companies
and cargo owners in an increasingly global competition where environmental performance is starting to become a relevant competitive advantage.

Implementation of the voluntary systems in practice shows various shortcomings in improving environmental performance through such largely
private informational governance arrangements. Cargo owners, which are
the powerful players in sea freight shipping, still rarely set strict environmental demands on shipping companies, as low costs and on-time delivery
prevail (Wuisan et al. 2012). As such, they do not frequently ask for transporting companies fulfilling green indices for shipping. But there are exceptions.
Liner shipping offers more possibilities than bulk shipping due to the lower
transport costs as a percentage of the final price and the higher public profile of consumer goods through liner shipping. Long-standing supply chain
relations and contracts between cargo owners and shipping companies show
more opportunities for the application of environmental performance indices and voluntary green shipping practices than short-term temporary contracts (Yang et al. 2013). Also, the large well-known cargo owners (e.g., IKEA,
Mattel, Nike, HP, and Walmart), mega-shipping companies (e.g., Maersk, APL,
NKY, OOCL, and Hapag-Lloyd), and well-known globally leading sea ports
(e.g., Rotterdam, Singapore, Hamburg, and New York) seem more willing to
implement and enforce voluntary standards than their smaller and less well-
known equivalents (Lai et al. 2011). In addition, the large number of intermediary actors (such as brokers, forwarders, operators, and managing companies)
complicate and limit the application of voluntary green performance indices

and systems. The limited public visibility and transparency of these voluntary systems also hamper public pressure on the environmental behavior of cargo owners and shipping companies (compared with, for instance, transparency and public pressure for sustainable fish). This can be illustrated by looking at five key examples: Table 6.2 shows, next to some basic information, that information about the outreach to consumers is lacking.

A more recent development is that some of the voluntary systems have turned compulsory, especially those that are (co)developed by the IMO or national governments. An example is the Energy Efficiency Development Index, although it is only compulsory for new ships built after its entering into force in 2013. This turn to mandatory systems may also happen for other voluntary systems in the future. A second development is a growing call (by shipowners, transport buyers, and international organizations such as IMO and WWF) for internationally accepted standards, a meta-standard, harmonization, and/or even a unified system. Confusion, incomparability, unfamiliarity with each of the systems, high administrative burdens, and difficulties in communicating environmental performance to the public are often-mentioned reasons for the meager proliferation and implementation of these informational governance instruments. Moreover, a further harmonization and integration of this too diverse field of voluntary systems and standards is believed to be essential for mitigating these shortcomings (see also Chapter 13 in this volume).

6.5 Governing Coastal and Marine Tourism

Tourism is one of the fastest growing economic sectors around the world, and in many countries it contributes to major increases in employment and economic growth. The most recent numbers from the United Nations World Tourism Organization (UNWTO) point to 1,135 million international travel arrivals in 2014, and forecasts indicate a 3%–4% growth in 2015 (UNWTO 2015, p. 11). Part of this tourism is related to the oceans, either in coastal areas through hospitality tourism in coastal hotels and resorts, beach activities, nature activities, docking stations for large cruise ships, marinas, or at open sea through cruises, sailing, diving, whale/dolphin watching, and so forth. With the intensification of marine tourism, the potential environmental impacts also increase, and thus the need for mitigation measures. Besides governmental policy and requirements, informational governance through voluntary environmental information provisioning, labels, benchmarking, sustainability rankings, and certification schemes, first introduced in Europe in the 1980s, is one of the strategies to reduce marine tourism environmental impacts. These voluntary systems can be consequential, especially in underregulated sectors such as cruise tourism.

A large variety of tourism-related environmental certification schemes, labels, and information provisioning have been developed over the last three decades, but most of them are related to the hospitality sector. Estimates range between 128 (Gössling and Buckley 2014) and 300 voluntary certification schemes (Lebe and Vrečko 2014). A small but increasing number of labels, information, and certification schemes are directly focused on the marine environment, such as those for cruises, beaches, marinas, cruise terminals, and touristic activities at sea. Quite a number of labels, information, and certification schemes are more general but also relevant for marine tourism, such as Green Globe and Green Key eco-labels (for hospitality on cruises) and the Carbon Disclosure Project (greenhouse gas emission reporting of cruise lines; see Lamers et al. 2015).

Voluntary informational governance is a very diverse, little structured, and not very transparent field in coastal and marine tourism. Labels, information, and certification schemes differ in various dimensions. They vary by the organization which has developed and implemented these labels: some are in the hands of private (for-profit) organizations/associations (such as the Voluntary Initiatives for Sustainability in Tourism (VISIT) in Europe), others are run by not-for profit NGOs (such as the Blue Flag program, run by the Foundation for Environmental Education; Creo and Fraboni 2011), while still others are fully public ones (such as the EU eco-label), with in-between all kinds of hybrid schemes (such as the UK Green Sea Initiative, which includes public and private actors).

Certification and information schemes differ also in their geographical spread, especially the place-based ones such as those focused on beaches, diving, and marinas. Some are truly global (e.g., Blue Flag, Green Globe, Global Reporting Initiative; Bonilla-Priego et al. 2014), others have a regional orientation (Certification for Sustainable Tourism in Latin America), and national ones also exist. Some cover a single activity such as a boat trip (whale/dolphin/seal watching; e.g., Taiwan Cetacean Society program; Chen 2011) or a dive (e.g., the STEP Dive Center Standard; PADI Environmental Achievement Award), others are directed at complex services (a full holiday including travel, overnight stay, and various activities) or infrastructure (e.g., a beach or marina; Botero et al. 2015). The diversity also relates to the coverage of the certification scheme or label. Some are single-issue labels, such as the carbon labels on travel (Gössling and Buckley 2014; e.g., the greenhouse gas emission standard of Shipping Efficiency), others cover a variety of environmental impacts ranging from energy and water use—via environmental management systems, purchasing and environmental training—to waste handling (such as the Clean Marine Program ratings of marinas, the Blue Flag program for beaches and marinas using 32 criteria, and Green Globe). Most of these (marine-related or general tourism) labels and certification systems are third-party certified, but some are second-party certified (such as the STEP Dive Center Standard and Go Eco Operator certification).

Besides labels or certifications, some programs also aim to benchmark marine tourism in a specific sector or provide recommendation lists. Friends of the Earth (FoE) annually reviews major cruise lines on their environmental performances (sewage treatment, air pollution, water quality compliance, and transparency) in their Cruise Ship Report Cards (FoE 2014); these reviews were initially done in cooperation with the Cruise Lines International Association, but since 2014 have been conducted independently. The U.S. Clean Beach Coalition of public and private actors provides an annual recommendation list of Blue Wave Beaches in the United States, as does the Marine Conservation Society with its Good Beach Guide. And Green Globe is a general ecotourism benchmarking system, also applicable for marine tourism.

How does this variety of voluntary systems work in improving environmental performance in coastal and marine tourism? Labels, information, and certification systems and benchmarks can impact individual consumer preferences as well as the behavior of institutional actors in the tourism value chain (investors, tour operators). Table 6.3 presents five early (but still existing) examples, showing outreach to either tourists or chain actors, or both. Various studies have been carried out to estimate to what extent tourists are sensitive to and (re)direct their behavior following environmental/sustainability certificates and schemes. In most tourism markets, tourists welcome such informational governance instruments but generally do not consider them of major importance in selecting tourist services and/or paying a premium; hence, there is limited market advantage (e.g., Chen 2011). Some studies identify a niche group of tourists that is guided in preferences by eco-labels (Blanco et al. 2009; ITB/IPK 2012), also as a kind of general quality assurance. For instance, the beaches on the US Blue Wave benchmark list attract more beach tourists than alternative beaches in the region.

Some institutional actors in the tourism value chain are also attracted by these informational governance systems. Tour operators and investors aim to fulfill environmental conditionalities of such voluntary programs to obtain a competitive advantage or prevent negative exclusion in markets, especially in highly competitive marine tourism markets. The Blue Flag label, for instance, does attract investment in additional hotel construction along labeled beaches (Blackman et al. 2014; Lucrezi et al. 2015). Some tour operators only select tourist facilities with specified labels. In addition, voluntary labels and certifications are also used to show civil society and other marine users the environmental advances made by the marine tourism sector, for instance, around marinas, or in conflicts between nature protection organizations and diving schools. Finally, these voluntary programs are sometimes applied out of intrinsic sustainability motivations of tourism facility/service operators.

But there are clear limitations. The rapid growth of tourism eco-labels and certification schemes endanger the effect of these schemes on customers and consumers, as their large number, unfamiliarity, confusingly similar names, unclear communication, and low credibility (especially of sector run programs; de Groot and Bush 2010) make their governing power limited (Lebe and Vrečko 2014). Also, the coverage of the labels (in terms of environmental criteria included) and the stringency of requested adaptation is the subject of frequent criticism: often requirements to obtain a label/certificate do not go beyond national laws.

6.6 Comparing Informational Governance Arrangements between Sectors

Informational governance draws on the governing strength of disclosing information about how, and to what extent, actors in their practices impact the marine environment, therewith pushing them toward sustainability change. Illustrated by informational governance reviews from three distinct sectors, it becomes clear that "regulatory" roles become available for a wide variety of actors. Positions in information gathering, processing, disclosing, communicating, verifying, certifying, and so forth, are not just bound to scientists, experts, and state authorities, as is most common in state-led governance and decision-making, but are open to and actively seized by a wide variety of actors. There are seemingly broad and ever-changing constellations of actors involved in informational governance: private companies within the sector; a wide range of other private companies, such as insurance companies, banks and investors, consultancies, and certification bodies; NGOs of different interest, size, focus, and operational scales (local to global); a diversity of state institutions (again from local to global levels); and scientific and expert institutions.

Some general differences in actors leading informational governance can be identified among the three sectors. In fisheries, NGOs clearly play a dominant role in agenda setting, designing informational instruments and implementing them. While informational governance by NGOs started with raising public awareness and changing retailing and consumption practices around dolphins and whales, the so-called seafood movement later moved toward protecting less attractive and mediagenic species, and addressed more complex themes as resource depletion and habitat destruction, especially through labeled products and wallet cards. In shipping, large for-profit private actors (such as cargo owners) and state actors (such as port authorities) are leading information governance at sea, although they are not always behind the design of these instruments. With the absence of wide public attention for environmental impacts of ocean transportation, the main emphasis is on pushing business partners up and down the chain toward more environmental awareness. Informational governance works especially where business relations are predictable and durable, as shown by the fact that more persistent modes of informational governance are found in liner shipping than in bulk shipping. In tourism, practices are more diverse and leaders/frontrunners in informational governance are not easily characterized. Trends in cruise tourism seem to be more or less in line with those in liner shipping, with big private companies (and global NGOs) focusing on affecting chain partners instead of addressing tourists directly. In localized tourism initiatives or single activities (beaches, diving, and whale watching), informational governance arrangements address individual tourists that engage in a particular practice.

Hence, the organization of the sector-specific value chain rather than the actual environmental impact of the marine practice determines the leading actors in informational governance of the marine environment. While in state-led marine governance the focus is on the ones closest to the problem (polluters/extractors), in (primarily private) informational governance this is not necessarily the case. The main focal actor for informational governance is often a publicly visible and/or leading actor in the value chain. In case of consumer products (e.g., fish on the shelf) or consumer experiences (e.g., ecotourism in a local/beach context), leverage points and key actors are often positioned close to consumers, using information to steer the practices consumers are engaged in (e.g., shopping, selecting a beach) toward sustainability. Informational governance in shipping and cruise tourism cannot make much use of public visibility and individual citizen/consumer choices. Here, informational governance works higher up in the value chain, between institutional actors. With these latter forms of informational governance, the leverage and driver is not so much related to price premiums or major new market shares, but rather part of safeguarding a social license to operate and preserving reputational capital, especially for major brands. Ecolabels, information and certification schemes, and sustainability ratings/benchmarks give account of a company's sustainability profile and hence its license to operate.

These forms of informational governance of the oceans are, of course, not without problems. Besides the standard environmental governance problems (lack of capacity, implementation deficit, marginal impact in changing practices), our review identified at least three additional weaknesses in informational modes of marine governance. First, more than conventional state-led forms of governance, informational governance is faced with data/information challenges such as environmental data deficiencies (especially in the global South and/or specific practices), data uncertainties, and data quality/verification. Second, for a number of environmental challenges in the three sectors, a danger of informational/labeling overflow is emerging. This overflow often goes together with competing informational systems (and competition between the interests and organizations that go behind these informational systems; Miller and Bush 2015) and little coordination or harmonization between these competing systems. The multiple informational systems for sustainable fish form a key example of this overflow, but there are others. This causes confusion and undermines the credibility and effectiveness of such informational instruments. Third, not all value chain actors related to marine environmental pollution or resource extraction practices are equally vulnerable for and reactive to informational governance arrangements. In quite a number of cases, companies or consumers in markets do not react upon informational governance instruments such as eco-labels, sustainability rankings, or information disclosure.

Hence, informational governance in our three sectors will not easily replace state-led governance for sustainability. More likely, informational governance of the oceans will complement, strengthen, and intertwine with state-led forms of marine governance, resulting in more complex polycentric and multiactor governance arrangements at sea. Informational governance of the oceans will especially emerge when it fills a void left by weak state environmental governance, by unambitious state environmental governance, and by extra-sovereignty spaces. Moreover, this void is likely to be filled successfully by sustainability advocates with *value chain power*: informational governance modes currently prevailing, at least in fisheries and shipping, seem to have a bias toward actors (NGOs, consumers, and private companies) from the global North.

6.7 Epilogue: Ecosystem-Based Governance through Information?

As has been widely stated and evidenced, state-based governance of marine resources has clear limitations in the current era. Increasingly, other actors perform on the marine environmental protection stage, using a wider variety of governance resources. The notion of informational governance captures the development that private actors use informational resources and processes to move marine practices in more sustainable directions. Our investigation of informational governance in three marine sectors (fisheries, shipping, and tourism) showed a wide diversity of such practices and arrangements with an equal diversity in impact. What has become clear from this review is that marine informational governance is sharply emerging and should no longer be considered a marginal or peripheral activity in protecting our marine environment. It is an important governance mode, which is here to stay, complementary to (rather than replacing) state-based marine governance.

In exploring informational governance at sea, this chapter focused on marine sectors and thus sectoral forms of marine governance. Sectoral forms of environmental governance are somewhat in contrast with the current call for a more integrated management of marine resources, where not so much individual practices or sectors are central foci of environmental governance, but rather the preservation of ecosystems. Ecosystem-based approaches to marine resources (e.g., Cury et al. 2005; Gilliland and Laffoley 2008; Ruckelshaus et al. 2008) take the desirable state of place-based complex ecosystems as their object of study and starting point of governance, rather than the impacts of social practices, sectors, or mobile polluters (e.g., cargo and cruise ships) or resources extractors (e.g., fishing vessels, sand or oil extracting companies). The boundaries of ecosystem-based approaches are typically based on scientific denominators derived from biogeography, oceanography, and the like, rather than on social characteristics of social practices, economic sectors, or companies. Regularly, skepticism has been expressed regarding the realism of such integrated and complex ecosystem-based approaches and models, and regarding defining reference points of optimum marine ecosystem metrics (Cury et al. 2005). But there seems to be wide acceptance that such complex ecosystem approaches are preferable to more single-species or single-use protection efforts of marine resources. Since the start of the new millennium, governments have been in the process of establishing marine protected areas (Halpern et al. 2010) and have emphasized marine spatial planning (Jay et al. 2013), both typical tools of ecosystem-based governance where the objective is place-based ecosystem protection.

In light of the rising importance of marine ecosystem-based management, the question is whether sector-based informational governance arrangements can be integrated to serve and support integrated ecosystem-based marine management. Could these sectoral informational arrangements in one way or the other be joined or combined to develop a more integrated informational governance approach toward marine ecosystem protection? This does not seem very likely. The informational governance arrangements for marine protection we studied differ fundamentally from ecosystem-based approaches in that these are (1) rather social system-based and oriented than ecosystem-based, (2) much broader in informational sources as they include not just scientific information but also other non-scientific information, and (3) less focused and referenced on place-based systems but rather center around (regulating and governing) mobile flows (of fish, ships, and tourists). Informational governance arrangements as discussed in this chapter could even be seen as the opposite of an ecosystem-based approach: a social system based approach, where governance is designed on the basis of the characteristics of social systems (fisheries, shipping, and tourism). As such, informational governance is not easily a basis, building block, or framework for designing and implementing ecosystem-based management approaches. What seems more likely is that, either within or outside the framework and contours of an ecosystem-based approach, sectoral informational governance arrangements and strategies are set to work to reduce environmental impacts.

References

Auld, G. and L. H. Gulbrandsen. 2010. Transparency in nonstate certification: Consequences for accountability and legitimacy. *Global Environmental Politics* 10: 97–119.

Beck, U. 1986. *Risikogesellschat. Auf dem Weg in eine andere Moderne.* Frankfurt: Subr-kamp.

Beck, U. 2009. *World at Risk.* Cambridge: Polity.

Blackman, A., M. A. Naranjo, J. Robalino, et al. 2014. Does tourism eco-certification pay? Costa Rica's Blue Flag program. *World Development* 58: 41–52.

Blanco, E., J. Rey-Macquieria, and L. Lozano. 2009. Economic incentives for tourism firms to undertake voluntary environmental management. *Tourism Management* 30: 112–122.

Bonilla-Priego, M. J., X. Font, and M. del Rosario Pacheco-Olivares. 2014. Corporate sustainability reporting index and baseline data for the cruise industry. *Tourism Management* 44: 149–160.

Botero, C.-M., A. T. Williams, and J. A. Cabrera. 2015. Advances in beach management in Latin America: Overview from certification schemes. In *Environmental Management and Governance: Advances in Coastal and Marine Resources,* edited by C. W. Finkl and C. Makowski, 33–63. Dordrecht: Springer.

Burn, G., T. Tyler, J. Zadkovich, et al. 2015. Legal issues in cross-border resource development. *Journal of World Energy Law & Business* 8: 154–172.

Bush, S. R., H. M. Toonen, P. Oosterveer, et al. 2013. The "devils triangle" of MSC certification: Balancing credibility, accessibility and continuous improvement. *Marine Policy* 37: 288–293.

Case, D. W. 2001. The law and economics of environmental information as regulation. *Environmental Law Reporter* 31: 10773–10789.

Castells, M. 1996/1997. *The Information Age: Economy, Society and Culture,* three volumes. Oxford: Blackwell.

Chen, C.-L. 2011. From catching to watching: Moving towards quality assurance of whale/dolphin watching tourism in Taiwan. *Marine Policy* 35: 10–17.

Christian, C., D. Ainley, M. Bailey, et al. 2013. A review of formal objections to Marine Stewardship Council fisheries certifications. *Biological Conservation* 161: 10–17.

Cohen, M. A. 2000. *Information as a Policy Instrument in Protecting the Environment; What Have We Learned?* Washington, DC: Environmental Defense Fund.

Convention on the Prevention of Marine Pollution by Dumping of Wastes and Other Matter. 29 December 1972. 1046 UNTS 120. As amended by the 1996 Protocol, Can TS 2006 No 5.

Creo, C. and C. Fraboni. 2011. Awards for the sustainable management of coastal tourism destinations: The example of the Blue Flag program. *Journal of Coastal Research* 61: 378–381.

Cressey, D. 2013. Fight over sustainable seafood labelling flares up. *Nature News Blog.* Accessed 11 July 2015. http://blogs.nature.com/news/2013/04/fight-over-sustainable-seafood-labelling-flares-up.html.

Crowder, L. and E. Norse. 2008. Essential ecological insights for marine ecosystem-based management and marine spatial planning. *Marine Policy* 32: 772–778.

Cury, P. M., C. Mullon, S. M. Garcia, et al. 2005. Viability theory for an ecosystem approach to fisheries. *ICES Journal of Marine Science* 62: 577–584.

Dayton, P. K., S. F. Thrush, M. T. Agardy, et al. 1995. Environmental effects of marine fishing. *Aquatic Conservation: Marine and Freshwater Ecosystems* 5: 205–232.

de Groot, J. and S. R. Bush. 2010. The potential for dive tourism led entrepreneurial marine protected areas in Curacao. *Marine Policy* 34: 1051–1059.

DeSombre, E. R. 2006. *Flagging Standards: Globalization and Environmental, Safety, and Labor Regulations at Sea.* Cambridge, MA: MIT Press.

de Vos, B. I. and S. R. Bush. 2011. Far more than market-based: Rethinking the impact of the Dutch Viswijzer (Good Fish Guide) on fisheries' governance. *Sociologia Ruralis* 51 (3): 284–303.

EMSA (European Maritime Safety Agency). 2007. *Study on Ships Producing Reduced Quantities of Ships Generated Waste: Present Situation and Future Opportunities to Encourage the Development of Cleaner Ships.* Lisbon: EMSA.

Esty, D. 2004. Environmental protection in the information age. *NYU Law Review* 79: 115–211.

FAO (Food and Agriculture Organization of the United Nations). 2014. *The State of World Fisheries and Aquaculture 2014: Opportunities and Challenges.* Rome: FAO.

FoE (Friends of the Earth). 2014. *2014 Cruise Ship Report Card.* Washington, DC: FOE.

Fung, A., M. Graham, and D. Weil. 2007. *Full Disclosure: The Perils and Promise of Transparency.* New York: Cambridge University Press.

Garibaldi, L. 2012. The FAO global capture production database: A six-decade effort to catch the trend. *Marine Policy* 36: 760–768.

Gilliland, P. M. and D. Laffoley. 2008. Key elements and steps in the process of developing ecosystem-based marine spatial planning. *Marine Policy* 32: 787–796.

Gössling, S. and R. Buckley. 2016. Carbon labels in tourism: Persuasive communication? *Journal of Cleaner Production* 111: 358–369.

Graham, M. 2002. *Democracy by Disclosure: The Rise of Technopopulism.* Washington, DC: Brookings Press.

Gray, T. 2005. *Participation in Fisheries Governance.* Dordrecht: Springer.

Greenpeace. 2013. 2013 Canned tuna sustainability ranking. Accessed 27 February 2015. http://www.greenpeace.org/canada/en/campaigns/ocean/Tuna/Get-involved/2013-canned-tuna-sustainability-ranking/.

Greenpeace. 2014. Protecting our oceans is everyone's business: Ranking supermarkets on seafood sustainability. Accessed 27 February 2015. http://www.greenpeace.org/canada/en/campaigns/ocean/Seafood/Resources/Reports/supermarket-ranking-2014/.

Gulbrandsen, L. H. 2009. The emergence and effectiveness of the Marine Stewardship Council. *Marine Policy* 33: 654–660.

Gupta, A. and M. Mason, eds. 2014. *Transparency in Global Environmental Governance: A Critical Perspective.* Cambridge, MA: MIT Press.

Halpern, B. S., S. E. Lester, and K. McLeod. 2010. Placing marine protected areas onto the ecosystem-based management seascape. *Proceedings of the National Academy of Sciences of the United States of America* 107: 18312–18317.

Halpern, B. S., S. Walbridge, K. A. Selkoe, et al. 2008. A global map of human impact on marine ecosystems. *Science* 319 (5865): 948–952.

Holm, P. 2002. Crossing the border: On the relationship between science and fishermen's knowledge in a resource management context. *MAST* 2: 5–33.

Howes, M. 2002. Reflexive modernization, the internet, and democratic environmental decision making. *Organization & Environment* 15: 328–331.

Iles, A. 2004. Making seafood sustainable: Merging consumption and citizenship in the United States. *Science and Public Policy* 31: 127–138.

ITB/IPK. 2012. World Travel Trends Report 2011/2012. Munich: IPK International.

Jacquet, J. L. and D. Pauly. 2007. The rise of seafood awareness campaigns in an era of collapsing fisheries. *Marine Policy* 31: 308–313.

Jacquet, J. L. and D. Pauly. 2008. Trade secrets: Renaming and mislabeling of seafood. *Marine Policy* 32: 309–318.

Jacquet, J. L., D. Pauly, D. Ainley, et al. 2010. Seafood stewardship in crisis. *Nature* 467 (7311): 28–29.

Jay, S., W. Flannery, K. Vince, et al. 2013. International progress in marine spatial planning. In *Vol. 27 of Ocean Yearbook*, edited by A. Chircop, S. Coffen-Smout, and M. McConnell, 171–212. Leiden: Brill Martinus Nijhoff.

Jivén, A. and K. Jivén. 1998. *Methods and Index Systems for Environmental Valuation of Sea Transports*. Department of Naval Architecture and Ocean Engineering Report No. X-98–101. Gothenburg: Chalmers University of Technology.

Karkkainen, B. C. 2001. Information as environmental regulation: TRI and performance benchmarking, precursors to a new paradigm? *Georgetown Law Journal* 89: 259–370.

Kleindorfer, P. R. and E. W. Orts. 1999. Informational regulation of environmental risks. *Risk Analysis* 18: 155–170.

Konar, S. and M. A. Cohen. 1997. Information as regulation: The effect of community right to know laws on toxic emissions. *Journal of Environmental Economics and Management* 32 (1): 109–124.

Konefal, J. 2013. Environmental movements, market-based approaches, and neoliberalization: A case study of the sustainable seafood movement. *Organization & Environment* 26: 336–352.

Kooiman, J. 2003. *Governing as Governance*. London: Sage.

Lai, K.-H., V. Y. H. Lun, C. W. Y. Wong, et al. 2011. Green shipping practices in the shipping industry: Conceptualization, adoption, and implications. *Resources, Conservation and Recycling* 55: 631–638.

Lamers, M., E. Eijgelaar, and B. Amelung. 2015. The environmental challenges of cruise tourism: Impacts and governance. In *The Routledge Handbook of Tourism and Sustainability*, edited by C. M. Hall, S. Gossling, and D. Scott, 430–439. London: Routledge.

Lebe, S. and I. Vrečko. 2014. Ecolabels and schemes: A requisitely holistic proof of tourism's social responsibility? *Systems Research and Behavioral Science* 32: 247–255.

Lewison, R. L., L. B. Crowder, A. J. Read, et al. 2004. Understanding impacts of fisheries bycatch on marine megafauna. *Trends in Ecology & Evolution* 19: 598–604.

Lucrezi, S., M. Saayman, and P. Van der Merwe. 2015. Managing beaches and beachgoers: Lessons from and for the blue flag award. *Tourism Management* 48: 211–230.

MARPOL (International Convention for the Prevention of Pollution from Ships). 2 November 1973. 1340 UNTS 184. As amended by the Protocol of 1978, 17 February 1978, 1340 UNTS 61 and the Protocol of 1997, 26 September 1997, Can TS 2010 no 14.

Miller, A. M. M. and S. R. Bush. 2015. Authority without credibility? Competition and conflict between ecolabels in tuna fisheries. *Journal of Cleaner Production* 107: 137–145.

Miller, D. D. and S. Mariani. 2010. Smoke, mirrors, and mislabeled cod: Poor transparency in the European seafood industry. *Frontiers in Ecology and the Environment* 8 (10): 517–521.

Mol, A. P. J. 2006. Environmental governance in the information age: The emergence of informational governance. *Environment and Planning C: Government and Policy* 24: 497–514.

Mol, A. P. J. 2008. *Environmental Reform in the Information Age. The Contours of Informational Governance.* Cambridge and New York: Cambridge University Press.

Mol, A. P. J. 2009. Environmental governance through information: China and Vietnam. *Singapore Journal of Tropical Geography* 30 (1): 114–129.

Mol, A. P. J. 2010. The future of transparency: Power, pitfalls and promises. *Global Environmental Politics* 10 (3): 132–143.

Mol, A. P. J. and G. Spaargaren. 2005. From additions and withdrawals to environmental flows reframing debates in the environmental social sciences. *Organization & Environment* 18 (1): 91–107.

Parkes, G., J. A. Young, S. F. Walmsley, et al. 2010. Behind the signs—A global review of fish sustainability information schemes. *Reviews in Fisheries Science* 18 (4): 344–356.

Pattberg, P. 2007. *Private Institutions and Global Governance. The New Politics of Environmental Sustainability.* Cheltenham: Edward Elgar.

Pauly, D., R. Watson, and J. Alder. 2005. Global trends in world fisheries: Impacts on marine ecosystems and food security. *Philosophical Transactions of the Royal Society B: Biological Sciences* 360 (1453): 5–12.

Roheim, C. A. 2009. An evaluation of sustainable seafood guides: Implications for environmental groups and the seafood industry. *Marine Resource Economics* 24 (3): 301–310.

Ruckelshaus, M., T. Klinger, N. Knowlton, et al. 2008. Marine ecosystem-based management in practice: Scientific and governance challenges. *BioScience* 58 (1): 53–63.

Seaman, T. 2009. Are sustainable seafood lists supposed to confuse? *IntraFish Media.* 25 February. http://www.intrafish.com/news/article1351738.ece.

Stewart, R. B. 2001. A new generation of environmental regulation? *Capital University Law Review* 29: 21–182.

Suárez-de Vivero, J. L. 2013. The extended continental shelf: A geographical perspective of the implementation of article 76 of UNCLOS. *Ocean & Coastal Management* 73: 113–126.

Svensson, E. and K. Andersson. 2012. Inventory and evaluation of environmental performance indices for shipping. Department of Shipping and Marine Technology Report No. R11:132. Gothenburg: Chalmers University of Technology.

Teisl, M. F., B. Roe, and R. L. Hicks. 2002. Can eco-labels tune a market? Evidence from dolphin-safe labeling. *Journal of Environmental Economics and Management* 43 (3): 339–359.

Tietenberg, T. 1998. Disclosure strategies for pollution control. *Environmental and Resource Economics* 11: 587–602.

Toonen, H. M. and H. J. Lindeboom. 2015. Dark green electricity comes from the sea: Capitalizing on ecological merits of offshore wind power? *Renewable and Sustainable Energy Reviews* 42: 1023–1033.

Toonen, H. M. and A. P. J. Mol. 2013. Putting sustainable fisheries on the map? Establishing no-take zones for North Sea plaice fisheries through MSC certification. *Marine Policy* 37: 294–304.

Tysiachniouk, M. 2012. *Transnational Governance through Private Authority: The Case of the Forest Stewardship Council Certification in Russia.* Wageningen: Wageningen University Academic Publishers.

UNWTO (United Nations World Tourism Organization). 2015. Annual Report 2014. Madrid: UNWTO.

Van Leeuwen, J. 2010. *Who Greens the Waves? Changing Authority in the Environmental Governance of Shipping and Offshore Oil and Gas Production.* Wageningen: Wageningen Academic Publishers.

Verweij, M. C., W. L. T. van Densen, and A. P. J. Mol. 2010. The tower of Babel: Different perceptions and controversies on change and status of north sea fish stocks in multi-stakeholder settings. *Marine Policy* 34: 522–533.

Wallace, B. P., C. Y. Kot, A. D. DiMatteo, et al. 2013. Impacts of fisheries bycatch on marine turtle populations worldwide: Toward conservation and research priorities. *Ecosphere* 4 (3): art40.

Ward, T. J. and B. Phillips. 2008. *Ecolabelling of Seafood: The Basic Concepts. Seafood Ecolabelling: Principles and Practice.* Oxford: Oxford University Press.

Washington, S. 2008. *Ecolabels and Marine Capture Fisheries: Current Practice and Emerging Issues.* Rome: FAO (GLOBEFISH Research Programme).

Wilson, D. C. 2009. *The Paradoxes of Transparency: Science and the Ecosystem Approach to Fisheries Management in Europe.* Amsterdam: Amsterdam University Press.

World Ocean Council. 2014a. *International Ocean Governance: Marine Planning Brief.* New York: World Ocean Council.

World Ocean Council. 2014b. Report of the World Ocean Council Business Forum: Ocean policy & planning, 28–30 September 2014. New York: World Ocean Council.

Worm, B., E. B. Barbier, N. Beaumont, et al. 2006. Impacts of biodiversity loss on ocean ecosystem services. *Science* 314: 787–790.

Worm, B., B. Davis, L. Kettemer, et al. 2013. Global catches, exploitation rates, and rebuilding options for sharks, *Marine Policy* 40: 194–204.

Wuisan, L., J. van Leeuwen, and C. S. A. van Koppen. 2012. Greening international shipping through private governance: A case study of the clean shipping project. *Marine Policy* 36: 163–173.

Yang, C.-S., C.-S. Lu, J. J. Haider, et al. 2013. The effect of green supply chain management on green performance and firm competitiveness in the context of container shipping in Taiwan. *Transportation Research Part E* 55: 55–73.

Young, O. R., G. Osherenko, J. Ekstrom, et al. 2007. Solving the crisis in ocean governance: Place-based management of marine ecosystems. *Environment* 49 (4): 21–32.

Zhang, L., A. P. J. Mol, and G. Z. He. 2016. Transparency and information disclosure in China's environmental governance. *Current Opinion in Environmental Sustainability* 18: 17–24.

7

Inducing Better Stakeholder Searches for Environmental Information Relevant to Coastal Conservation

Diana L. Ascher and William Ascher

CONTENTS

7.1 Introduction

Sound knowledge is crucial for good coastal policy and management decision-making, as well as to guide farsighted practices by relevant resource users. We have in mind both group and individual decisions: (1) to support or oppose policy initiatives impacting coastal conservation, (2) to manage resources in particular ways, (3) to comply with conservation regulations, and (4) to engage in voluntary conservation actions. In many contexts, marvelous opportunities to protect or improve coastal systems exist, but they are neglected because their benefits are underappreciated compared to their costs. And, of course, long-term risks also exist, for example, situating oil facilities nearshore or off-shore, applying fertilizer to farmland near estuaries, and resisting or defying fishing moratoria; when such risks are underappreciated, significant damage can be done by misguided policies and practices.

Experts have recognized for decades that the complexity of coastal systems—involving land and sea effects, multiple affected industries, and multiple regulatory jurisdictions—necessitates integrated coastal management (ICM). (These challenges are explored in detail in Mercer Clarke 2010.) The need for integration requires that knowledge of coastal systems must not only be generated, but also be incorporated sufficiently in decisions by resource users, resource managers, and policy makers. In light of the highly polarized opinion camps on coastal conservation issues, reliance on overly simplistic information gathering and interpretation, especially when it leads to confirming existing narrowness and biases, creates a significant barrier to leaders who want to ensure that stakeholders have adequate knowledge about complex issues. Therefore, it is desirable for such information providers to induce stakeholders to engage in more active, systematic information seeking, as opposed to rudimentary information acquisition—or not seeking relevant information at all.

Yet, in many contexts over the past several decades, knowledge use has become more problematic than knowledge generation. As we shall argue, the use of knowledge to make sound decisions based on ICM inputs is challenging. Although remarkable progress has been made in generating information relevant to environmental decision-making, little attention has been paid to the creation of supplemental information to guide stakeholders in the *use* of this knowledge. Similarly, though challenges in the generation of sound environmental information persist, it is clear that the uptake of sound environmental information has not kept pace with its supply. Citizens, resource users, and policy makers too often rely on partial, inadequate, or inappropriate environmental information. To overcome the difficult challenges of integrated coastal management, the technical knowledge that often is *siloed* within esoteric groups or insulated fields must be synthesized, translated, and made actionable for the community at large. This is important particularly for selecting from among policies and practices that pose different types of risks residing in what seem like different areas of concern. Cass Sunstein (2002) documents environmental policy failures that occur due to misjudging or neglecting the relative risks of alternative policies. For example, declining agricultural yields may seem like an isolated challenge to farmers. However, to other stakeholders, such as fishers, this challenge presents a significant risk for long-term yields because of the potential for algal blooms and eutrophication of coastal waters due to increased fertilizer use (Anderson et al. 2002).

There is no shortage of possible reasons for the limited uptake of sound, actionable coastal conservation information by stakeholders. Unsound environmental messages can be retransmitted easily via social media networks, creating greater salience for members of the network than sound information that better represents the complex issues relevant to coastal conservation. Assessing the authenticity and authority of information sources can be challenging, raising doubt as to stakeholders' ability to identify valid information about coastal policies and practices. Similarly, overly technical information is daunting to stakeholders who lack the esoteric vocabulary that comes easily to those working directly on coastal conservation within governments, nongovernmental organizations (NGOs), and coastal management groups. In addition to the challenges posed by technical terminology and a lack of confidence in the ability to discern credible, authoritative information, the general overabundance of environmental information can cause stakeholders to develop an inaccurate judgment of the capacity for individual action to make a significant difference in the long-term sustainability of coastal conservation efforts. Further, a lack of differentiation from other messages concerning the environment, such as global climate change, makes coastal conservation knowledge vulnerable to issue fatigue.

These problems have been addressed largely with efforts to package and deliver sound environmental information more effectively. While such efforts are useful—particularly with regard to education and research in the technical arena—we maintain that the primary problems challenging the salience of sound environmental messaging and the ability of coastal conservation information to motivate action must be addressed from the other side, that is, by devising approaches to induce more effective acquisition and use of coastal conservation knowledge by stakeholders. Therefore, understanding how stakeholders acquire, interpret, and make decisions based on coastal conservation knowledge is instrumental to any attempt to improve coastal conservation knowledge transfer. This chapter offers approaches to enhance the salience of sound coastal conservation information and to motivate stakeholders to take action based on this information. Such approaches are not intended to supplant marketing and communications strategies to tailor information content and format for particular audiences; rather, our focus pertains to strategies informed by social interaction, identity formation, and triggers that motivate the use of sound coastal conservation knowledge, whether it is acquired actively or encountered passively.

It is important to note that the approaches may target one or more of four somewhat overlapping arenas in which acquisition and use of coastal conservation knowledge may be relevant to stakeholders:

- Individual action and state of mind: for example, complying with or defying regulations; adopting best conservation practices in farming, fishing, and so on; overcoming the anxiety arising from feelings of lack of mastery

- Peer interaction: for example, explaining ecosystem interactions, expressing views on what ought to be done regarding coastal management
- Collective action of nongovernmental stakeholders with common interests: for example, farmers jointly mounting a lobbying effort for or against a conservation proposal
- Government actions: funding research, creating regulations, invoking regulations in particular cases, assessing effectiveness.

Stakeholders may straddle various arenas, or information contexts. For example, sometimes stakeholders with common interests are also bound together through simple friendship or family ties, or membership in social or civic clubs that are not directly related to collective action with respect to resource exploitation. Sometimes stakeholders with personal or group interests also serve on governmental or quasi-governmental planning commissions, water district boards, and so on. Despite such overlaps, distinguishing among these arenas is important because it clarifies the multiplicity of instrumental uses of the knowledge (select a position, strengthen the argument for that position within the collective action group, strengthen the argument of that position vis-à-vis the government) and a host of other motives, including fulfilling basic psychological impulses, expressing value preferences, gaining intellectual mastery, and achieving personal advancement.

7.2 Knowledge for Sound Coastal Conservation Decision-Making

What does it mean to assert that a stakeholder has adequate knowledge to make sound decisions about activities that affect coastal conservation? This is the epistemological facet of the problem. Stakeholders must recognize that information is relevant and useful in order for them to decide to incorporate the information into their understanding of an issue. Relevance and utility of information are assessed in a variety of ways, including source trustworthiness, social consensus, and whether the new information makes sense in the context of the stakeholder's beliefs and value system.

Regarding the content facet of the problem, it is important to note that the standard scientific knowledge that is typically the heart of the ICM knowledge base has to be reinforced by the oft-neglected knowledge of the priorities and outlooks of other actors. What do *they* want; how do *they* see the world? For example, knowledge of current circumstances and causal patterns should encompass the objectives, intensity, and outlooks of both allies and opponents. This includes beliefs about causal patterns. Thus, if opponents of a conservation plan see a causal link between the plan's elements and serious declines in their yields, they may conclude that the plan would destroy their businesses. The plan's proponents, however, may be surprised by this conclusion if they had believed it was common knowledge that the effects on yields would be minimal. In short, the actual causal patterns are only one aspect of conditioning factors that one must assess; the perceptions held by others are essential, whether accurate or not. These viewpoints often are difficult to understand without research, especially because opposing groups may have significantly different value systems from which their issue-oriented positions derive. More often than not, advocates of one perspective assume opponents are ignorant, naïve, misinformed, or mal-intentioned, rather than consider the possibility that opponents are operating with a different worldview.

To comprehend the multiple perspectives of other stakeholders and the motivations underlying their actions requires a sufficiently comprehensive mapping of priorities, and sufficiently insightful understanding of *others'* understandings of the effects of alternative policies and practices. In polarized policy debates, with positions often expressed with hyperbole and demands couched in extreme terms for the sake of negotiation, nuanced information about the preferences and outlooks of others rarely can be acquired without active information search. This greatly increases the knowledge burden for those engaging in coastal conservation policy issues and, consequently, the need for more comprehensive information searches.

To address the challenges of inducing more useful information searches, we first present an overview of knowledge transfer and information behavior. Next, we identify opportunities for intervention in the knowledge-acquisition process that can be leveraged to enhance the salience of sound coastal conservation information and motivate informed use of this knowledge by stakeholders. Finally, we suggest categories of strategies to enhance coastal conservation stakeholders' decision-making. The chapter concludes with implications of these approaches and suggestions for future research.

7.3 Knowledge Transfer, Acquisition, and Information Behavior

If the aim is to improve stakeholder use of sound information, several aspects of the knowledge-acquisition process must be understood and leveraged to ensure stakeholders have adequate knowledge for coastal conservation decision-making. Decision-making and information behavior are interconnected in both theory and practice. Information studies researchers have tended to focus on the stages and manners in which people use systems and networks to find and/or encounter information that affects their decision-making. The emphases of these studies include access, ease of use, quality, quantity, relevance, and speed, with information scientists emphasizing how systems help or hinder the information-seeking process, and information studies researchers tending to focus on the information context and the societal forces that can threaten access and use. Behavioral and decision scientists, on the other hand, tend to focus on how and why various biases prevent people from using information to generate the best outcomes according to rational economic theory. Researchers in both broad domains (as well as in psychology, sociology, and public policy) have explored the roles of uncertainty and information need as motivators of information search.

Most models of information-seeking behavior are based on the premises that (1) individuals are engaged in an active search task to reduce anxiety arising from the recognition of a knowledge deficit, (2) the search process has a discernible start and terminus, (3) knowledge acquisition is equivalent to understanding, and (4) information retrieval systems are unbiased. For marvelous detail on information behavior models, see Case (2012). Information studies research over the past half-century has yielded several useful models of information behavior: Ellis's (1984) behavioral model of information search strategies, Kuhlthau's (1988, 1991) information search process, and Wilson's (1997, 1999) problem-solving model, as well as contributions from Bates (1989), Belkin (1996), Choo et al. (1998, 1999), Dervin (1998), Ingwersen (1984, 1996, 2001), Krikelas (1993), Leckie and Pettigrew (1997), Leckie et al. (1996), Marchionini (1995), Savolainen (2007), Sonnenwald (1999), Sonnenwald et al. (2001), and Spink (1997), among others.

Most of these models are predicated on the notion of the needy information seeker engaged in a goal-directed search using context-agnostic information-retrieval systems that, when properly configured, transfer knowledge from information generator to information seeker. However, an examination of the constraints of the most highly regarded models of information-seeking behavior yields insight into how coastal conservation knowledge transfer efforts can be enhanced.

Four important aspects of knowledge transfer are missing from several of these models: (1) information may be sought, but also it may be encountered, raising questions of intentionality; (2) knowledge acquisition is a dynamic, recursive, nonlinear process, raising issues of salience and classification of information; (3) information acquisition does not ensure knowledge, raising epistemological concerns; and (4) information retrieval systems manifest the biases and assumptions inherent in their algorithms, raising apprehension about objectivity. We briefly address each of these concerns in the context of coastal conservation knowledge transfer.

7.4 Seeking and Encountering Information

While we know that people seek information through a variety of channels, the advice of family members and friends remains one of the primary sources of opinion-forming sustainability information for many people. For example, in the United States, an assessment of the conditions related to the adoption of sustainable agricultural practices concludes:

> It has long been known that information sources besides agriculture professionals, such as mass media and family and friends, are vitally important in helping a farmer become aware of new agricultural techniques. ... More recently, farmers have reported that their most utilized sources of information are chemical and fertilizer dealers, followed by family and friends, and media publications. Professional sources of information, such as the extension service and USDA [US Department of Agriculture] personnel, have been ranked lower in importance. (Fazio et al. 2005, p. 27)

We also know that the Internet has expanded people's search options—there were an average of 5.74 billion Google searches conducted per day in 2014 (Statistic Brain Research Institute 2015)—both in terms of information retrieval mechanisms and social networks. Further, traditional media channels now incorporate new media messages as sources of goals, trends, conditioning factors, projections, and preferred alternatives, drawing attention in popular discourse. Aside from turning to these information channels, people encounter information—serendipitously and/or incidentally—throughout their daily activities (Rice et al. 2001). Understanding the contexts of encountered information and how such information is incorporated into people's stances on hotly debated issues is at least as important as generating sound coastal conservation technical information. Therefore, those concerned with coastal conservation knowledge transfer must consider both passive information encountering, as well as active, systematic information-seeking behavior. This is particularly important, because people tend to evaluate encountered information using heuristics, which we explain after a brief discussion of intentionality.

The balance of passive and active information behavior tends to correlate with the degree of intentionality of the information recipient. By definition, active information seekers intend to locate and interpret knowledge about a topic and use various resources to do so; passive information recipients encounter knowledge about a topic without undertaking an intentional search for that information. In both cases, the judgments about whether and how to classify and use the acquired knowledge are governed by several factors, including the information recipient's (1) background and beliefs, (2) cognitive ability and load, and (3) estimation of the usefulness of the information.

The intersection of information behavior and information literacy is, perhaps, the most important area on which to focus in efforts to improve stakeholder decision-making based on sound coastal conservation knowledge. As Williamson and Asla (2009) observed in their study of people in the "Fourth Age," information literacy (defined as the ability to recognize a need for information and take action to acquire it) is very often the result of engagement in strong information networks. In addition, Williamson and Asla concluded that, in some circumstances, the knowledge gained as a result of active information-seeking behavior is less useful than the information people passively encountered through interactions within social and professional networks.

Table 7.1 presents a matrix of active versus passive information seeking and the process by which information recipients classify knowledge for use. As noted above, adequate knowledge for coastal conservation decision-making requires either:

1. Active, systematic knowledge acquisition to become informed about the assumptions, perspectives, and motivations that create the information context in which a stakeholder makes decisions, or

2. Passive, heuristic knowledge classification that allows for rapid assimilation of new information into the stakeholder's perspective on the issue at hand.

Often, the relevance, veracity, completeness, and usefulness of information are not evaluated through thorough analysis; instead, they are evaluated according to analytical shortcuts that go by the label of "heuristics." For example, an environmental activist might see a particular pro-economic-growth leader on television advocating for incentives for oil companies and have the heuristic response of filing the messages under "total nonsense." This heuristic—anything that leader says is total nonsense—is not universally accurate, but it saves the activist the time and effort of analyzing the leader's messages and determining their value relative to all the other

TABLE 7.1

Intentionality and Information Behavior

	Systematic	Heuristic
Active	Explore multiple perspectives Evaluate authenticity and authority Assess dynamic context • Goals • Trends • Conditioning factors • Projections • Preferred alternatives	Classification according to individual or group identity • Confirmation bias • Assimilation bias
Passive	Opportunity for intervention	Classification according to individual or group identity • Confirmation bias • Assimilation bias

information about corporations, the environment, and climate change that he/she seeks and encounters.

It is crucial to understand that under conditions of uncertainty, no judgment can be fully comprehensive. Herbert Simon's 1978 Nobel Prize in Economics was based on his twin insights of (1) *bounded rationality*, whereby future events and conditions depend on an unbounded number of possible conditioning factors; hence, rationality (in the sense of selecting the definitively known optimal decision) is always bounded; and (2) *satisficing*, whereby given the impossibility of knowing everything, and given constraints on resources, people tend to end their searches when they deem the results are "good enough," where "good enough" is a function of expected effectiveness and the individual's assessment of the trade-offs involved in expending the effort (Simon 1959). The most prominent proponents of the *heuristics* and *biases* paradigm embrace Simon's insights with the premise that reaching judgments, whether through deliberate search or intuition, cannot entail a fully comprehensive search (Kahneman 2003). Kahneman and Frederick (2002) propose that because the case at hand cannot be subjected to intensive scrutiny given the limitations of time and other resources, the heuristic process entails "attribute substitution," whereby characteristics of the current case are substituted with attributes of prior cases or generalizations regarding those cases.

Thus, when Todorov, Chaiken, and Henderson contrast heuristic and systematic information processing related to risk assessment, what they really mean (or at least ought to mean) is that some processing is rudimentary while other processing is *more, but never fully*, systematic. They try to clarify the distinction by noting that "[i]n a systematic mode, people consider all relevant pieces of information, elaborate on these pieces of information, and form a judgment based on these elaborations" (Todorov et al. 2002, p. 196). They contrast this with the heuristic mode:

> However, even if people are not sufficiently motivated or do not have sufficient cognitive resources, they can engage in superficial or heuristic processing of available information … people consider a few informational cues—or even a single informational cue—and form a judgment based on these cues. For instance, such cues may be the source of the message or the length of the message. (Todorov et al. 2002, p. 196)

The premise that people *can* consider *all* pieces of relevant information flies in the face of the bounded rationality concept. In fact, although Todorov, Chaiken, and Henderson list reliance on the source of a message as a possible heuristic action, even highly systematic information processing depends to a certain extent on the shortcut acceptance of information as credible based on a belief that the source is regarded to be expert, truthful, or both, rather than on the basis of corroborating or validating evidence. The useful distinction between these forms of information processing is that some information processing relies on a few new cues, while other information processing relies on a richer set of cues. In other words, heuristics can be rudimentary or sophisticated.

Moreover, Kahneman (2003, p. 697) points out that heuristics can be deliberate or intuitive. Deliberate reliance on heuristic classification of information can be a conscious choice selected to increase efficiency of knowledge acquisition. In Table 7.1, deliberate heuristic judgment would fall in the upper-right quadrant, active heuristic classification according to the decision maker's acknowledged values, beliefs, and identity. Intuitive heuristics, on the other hand, are used "automatically and rapidly" (Kahneman 2003, p. 697) and occur within particular cultural, political, or social milieus that do not encourage teasing out the layers of complexity that result in opposing viewpoints. They would fall in the lower-right quadrant of Table 7.1; people making decisions based on passive heuristic information acquisition may not even be cognizant of the heuristic judgments entailed in the decision-making process. For example, the rapid, automatic assumption that a highly respected professor's findings are valid presumes that she has employed an appropriate paradigm. In most countries today, full professors at distinguished universities are regarded broadly as likely to be highly expert in their areas of specialization. This was less so, for example, in the United States in the 1960s and 1970s.

Therefore, biased searches can derive from passive heuristic judgments that entail attribute substitution, that is, replacing a more complex, analytically challenging attribute such as the current issue at hand with a more easily understandable attribute, such as an expert's view or the outcome of a past case. This would cut down on the evaluative work of information classification during a search. By substituting attributes, for example, the decision maker assumes the professor's findings are valid and can allocate mental, physical, and financial resources to a more extensive exploration of other information related to the decision (Schulz-Hardt et al. 2000). Thus, accepting the opinion of the authoritative source rests on the representativeness heuristic: the individual believes the expert's opinion is similar to the population of the expert's past opinions; because these opinions are regarded as correct, the expert's opinion on the current matter is assumed to be correct, as well.

The ambivalent treatment of heuristics in the literature, as a source of bias but also as "efficient cognitive processes" (Gigerenzer and Gaissmaier 2011, p. 451), raises the question of whether people can be equipped with guidance to employ sound and efficient search heuristics actively, yet remain cognizant of behavioral tendencies that can undermine systematic or active heuristic decision-making. Gigerenzer and Gaissmaier note that

> [b]ecause using heuristics saves effort, the classical view has been that heuristic decisions imply greater errors than do "rational" decisions as defined by logic or statistical models. However, for many decisions, the assumptions of rational models are not met, and it is an empirical rather than an a priori issue how well cognitive heuristics function in an uncertain world.

Going back to Todorov, Chaiken, and Henderson's juxtaposition of heuristic versus systematic searches, we must conclude that if a relatively quick search is the only search an individual is willing to undertake, guidance on how to make the most of it could be very helpful.

7.5 Overcoming Obstacles to Inducing More Effective Searches

While individuals or organizations may be *capable* of engaging in more systematic information-seeking behavior, different scenarios dictate the extent and nature of further search activity. First, some may be unaware of the risks and/or opportunities involved with the issue, or find the issue to be of such low salience to their welfare or deference values—Lasswell's distinction between valued material outcomes and valued relationships, respectively, which we detail in Section 7.7.3—that no search is worth the effort. Second, some may regard the issue as salient, but despair that a search would not provide useful guidance in making decisions, either because they (a) believe constructive understanding is beyond their grasp or (b) do not believe that gaining the knowledge would help them take effective action. Finally, some may have views based on prior knowledge acquisition such that they are confident they can make sound decisions without seeking additional information. None of these scenarios induces stakeholders to believe that further search would have a reasonable chance of improving the instrumental effectiveness of their actions or their deference rewards. Each of these implies thresholds of salience, confidence, or both. These thresholds, which differ for every person and in every unique context, are set according to the potential information seeker's assessment of the expected intelligibility and usefulness of additional information, the trustworthiness of the information source, the gravity of the decision, and the belief that the process by which the information was acquired will stand up to scrutiny. In other words, a person sets these thresholds, often unconsciously, based on his or her belief that the search effort is commensurate with the level of attention that the issue at hand deserves. This complex decision-making protocol exposes several facets of information behavior that are not addressed by Zipf's principle of least effort, which asserts information seekers tend to use the most convenient, minimally demanding search method until just barely acceptable information is acquired (Zipf 1949).

Those interested in inducing more effective information-seeking behavior with the aim of helping stakeholders base their decisions on sound environmental (or any other kind of) knowledge should note that any stakeholder is not merely a needy, lazy information seeker. Rather, every individual employs a multifaceted decision-making process based on the individual's assessment of how each of the factors (confidence, salience, utility, gravity) affects to his or her ability to improve directly instrumental effects and deference rewards. For example, if an influential farmer is considering a request by a member of his or her political party to persuade others to acquiesce to a policy initiative that tightens run-off regulations, he or she may conduct a search by reading the government's policy briefs, talking with the agricultural extension agent, visiting websites about the risks of fertilizer and pesticide run-off, skimming Twitter feeds, reading the bulletins of farmers' associations, or consulting any number of other information sources. Some of these searches may seem too time-consuming or unlikely to provide understandable results. The farmer may, in fact, do no search, relying on his or her existing beliefs; otherwise, he or she not only must select sources of information, but also must decide how much effort to put into the search.

The amount of effort a decision maker is willing to expend also is influenced by the degree and nature of affect. The degree of affect associated with commitments to influence coastal policies or practices is part of the implicit or explicit benefit-cost calculus that determines whether a particular search activity is worthwhile for that individual. While the predominant focus of efforts to increase commitment to conservation has focused on the positive affects associated with love for nature or for future generations, it is important to recognize the mobilizing strength of appealing to darker impulses. The focus on emotion and affect in the information studies field largely has concerned feelings that would deter otherwise-motivated searches, such as anxiety and intolerance of uncertainty (Wilson, 1997, p. 555), lack of confidence, frustration, doubt, pessimism, or disappointment (Kuhlthau, 1991, p. 367). We propose the alternative of focusing on fundamental drives that may heighten basic motivation and overcome these feelings, resorting to the classical distinctions among raw impulse, reason, and conscience. (These categories were defined in the field of political psychology in the more traditional Freudian language of id, ego, and superego (Lasswell 1932)). This slice of the mental process is a useful basis for search-promotion strategies because appeals can be targeted systematically to each drive or combination thereof. Aside from the obvious instrumental goals pursued through reason, appeals can be directed to positive impulses such as camaraderie or to negative impulses like aggression; there can be appeals to conscience, with the potential flexibility of specifying different norms as ethical. Such appeals change the least-effort calculus in terms of the value individuals and groups place on their identification as environmentalists, good citizens, winners, or ethical people, respectively.

7.6 Strategies for Mobilizing Motivations

Under the assumption that intentional searches are goal-directed, the first aspect of constructing potential strategies is to offer a map of possible goals held by individuals for whom we would hope to induce better information-seeking behavior. A useful organizing principle is the distinction between directly instrumental motives related to advancing material interests

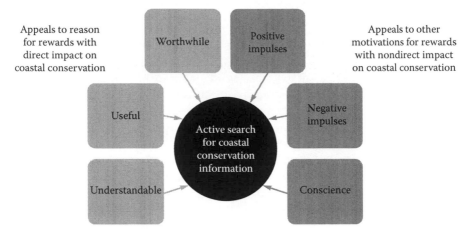

FIGURE 7.1
Appeals to induce active search for coastal conservation information.

through activities or policies affecting coastal systems, and motives that have no such direct connection. The breadth of potential strategies to motivate searches rests on the fact that while the objectives of stakeholders may pertain directly to rewards from achieving preferred outcomes of coastal policy or practices, others' objectives do not. Those who develop coastal conservation information and strategies to induce the acquisition of this information are likely to focus intently on motivating stakeholders to choose sound policies and practices, yet other, indirect motivations may be even more useful in particular contexts. In other words, opportunities to encourage improved information-seeking behavior arise not only from efforts to increase the reach and salience of targeted coastal conservation messaging, but also from appealing to motivations associated with achieving objectives that may or may not have anything to do with coastal conservation at all. The simple diagram in Figure 7.1 conveys the implications of the two kinds of appeals that can act as motivation for active information search related to directly instrumental and nondirectly instrumental objectives.

7.7 Strategies

7.7.1 Category 1: Supplemental Information

7.7.1.1 Supplemental Information on How Core Information Can Be Used

Focusing first on the left-hand side of the diagram in Figure 7.1, those appeals to reason that have direct impact on coastal conservation, we make the perhaps obvious point is that for a search to be motivated by the desire to choose optimal policies or practices, the individual must have sufficient confidence that knowledge exists and that it can be accessed readily, understood adequately, and used effectively. The individual must also have sufficient confidence that an instrumentally effective search stemming from an appeal to reason is worth the time and other resources required for the search. Although good effort has been put into organizing the knowledge needed for managers to engage in ICM (e.g., Perry et al. 1999; Berkes et al. 2007; Lertzman 2009), or its cousin, adaptive management (e.g., Allan and Curtis 2003; Lawrence and Bennett 2002), to make directly applicable knowledge more understandable and usable for lay stakeholders, another class of knowledge is important: *how to use* the direct-impact knowledge. Such *utilization information*, beyond the technical knowledge of the projected consequences of policies and resource practices, may be needed to instill this confidence in the decision maker. One type of utilization knowledge is epistemological: whether and how the knowledge can be understood. The other type of utilization knowledge is sociopolitical: how the knowledge can further the objectives of the individual, group, or organization. For example, in addition to understanding that agricultural run-off may erode nearshore ecosystems, the motivation to master this knowledge may require confidence in its usefulness, perhaps to invoke scientific projections in support of a lawsuit demanding a stronger conservation effort. There may be opportunities, such as workshops, write-ups of past cases, and so on, to help stakeholders use more helpful heuristics as bases for judgments of relevance and usefulness of information. It should be noted that decision aids abound to assist resource managers (Holling 1978; Pearson et al. 2010), but few exist to assist lay stakeholders.

Another approach entails organized interactions between experts and lay stakeholders. A host of formats, many pioneered in Europe, such as citizen advisory committees, citizen juries, consensus panels, science shops, study circles, and joint fact-finding commissions, are among the venues that can support such expert-novice engagement. These are all variants of exchanges in which stakeholders clarify their objectives and concerns, work with experts to project possible outcomes of policies, and help orient scientific research and knowledge dissemination (Rowe and Frewer 2000; Ascher et al., 2010, pp. 195–196). Such strategies translate well to online discussion groups, webinars, podcasts, and other technologically facilitated interaction. In all cases, however, the stakeholder must know the opportunity for interaction exists and is accessible, and that the information that may be gained through such interaction is understandable, actionable, useful, and worth the effort to acquire. These approaches have the advantage of both conveying that relevant knowledge can be grasped, and presenting stakeholders with current information to offset searches that return obsolete information. However, these approaches can run into opposition if some participants believe that the formats are manipulated to gain compliance for particular policies. Even so, strong efforts by participating groups can be reassuring if their leaders have sufficient knowledge and influence regarding the balance and credibility of the experts brought into the deliberations.

7.7.1.2 Supplemental Information on Calibrated Uncertainty

Although ICM theorists and practitioners have endorsed the notion that uncertainty ought to be embraced rather than ignored or used as a pretext for inaction, it is still important to overcome the dual dangers that information about projections of future consequences of policies or practices is either certain—which could cut off the search prematurely—or totally uncertain—which could discourage the search entirely.

7.7.2 Category 2: Mitigating Heuristic Bias in Information Search

Numerous strategies may be suitable for *de-biasing*, or mitigating human judgment biases, as well as the *filter bubble* that arises from search algorithms that tailor search engine results based on a searcher's prior online activity:

- Nisbett et al. (1982, pp. 448–451) suggest that clarifying the distinctiveness of the circumstances in which the most available or apparently representative cases occurred could help decision makers weigh the prevalence of accounts described in search results more appropriately. Thus, calling attention to particular circumstances under which relevant outcomes occurred may counteract the availability and representativeness biases. Clarifying the circumstances in which presumed experts produced their assessments, and conveying that changed circumstances bring these assessments into question, might induce active seeking for updated information.
- The availability bias can be offset further by publicizing summaries of a broader range of circumstances, events, or policies, to focus attention on other cases in addition to the most prominent one(s). For example, to offset the presumption that the consequences of the next initiative to enact a fishing moratorium will be the same as the latest or most painful moratorium, the local newspaper could run a feature on "The History of Fishing Bans—Balancing Access and Sustainability." The tendency to presume that the pending issue is similar to a narrow set of previous cases could be addressed by the same tactic of giving prominence to many different cases. Of course, this becomes more difficult as Internet users customize their search behavior and tools in ways that fuel the confirmation bias.
- Nisbett et al. (1982, p. 447) and Fischhoff (2002, p. 746) suggest that specifying the assumptions underlying the analyses of credible sources may reveal their possible shortcomings. This could also

clarify which assumptions may no longer hold. Further, explicit description of tendencies of humans and algorithms to focus attention on information that confirms existing beliefs might prompt stakeholders to investigate alternate perspectives.

- Wilson et al. (2002, p. 197) suggest that conveying a full range of possible outcomes can offset overconfidence in the most obvious possible outcome by presenting pathways and explanations that could lead to other results. This is a means of conveying uncertainty without implying that the uncertainty is so crushing that further information seeking would be futile.

- Specifying how the current issue differs from some prior issues can make the set of representative cases more appropriate. This may require discipline to avoid overpublicizing the most striking or exciting cases; by countering salient generalizations with strategic promotion of specific, distinctive cases to serve as representative cases, calling attention to differences may induce stakeholders to undertake deeper information searches.

- The presumption that the view of the individual or institution perceived as most authoritative ought to be accepted unquestionably may be offset by publicizing the views of other individuals or organizations of comparable repute. For example, the views of major environmental groups such as the Natural Resources Defense Council (NRDC), the Environmental Defense Fund, the Sierra Club, and Greenpeace on issues such as hydraulic fracturing or offshore drilling, or on the advisability of collaborating with the corporate sector, vary considerably. Exposing pro-environment individuals to this range of *pro-environment* views would clarify that no single source has a monopoly on pro-environmental expertise or commitment.

7.7.3 Category 3: Social Strategies for Strengthening Search Incentives

The right side of Figure 7.1, depicting appeals leading to nondirect impacts, suggests that searches can be triggered by any of a range of rewards expected as a result of undertaking the search, regardless of whether or not the search ultimately provides directly instrumental coastal conservation benefits resulting from particular policies or practices. For example, while a fisher's decision to learn whether it is in her interest to abide by a fishing moratorium is an obvious example of the directly instrumental category, the alternative, nondirectly instrumental category would be exemplified by a parks and wilderness society member's effort to gain greater mastery of coastal ecology in order to earn the respect of other chapter members or simply to feel good about the increased mastery. We could hope that this nondirectly instrumental motivation will lead to knowledge acquisition that is deployed soundly in selecting directly relevant actions.

Rather than fastening on one or a few motivations, such as mastery per se or respect from peers, a fairly comprehensive map is necessary to depict the enormous variety of motivations that may be mobilized effectively across contexts. Such a map yields a host of possible appeals with the potential to induce better searches. Even so, for a particular context, these potential motivations can be identified systematically. We propose the use of the "valued outcomes" categories of the policy sciences framework (Lasswell and McDougal, 1991, pp. 35–38) as an effective tool for assisting in the identification of both potential motivations involving expected rewards from policies and practices, as well as the rewards that do not derive from these policies and practices. The categories and illustrative examples are displayed in Table 7.2.

Lasswell and Kaplan (1950, pp. 55–56) distinguish two broader motivational categories: "welfare values" ("those whose possession to a certain degree is a necessary condition for the maintenance of the physical activity of the person") and "deference values" ("those that consist in being taken into consideration [in the acts of others and of the self]"). As a first approximation, the welfare values pertain largely to instrumental motives. The deference values, especially power, can be deployed to pursue welfare values, but the key point is that they may be valued in and of themselves, apart from advancing the material interests involved in coastal management. Because each of these motivational categories listed in Table 7.2 is plausible, it is

TABLE 7.2

Categories of Valued Outcomes

Category	Value	Example
Welfare values	Enlightenment	Greater sense of understanding and mastery
	Wealth	Greater sustainable yields; higher ecotourism revenues
	Wellbeing	Lower health risks; reduction of anxiety stemming from feelings of lack of mastery
	Skill	Higher status within an organization; greater success for the organization
Deference values	Power	Influence within an organization; group's success vis-à-vis other groups
	Respect	Status among peers and within an organization
	Affection	Friendship among peers and within an organization
	Rectitude	Being a responsible citizen or group representative

worth exploring which motives are most compelling on a case-by-case basis. The virtue of both instrumental and noninstrumental objectives presents a challenge for those who are trying to motivate more effective searches. Nevertheless, we present several strategies appropriate when particular motivations are known to be potent.

7.7.3.1 Strategies for Conservation Groups

One approach available to nongovernmental conservation organizations is to decentralize decision-making such that the members of local branches or chapters (a) feel a responsibility to be well-informed in order to take positions, and (b) reward local members with respect and power insofar as their mastery is known and regarded as an asset to the organization. In the United States, the Sierra Club, with its fairly small chapters and its penchant for giving a host of local and national awards every year, is a notable example of this approach. In contrast, the NRDC, with well over a million U.S. members, a lack of chapters, and its trumpeting of its "expertise of more than 350 lawyers, scientists and other professionals" (NRDC 2014), provides little incentive for members to engage in their own searches on environmental issues. As one member communicated in confidence, "[t]he NRDC has very smart people. If they say that fracking is a bad idea, it's a bad idea." Why search for knowledge about the relative risks of fracking versus continued reliance on coal, with its greater burden of greenhouse gas emissions and conventional pollution than natural gas, when smart people have already determined the best course of action? And how can one debate the fracking advocate who cites statistics on emphysema from particulates, except to say that smart people oppose fracking? Although it is unclear whether the NRDC leadership's actual position is truly so definitive, or whether it is a negotiating stance vis-à-vis the government and the energy industry, the heuristic shortcut of taking its position as the last word in expert analysis is clearly problematic.

Another strategy for conservation NGOs is to co-finance and collaborate in designing research projects along with the relevant industrial groups. Busenberg (1999) has demonstrated the greater credibility of collaborative research and guideline development regarding oil spills. This approach has been shown to reduce confusion over the authenticity and authority of information sources in a manner akin to governmental checks and balances.

7.7.3.2 Strategies for Government

As mentioned previously, governments have the potential to organize interactions between lay stakeholders and experts. They should also consider increasing stakeholder affect by stimulating face-to-face debates among citizens with opposing views. Although rivalry seems to have no place in the idealized vision of rational debate, the reality is that conservation debates tend to be acrimonious. Pro-conservation activists are criticized either as naïve, accepting the doomsday scenarios of radical environmental activists, or as elitists, uncaring about the economic burdens on less well-off citizens imposed by stringent conservation regulations. Opponents of stronger conservation measures are criticized as selfish, shortsighted, and naïve in ignoring the consequences of weak conservation efforts. Therefore, the raw-impulse motive of feeling superior to those with opposing views—afforded by being able to marshal more evidence to win the debate—can be a powerful motivation, which dovetails with the rational need for gratifying affiliations and identifications, as well as the need to be a conscientious citizen. At the same time, face-to-face debates, such as town meetings and open hearings, provide each side insight into the perspectives of others.

7.8 Conclusion

While those charged with providing coastal conservation information for stakeholders have focused on generating and organizing core technical information, much more needs to be done to strengthen the incentives for stakeholders to search for, make sense of, and make decisions based on the knowledge needed for sound resource practices and stances toward conservation policies. Both to increase the motivation to seek information actively on the effects of coastal policies and practices and to make the acquired knowledge more useful, the generation of information must be broadened to encompass how the core information can be supplemented by information to mitigate bias in the decision-making process, as well as to present alternate perspectives to provide opportunities for greater understanding of the range of relevant viewpoints. Regarding social pressures, the basis for optimism is that so many potential appeals can be made. The more general point is that organizations convey the roles that they expect their members to play. If the role is to engage in meaningful debate with policy adversaries or to persuade other resource users that sustainable practices are imperative, it is incumbent upon the organization to emphasize and reward this role. As we have described, inducing better searches for coastal conservation information can be accomplished through appeals to reason that trigger stakeholders' consideration of welfare values concerning direct effects on coastal conservation, as well as by leveraging other motivations, which appeal to deference values to nudge people to conduct more thorough searches and to take action that has nondirect effects on coastal conservation.

In short, insights from information studies and the psychology of motivating more sound information-seeking behavior can be useful to guide prototypes and focus groups to determine which strategies are promising in particular contexts. What it cannot do is specify a general strategy that will hold in every case, as contextual factors must inform the development of strategies to motivate better knowledge practices.

― ───

References

Allan, C. and A. Curtis. 2003. Learning to implement adaptive management. *Natural Resource Management* 6: 23–28.

Anderson, D., P. Gilbert, and J. Burkholder. 2002. Harmful algal blooms and eutrophication: Nutrient sources, composition, and consequences. *Estuaries* 25 (4): 704–726.

Ascher, W., T. Steelman, and R. Healy. 2010. *Knowledge and Environmental Policy: Re-Imagining the Boundaries of Science and Politics.* Cambridge, MA: MIT Press.

Bates, M. 1989. The design of browsing and berrypicking techniques for the online search interface. *Online Review* 13 (5): 407–424.

Belkin, N. 1996. Intelligent information retrieval: Whose intelligence? In *Proceedings des 5. Internationalen Symposiums für Informationswissenschaft (ISI '96)*, 25–31. Konstanz: Universtaetsverlag Konstanz.

Berkes, F., M. Kislalioglu Berkes, et al. 2007. Collaborative integrated management in Canada's north: The role of local and traditional knowledge and community-based monitoring. *Coastal Management* 35: 143–162.

Busenberg, G. 1999. Collaborative and adversarial analysis in environmental policy. *Policy Sciences* 32 (1): 1–11.

Case, D. O. 2012. *Looking for Information: A Survey of Research on Information Seeking, Needs, and Behavior.* Bingley: Emerald Group.

Choo, C. W., B. Detlor, and D. Turnbull. 1998. A behavioral model of information seeking on the web: Preliminary results of a study of how managers and IT specialists use the web. Paper presented at the American Society of Information Science, Pittsburgh, PA.

Choo, C. W., B. Detlor, and D. Turnbull. 1999. Information seeking on the web: An integrated model of browsing and searching. In *Proceedings of the 62nd Annual Meeting of the American Society for Information Science, Washington, DC. Knowledge: Creation, Organization and Use,* edited by L. Woods, 3–16. Medford, NJ: Information Today.

Dervin, B. 1998. Sense-making theory and practice: An overview of user interests in knowledge seeking and use. *Journal of Knowledge Management* 2 (2): 36–46.

Ellis, D. 1984. The effectiveness of information retrieval systems: The need for improved explanatory frameworks. *Social Science Information Studies* 4: 261–272.

Fazio, R., J. M. Rodriguez Baide, and J. Molnar. 2005. Barriers to the adoption of sustainable agricultural practices: Working farmer and change agent perspectives: Final report. Auburn, AL: Auburn University, Department of Agricultural Economics and Rural Sociology.

Fischhoff, B. 2002. Heuristics and biases in application. In *Heuristics and Biases: The Psychology of Intuitive Judgment,* edited by T. Gilovich, D. Griffin, and D. Kahneman, 730–748. Cambridge: Cambridge University Press.

Gigerenzer, G. and W. Gaissmaier. 2011. Heuristic decision making. *Annual Review of Psychology* 62: 451–82.

Holling, C. S., ed. 1978. *Adaptive Environmental Management and Assessment.* Chichester: Wiley.

Ingwersen, P. 1984. Psychological aspects of information retrieval. *Social Science Information Studies* 4 (2/3): 83–89.

Ingwersen, P. 1996. Cognitive perspectives of information retrieval interaction: Elements of a cognitive IR theory. *Journal of Documentation* 52 (1): 3–50.

Ingwersen, P. 2001. Users in context. In *Lectures on Information Retrieval: Third European Summer-School 2000, Varenna, Italy, September 11–15, 2000, Revised Lectures,* edited by M. Agosti, F. Crestani, and G. Pasi, 178–200. Berlin: Springer-Verlag.

Kahneman, D. 2003. A perspective on judgment and choice: Mapping bounded rationality. *American Psychologist* 58 (9): 697–720.

Kahneman, D. and S. Frederick. 2002. Representativeness revisited: Attribute substitution in intuitive judgment. In *Heuristics and Biases: The Psychology of Intuitive Judgment,* edited by T. Gilovich, D. Griffin, and D. Kahneman, 49–81. Cambridge: Cambridge University Press.

Krikelas, J. 1993. Information-seeking behavior: Patterns and concepts. *Drexel Library Quarterly* 19 (2): 5–20.

Kuhlthau, C. 1988. Developing a model of the library search process: Cognitive and affective aspects. *Reference Quarterly* 28: 232–242.

Kuhlthau, C. 1991. Inside the search process: Information seeking from the user's perspective. *Journal of the American Society for Information Science* 42 (5): 361–371.

Lasswell, H. D. 1932. The triple-appeal principle: A contribution of psychoanalysis to political and social science. *American Journal of Sociology* 37: 523–538.

Lasswell, H. D. and A. Kaplan. 1959. *Power and Society.* New Haven: Yale University Press.

Lasswell, H. D. and M. McDougal. 1991. *Jurisprudence for a Free Society.* Dordrecht: Kluwer.

Lawrence, P. and J. Bennett. 2002. Improved planning and management in coastal environments using an adaptive management framework. *Water: The Journal of the Australian Water Association* 29 (6): 24–27.

Leckie, G. and K. Pettigrew. 1997. A general model of the information seeking of professionals: Role theory through the back door? In *ISIC '96 Proceedings of an International Conference on Information Seeking in Context,* 99–110. London: Taylor Graham Publishing.

Leckie, G., K. Pettigrew, and C. Sylvain. 1996. Modeling the information seeking of professionals: A general model derived from research on engineers, health care professionals, and lawyers. *The Library Quarterly* 66 (2): 161–193.

Lertzman, K. 2009. The paradigm of management, management systems, and resource stewardship. *Journal of Ethnobiology* 29: 339–358.

Marchionini, G. 1995. *Information Seeking in Electronic Environments.* New York: Cambridge University Press.

Mercer Clarke, C. S. L. 2010. Rethinking responses to coastal problems: An analysis of the opportunities and constraints for Canada. PhD diss., Dalhousie University.

Nisbett, R., D. Krantz, C. Jepson, et al. 1982. Improving inductive inference. In *Judgment under Uncertainty: Heuristics and Biases,* edited by D. Kahneman, P. Slovic, and A. Tversky, 445–459. Cambridge: Cambridge University Press.

NRDC (Natural Resources Defense Council). 2014. About NRDC. http://www.nrdc.org/about/.

Pearson, L., A. Coggan, W. Proctor, et al. 2010. A sustainable decision support framework for urban water management. *Water Resources Management* 24 (2): 363–376.

Perry, R. I., C. Walters, and J. Boutillier. 1999. A framework for providing scientific advice for the management of new and developing invertebrate fisheries. *Reviews in Fish Biology and Fisheries* 9 (2): 125–150.

Rice, R., M. McCreadie, and S. Chang. 2001. *Accessing and Browsing Information and Communication: An Interdisciplinary Approach*. Cambridge, MA: MIT Press.

Rowe, G. and L. Frewer. 2000. Public participation methods: A framework for evaluation. *Science, Technology, & Human Values* 25 (1): 3–29.

Savolainen, R. 2007. Information behavior and information practice: Reviewing the "umbrella concepts" of information-seeking studies. *The Library Quarterly* 77 (2): 109–113.

Schulz-Hardt, S., D. Frey, C. Lüthgens, et al. 2000. Biased information search in group decision making. *Journal of Personality and Social Psychology* 78 (4): 655–669.

Simon, H. 1959. Theories of decision-making in economics and behavioral science. *American Economic Review* 49 (3): 253–283.

Sonnenwald, D. H. 1999. Evolving perspectives of human information behaviour: Contexts, situations, social networks and information horizons. In *Exploring the Contexts of Information Behaviour*, edited by T. Wilson and D. Allen, 176–190. London: Taylor Graham.

Sonnenwald, D. H., B. M. Wildemuth, and G. Harmon. 2001. A research method using the concept of information horizons: An example from a study of lower socio-economic students' information seeking behavior. *The New Review of Information Behavior Research* 2: 65–86.

Spink, A. 1997. Information science: A third feedback framework. *Journal of the American Society for Information Science* 48: 728–740.

Statistic Brain Research Institute. 2015. Google annual search statistics, Comscore. http://www.statisticbrain.com/google-searches/.

Sunstein, C. 2002. *Risk and Reason: Safety, Law, and the Environment*. Cambridge: Cambridge University Press.

Todorov, A., S. Chaiken, and M. Henderson. 2002. The heuristic-systematic model of social information processing. In *The Persuasion Handbook: Developments in Theory and Practice*, edited by J. Dillard and M. Pfau, 195–211. Thousand Oaks, CA: Sage.

Williamson, K. and T. Asla. 2009. Information behavior of people in the fourth age: Implications for the conceptualization of information literacy. *Library and Information Science Research* 31 (2): 76–83.

Wilson, T. 1997. Information behaviour: An interdisciplinary perspective. *Information Processing & Management* 33 (4): 551–572.

Wilson, T. 1999. Models in information behaviour research. *Journal of Documentation* 55: 249–270.

Wilson, T., D. Centerbar, and N. Brekke. 2002. Mental contamination and the debiasing problem. In *Heuristics and Biases: The Psychology of Intuitive Judgment*, edited by T. Gilovich, D. Griffin, and D. Kahneman, 185–200. Cambridge: Cambridge University Press.

Zipf, G. K. 1949. *Human Behavior and the Principle of Least Effort*. Cambridge, MA: Addison-Wesley.

8

When Scientific Uncertainty Is in the Eye of the Beholder: Using Network Analysis to Understand the Building of Trust in Science

Troy W. Hartley

CONTENTS

8.1 Introduction

Former U.S. Secretary of Defense Donald Rumsfeld is commonly quoted for his views on decision-making under conditions of uncertainty; in a 12 February 2002 news briefing, he said,

> there are known knowns; there are things we know we know. We also know there are known unknowns; that is to say we know there are some

things we do not know. But there are also unknown unknowns—the ones we don't know we don't know. (Rumsfeld 2002)

In all aspects of coastal and ocean policy and management, there are uncertainties and decisions that have to be made in spite of those uncertainties. The scientific understanding of *how* institutions, organizations, and individuals perceive and grapple with those uncertainties has been growing, including in the field of fisheries science and management.

Scientific uncertainty arises from many different sources in coastal, marine, and fisheries planning and management. For example, there can be varying degrees of confidence in the quality of our measurement instruments, data, and data-analysis methods. Fisheries science applies extensive at-sea sampling and sophisticated stock assessment modeling techniques to estimate the biomass and population structures of fish species, which in turn inform fishing management rules. In fact, there are many fish stocks for which simply too few data are available to conduct the most robust stock assessment methods. Increasingly, fisheries scientists are recognizing the variability in environmental conditions and alterations in fish stock status due to environmental conditions from systematic changes such as climate change. In Rumsfeld-ese, these are known unknowns that science is increasingly learning about and developing strategies to manage, although it is likely that unknown unknowns will surprise scientists as they continue to study the natural sciences relevant to fisheries. To account for these types of uncertainties, scientists may, for example, apply statistical analyses, present confidence intervals, conduct various sensitivity analyses, employ multiple model strategies, and report a range of potential outcomes from a given management decision.

Further, human dimensions may introduce additional uncertainties, and while we have been studying the human dimensions of uncertainties in fields of risk perception, decision sciences, communication, and others—mostly within the context of public health, environmental pollution, and other hazards—the application of risk perception and decision and communication sciences is a relatively new area of study for fisheries, coastal, and marine management. An example of the human dimensions surrounding uncertainty would be the imprecision that can exist in the implementation of management actions. We do not know the likelihood of compliance with rules and how fishers, local policy implementers, or other stakeholders actually might behave given the incentives created by the management decision. For fisheries, coastal, and ocean managers in democratic societies, there are many stakeholders with different interests and values in the resource management objectives, and typically there is no single consensus objective. Thus, uncertainty arises when picking a policy objective among several value-laden preferences. Still further, as society expands the scope and boundaries of coastal, marine, and fisheries resource management beyond the socioeconomic objectives of fish stock and fisheries, integrated management seeks a balance with maritime trade, offshore energy, recreation, and many other competing uses. Additional uncertainties arise in our efficacy in achieving these multisector goals and the contribution that fisheries management might make to realize other sectors' objectives. There are far more unknown unknowns in the human dimensions surrounding uncertainty.

The balance of knowing and not-knowing about coastal and marine resources and management has impacts on decision-making, including the attitudes and perceptions of stakeholders, which has been particularly critical in fisheries management in the United States. Fisheries science and management have seen their perceived credibility undermined as new knowledge and sophisticated modeling of fisheries stock assessments have been introduced. In other words, fisheries science has experienced Rumsfeld's unknown unknowns that become known and has produced surprises for scientists, managers, and fishers.

After reviewing how scientific uncertainty has been handled in U.S. fisheries management, this chapter considers whether fisheries science and strategies for addressing uncertainty have built or undermined credibility and trust in the science. Governance network analysis research is presented from two cases of fisheries management as a tool to explore information flow and the use of that information in an integrated management context in the United States and to further assess how to build credibility and trust in science. The findings have ramifications for how we orchestrate professional networks and deploy boundary-spanning organizations and individuals to constructively manage scientific uncertainty in coastal and marine resource management. Reflection on the growing and promising use of network analysis concludes the chapter.

8.2 U.S. Fisheries Management and Uncertainty

Fisheries scientists have identified a consistent overestimation of fish stock size and underestimation of fishing mortality and have applied an ad hoc downward adjustment to stock biomass estimates to account for these biases (Legault 2009). There is also a substantial probability that the stocks are classified as overfished, when in fact they are not (National Research Council 2014). Uncertainty is plentiful in U.S. fisheries management. Some fisheries scientists have argued that a focus on uncertainty can lead to management paralysis (Rosenberg 2007), while others have argued that underemphasizing uncertainty is riskier because it can cause long-term harm to the underlying credibility of the science (Keepin and Wynne 1984; Kloprogge and van der Sluijs 2006; van der Sluijs et al. 2008). The International Council for the Exploration of the Seas (ICES) Working Group on Fisheries Systems considered the social implications of underemphasizing and overemphasizing uncertainty. It recommended addressing uncertainty in a transparent manner, early, and continuously in the fisheries decision-making process (Dankel et al. 2012). While uncertainties remain, a lack of transparency surrounding these uncertainties can undermine the credibility of the science (Röckmann et al. 2012).

In the 2014 National Research Council (NRC) report on the effectiveness of fisheries stock rebuilding in the United States, the ramifications of how uncertainty is treated in decision-making was clear in several cases (NRC 2014). In Gulf of Maine cod, the application of new stock assessment models in 2010 showed a significant stock assessment bias in the 2008 assessment. Whereas managers and fishers originally thought that overfishing was not occurring and that the stock was on track to be completely rebuilt by its legally mandated date of 2014, the stock in fact showed a huge, unexpected decline in biomass. Completely halting cod fishing in 2011 and 2012 would not lead to rebuilding by 2014, and management measures under consideration in 2011 and 2012 would lead to reductions in groundfish revenue in New Hampshire (91% reduction), Maine (54%), and Massachusetts (21%) (U.S. Department of Commerce 2012; Gulf of Maine Cod Working Group 2012). Fishing communities and stakeholders have called these rapid, unexpected reversals in a stock status a *yo-yo* or *whip-saw* effect, which can lead to rapid, negative socioeconomic impacts on fishing communities and can undermine trust in fisheries science (NRC 2014). In New England, significant and rapid reductions in fishing effort from one year to the next have occurred for Georges Bank yellowtail, Gulf of Maine cod, witch flounder, pollock, Georges Bank cod, Georges Bank winter flounder, and plaice (Nies 2012). The yo-yo changes in stock status result from the continuous improvements and updating of stock-assessment models, followed by a retrospective analysis of stock health in previous years, and revisions to previous years' stock assessments.

Looking nationwide at stock rebuilding in the United States, the NRC (2014) reached several conclusions related to the treatment of uncertainty, including that the treatment of uncertainty varied across fisheries management plans and regions. The NRC noted that the treatment of uncertainty is not integrated across ecological, economic, and social dimensions, and it was not clear whether the appropriate level of precaution was being applied in fisheries management. The socioeconomic impacts of these findings on fishing communities could be substantial, and thus the fishing stakeholders are given motivations to view fisheries science with suspicion. In fact, fishers' mistrust and suspicion of scientists and managers in New England is well documented (e.g., Hartley and Robertson 2008; King 1999; Conway and Pomeroy 2006; Jones et al. 2007). In the early 2000s, for example, fishers noticed a problem with the trawl gear on the federal government's research vessel, which is used to conduct fish-sampling trawls and to provide data for the stock assessments that inform fisheries management decisions about catch limits. The cables attached to each side of the trawl were not of equal length, and thus it was likely dragging the net unevenly through the water. Fishers who discovered the problem were concerned that it would lead to the undersampling of fish and thus greater fishing restrictions than were warranted. This discovery led to a heated public debate surrounding "trawlgate" and about the adequacy and accuracy of the fisheries data (Kaplan and McCay 2004). As one fisher noted, "The biggest problem is not the [uneven] trawl wire. The biggest problem is that nobody there [in government] knew it" (Cook and Daley 2003).

8.2.1 Advancements in Treating Uncertainty in Fisheries Decision-Making

The ICES Working Group on Fisheries Systems recommended addressing uncertainty in a transparent manner, early, and continuously in the fisheries decision-making process in order to overcome mistrust of science (Dankel et al. 2012). Complicating the challenge of building trust in science and addressing uncertainty, risk communication scientists have documented that the general public looks at uncertainty very differently than scientists. The public can interpret uncertainty as indicating a poor understanding of the topic and those who communicate about uncertainty as untrustworthy, invoking confusion and anger (Johnson and Slovic 1995). In contrast, scientists view estimation and a rigorous discussion of uncertainty as reflecting a deeper understanding and more credible science. These perceived differences between fishers and scientists were observed in the stock assessment yo-yo effect documented by the NRC study (NRC 2014). Attitudes toward science—for example, science as debate versus science as the search of truth—have also been seen to contribute to how individuals respond to the uncertainty surrounding climate change; communicating higher uncertainty can be persuasive for some while delegitimizing for others (Rabinovich and Morton 2012).

In a comprehensive case study of the use of scientific information by resource managers in ICES, Wilson (2009) identified the significance of salience, credibility, and legitimacy to ensure that scientific information, uncertainties and all, was used in management. Salience is the usefulness of information to the problem at hand. Credibility reflects whether stakeholders perceive fisheries science and the methods of stock assessments as, for example, meeting a standard of plausibility and adequacy; and legitimacy refers to whether stakeholders perceive the output of the stock assessment process as unbiased and meeting the standards of fairness (Wilson 2009).

8.2.2 Building Credibility and Trust in Science

Communicative action theories of social institutions illustrate the significant role of iterative communication within safe and secure public spheres in developing mutual understanding and trust and in promoting acceptance of information, values, interests, and common objectives—along with uncertainty (Habermas 1984, 1987). However, the generation of scientific information and the production of management outputs to solve problems can occur in their own separate spheres and thus are not always conducive to iterative communication (Williams and Matheny 1995). Iterative communication occurs over time, includes multiple communication exchanges between individuals, provides time for reflection and asking clarifying questions, and builds a deeper understanding of each other's perspectives. Few professional opportunities exist for scientists, managers, and other fisheries stakeholders to engage in long-term, iterative dialogue in order to establish the salience of scientific information, the credibility of the science, and the perceived legitimacy of the scientific process among nonscientist stakeholders.

Fisheries management in the United States is a multistakeholder enterprise involving the commercial fishing industry; recreational fishing interests; conservation nongovernmental organizations; municipal, state, and federal governments; additional water-dependent industry sectors; and other interested parties. Fisheries management has been criticized as slow, co-opted, and ineffective because of this structure (Heinz 2000; Okey 2003; Rosenberg 2003), although there have been claims of considerable successes (Witherell 2004; Hilborn 2007). Further, while many stakeholders are at the table, stakeholder engagement often leads to disillusionment, frustration, and conflict—rather than mutual understanding and trust—in these complex natural resource and fisheries management contexts (Butler et al. 2001; Reed 2008).

New strategies in fisheries science and management seek to enhance the credibility among stakeholders of fisheries science with its uncertainties (see a description of these strategies in Table 8.1). To varying degrees, pedigree analysis, uncertainty matrix, extended peer review, integrating local ecological knowledge, participatory modeling, collaborative research, and other new approaches engage stakeholders in iterative communication and aim to develop greater mutual understanding.

Given the large number of diverse stakeholders, fisheries management has been conceptualized as a governance network (Hartley 2010; Gibbs 2008). A governance network is a group of individuals or organizations working together toward a common outcome—for example, the development of a fishery management plan with varying degrees of nonhierarchical and self-organizing features. The networks employ communication and coordination tactics, such as regular meetings, formal communication procedures, coordinating staff or leaders, defined decision-making procedures, and division of labor or responsibilities and expectations (Agranoff 2007). Networks have been shown to enable considerable trust-building, the development of mutual understanding, and innovation (Bodin and Crona 2009; Stojanovic et al. 2009). A multiparty network gives its members access to each other and to decision-making, and with that access comes the opportunity to influence each other and the outcome through iterative communication and greater mutual understanding (Verschuren and Arts 2004; Betsill and Corell 2008). This conceptual link between addressing uncertainty and networks provides a new analytical strategy for understanding how the structure and function

TABLE 8.1

Management Approaches for Addressing Uncertainty in Fisheries Science

Technique	Description	References
Pedigree analysis	Multicriteria, qualitative characterization of the origins and status of information and data.	Dankel et al. 2012
Uncertainty matrix	Classification method whereby a panel of experts numerically rates the nature and scale of the uncertainty on several defined parameters.	Walker et al. 2003
Extended peer review	Involving multiple disciplines and stakeholder perspectives on a peer-review panel.	Wilson 2009; Dankel et al. 2012
Incorporating traditional or local ecological knowledge	Participatory approaches and incorporation of traditional or local knowledge; e.g., Q-Method is based on the conceptual framework of factor analysis, seeking correlations between variables. The Q-Method is concerned with individuals' viewpoints, seeking shared views or correlations across a sample of individuals and clarification on points of agreement and disagreement.	Danielson, Webler, and Tuler 2010; Carr and Heyman 2012
Participatory modeling	Facilitated, structured dialogue about uncertainty and the quality of the state of knowledge among scientists and stakeholders to enhance scientific understanding.	Parma et al. 2003; Bentley and Stokes 2009; Röckmann et al. 2012
Collaborative research	Joint development, design, and implementation of scientific monitoring or research activities. Collaborative at all stages of the scientific process from developing questions through design and implementation to communicating findings and results.	Conway and Pomeroy 2006; Hartley and Robertson 2006, 2008; Johnson and van Densen 2007

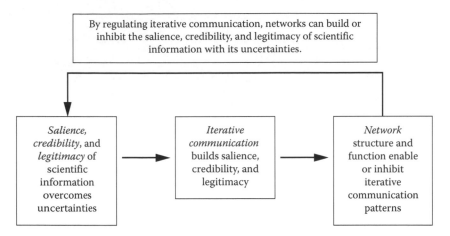

FIGURE 8.1
Conceptual framework: networks and uncertainty.

of fisheries management networks may enable or inhibit the development of salience, credibility, legitimacy, and trust of science and information in the face of uncertainty. The conceptual framework is illustrated in Figure 8.1. We employ network analysis methods to begin to assess these relationships by examining the potential for fisheries management networks to foster iterative communication across sectors and boundaries and establish the salience and credibility of the scientific information.

8.3 Governance Network Analysis

Governance networks have structural and functional characteristics, that is, network map shapes and roles that individuals play to enable the network to move information and resources. The structural patterns and functional roles of individuals reflect the operations of a network. Communication patterns in networks can develop mutual understanding, organizational learning, trust, and other features critical to the development of the credibility and legitimacy of scientific information (Manring 2007; Bodin and Crona 2009). Network structure can also indicate the influence of information and ideas and the potential to create innovation by connecting diverse expertise (Wyborn and Bixler 2013; Stojanovic et al. 2009).

Network structures can indicate the overall group behavior and guide functions of subgroups and individual members. For example, high density, reflecting many links between many members, can enable a sense of belonging and group identity (Coleman 1990). Density measures of a network map can indicate the level of trust and the potential for controlling behavior among the group members (Bodin et al. 2006). Bridgers are individuals linking otherwise less connected groups or individuals, and they provide access to diverse resources that can enable more adaptive and creative behaviors than high-density networks with many tight links (Newman and Dale 2005). Bridgers can synthesize a larger pool of knowledge, learn about the organizational cultures and interests of subgroups, and have advantages in being able to identify key individuals to connect with and the most appropriate strategies for establishing connections (Burt 2003). Bridgers can also serve an inhibiting role, however, particularly if they are overwhelmed by their bridging role; or they can hoard information and thus function as gatekeepers (Long et al. 2013; Cranefield and Yoong 2007).

Further, a network member's position in the network and his/her communication links are indicators of the member's role in managing information flow and his/her access to key decision makers. Friedkin (1983) coined the term "horizon of observability" as the threshold where information becomes inaccessible. Information, knowledge, or other types of resources that are within two communication links of another network member are readily available for decision-making. If members are two to three links away from each other, their information, knowledge, or other resources are generally known to the network members; however, the members are less conscious of the specific content and its availability. Members who are three links or greater away are incrementally moving beyond the horizon of observability, and their knowledge is beyond easy access to others in the network. Nonetheless, not all network links are equal, and there can be tremendous value in relatively infrequent network ties, also called *weak ties* (Granovetter 1983); for example, securing and transmitting financial resources can be an important but less frequent exchange than information sharing.

Network analysis quantitatively measures characteristics of connections (links) between individuals or groups (nodes), typically with surveys and interview instruments. Software—for example, InFlow and UNICET—is readily available to graphically map a network and to analyze its connectivity. Common measures of a professional network include communication frequency, although additional link characteristics associated with willingness to use information from the source in decision-making and other relational characteristics can be measured. Demographic measures of the nodes are collected to assess correlations with roles that individuals may play. Path lengths or degrees of separation among individuals reach across the network, and an individual's centrality measures of betweenness and closeness illustrate their roles in bridging between otherwise unconnected subgroups of information and dissemination of information.

Network analyses are presented from two cases: (1) a large regional fisheries-management process for Atlantic herring in the Gulf of Maine composed of scientists, industry, state, and federal governments in both the United States and Canada, regional fisheries management councils for the northeast, mid-Atlantic and the Atlantic states, and other stakeholder groups; and (2) a county-level fish land-use/habitat planning and development network in Virginia and the bordering county in Maryland in the United States. The first case assessed the communication network structure and function, including the iterative communication among stakeholder groups and the connectivity of scientific information sources, while the second examined communication and use of information from sources, that is, information that was seen as sufficiently trusted, salient, and credible for use in management decision-making. The cases contain multiple stakeholders and considerable scientific uncertainty and information that inform resource management decisions at a local, state, and federal level.

8.4 Case Study 1: Atlantic Herring Fisheries Management

The small, oily, schooling pelagic fish, Atlantic herring (*Clupea harengus*), is a key member of a complex, multispecies ecosystem and economy of the Northwest Atlantic, which is distributed along the North American Atlantic coast from Cape Hatteras, North Carolina, to the Canadian Maritime provinces. Atlantic herring serve as a foundation forage food for marine mammals, birds, sharks, and over 20 other fish species, which contribute to an ecotourism industry in whale and dolphin watching. Atlantic herring eat zooplankton and connect lower and upper trophic levels in the large Northwest Atlantic food web. There is a directed purse seine fishery, and Atlantic herring are critical bait for the lucrative lobster fishery. Atlantic herring support a breadth of economic and community activity (New England Fisheries Management Council 2003).

In the Northwest Atlantic, herring are not currently in an overfished state, and overfishing is not occurring—two important management classifications that trigger additional management actions atop the regular fishery management plan development. However, the potential for localized depletion and negative economic impacts on other fisheries, particularly lobster, and other economic sectors have contributed to a sequence of management plans and amendments in recent years. While more recent fisheries management plan (FMP) actions have been taken, the network analysis was conducted on a joint federal-state action undertaken in the mid-2000s. Federal Amendment 1 was completed by the New England Fishery Management Council in May 2006, while State Amendment 2 was issued in August 2006 (U.S. Department of Commerce 2007; Atlantic States Marine Fisheries Commission 2006).

8.4.1 Science Underpinning the Fisheries Management Plans

Stock assessments of the fish population size and distribution are the scientific foundation for management deliberations. Atlantic herring are managed as one stock in the Northwest Atlantic, that is, the Gulf of Maine and Georges Bank. Nonetheless, there is evidence of three distinct stocks in the region with different spawning times, locations, and biological characteristics; however, the lack of quantitative data on relative stock sizes for each distinct stock has led to difficulties in assessing individual stock status (Atlantic States Marine Fisheries Commission 2006; Overholtz et al. 2006).

Several sources of data on Atlantic herring stocks have been assembled, the primary being the U.S. and Canadian federal governments' trawl surveys. Further, for over 40 years, the U.S. federal government has conducted annual acoustic surveys offshore on Georges Bank and Nantucket Shoals. In addition, the herring fishing industry and scientists at a private research institution conduct collaborative acoustic survey research inshore on Atlantic herring spawning beds. Canadian and U.S. bottom-trawl surveys have been used to model herring stock population trends and abundance over time, although challenges are acknowledged; for example, environmental factors, altered herring behavior, and changes in survey gear or timing have been associated with significant annual variability. In part, to address some of these limitations, effort has been put into developing the acoustic survey designs for herring. In the United States, an acoustic research and monitoring survey was established in 1998 by the National Marine Fisheries Service (NMFS) to assess prespawning herring offshore on Georges Bank, followed in 1999 by an industry–science Collaborative Acoustic Stock Survey (CASS) covering inshore spawning components of the stock along the Maine–New Hampshire–Massachusetts coast.

A scientific panel converts survey data and analysis conducted by U.S. and Canadian government fisheries scientists into an overall stock assessment of abundance, geographic and temporal distribution, biomass, and scientific advice on quotas. Given the transboundary nature of Atlantic herring and the fishery, joint U.S. and Canadian stock assessment processes have been established. Since 1998, the Transboundary Resources Assessment Committee (TRAC) has reviewed stock assessments and the projections necessary to support management activities for shared resources across the United States–Canada boundary in the Gulf of Maine and Georges Bank. These assessments provide advice to federal and state resource managers on the status of fish stocks and the likely consequences of management alternatives. The TRAC cochairs (one Canadian government fisheries appointee and one U.S. government fisheries appointee) identify coexperts (one each from the Canadian and U.S. government fisheries agencies) responsible for coordinating data preparation, leading the analysis, facilitating the production and presentation of the working paper, and inviting independent peer review. TRAC drafts scientific consensus stock assessment reports and presents the results to U.S. and Canadian fisheries managers (DFO 2009).

TRAC produced reports in 2003 (Overholtz et al. 2004) and 2006 (O'Boyle and Overholtz 2006) referencing several sources, including U.S. winter, spring, and autumn bottom-trawl surveys, Canadian winter bottom-trawl surveys, U.S. and Canadian larval herring surveys (United States 1971–1994; Canada 1987–1995), the U.S. acoustic surveys on Georges Bank, and CASS (Overholtz et al. 2004). The fishery-dependent data from CASS have not been used regularly, although before the 2005 peer review, CASS data were cited by Overholtz et al. (2004) and in the NMFS Stock Assessment and Fishery Evaluation (SAFE) reports for Atlantic herring (Northeast Consortium 2006). Since the 2005 CASS peer review, the survey could not be considered a consistent time series of stock assessment data, and the data have been used more qualitatively by TRAC (Northeast Consortium 2006).

8.4.2 Fisheries Management Plan Development

These stock assessments and other relevant science about Atlantic herring feed into the deliberations of the fisheries management Plan Development Teams (PDT) charged with developing a fisheries management plan. PDTs consist of scientists and staff from the U.S. federal fisheries agency (NMFS) and staff from the regional fisheries management councils, state fisheries agencies, and research institutions. PDTs review stock assessment and other scientific findings before drafting regulatory measures and developing proposals for the species-specific oversight committee, which is a subset of the council members. Advisory panels are formed for each fishery among recreational and commercial fishers, charter boat operators, buyers, sellers, consumers, and other knowledgeable and interested stakeholder groups to provide advice and input to their respective PDTs, an oversight committee, and the councils. The oversight committee presents management strategies and measures to the full regional fisheries management council (for herring, the Northeast Fishery Management Council) for the approval and formation of a final FMP, which is then presented to NMFS for approval.

In addition to the federal fisheries management process, individual states in the United States are responsible for managing fisheries in state waters (within three miles of shore), although they must be consistent with federal rules. Established in 1942, the Atlantic States Marine Fisheries Commission (ASMFC) has three commissioners from each of the 15 Atlantic coast states from Florida to Maine, specifically the director of each state's marine fisheries agency, a state legislator for each state, and an appointed knowledgeable and interested individual. Each state has a single vote. The ASMFC adopts FMPs for coastal fisheries although it has limited regulatory authority, and it works cooperatively with lead-state regulatory agencies on interstate fisheries management, research, and statistical analysis; fisheries science; habitat conservation; and law enforcement.

8.4.3 Methods: Atlantic Herring Management

A questionnaire was administered via the web, telephone, or as hard copies among 249 participants identified in the public records as participating in the Atlantic herring FMP process and confirmed as participants by key informants. The questionnaire asked respondents to indicate the frequency with which they communicated with each listed individual on a scale from 0 to 5, with 0 meaning never; 1, yearly; 2, quarterly; 3, monthly; 4, weekly; and 5, daily. A 1–5 scaled frequency of communication is a common network analysis measure (Scott 2000; Monge and Contractor 2003).

Further demographic information was gathered, including age, years of experience on the job, educational achievement, and discipline or profession. The individuals participating in the CASS and TRAC stock assessment processes were identified based on a public record so that they could be located in the network maps. Standard survey data collection procedures and quality control standards were used in the design and administration of the questionnaire (Dillman 1999).

Data on the links (i.e., scaled frequency measures) and the nodes (i.e., demographic identifiers for the individuals) were entered into a database for importation into the InFlow software to generate network maps and to run the network-connectivity measures. An algorithm from mathematical graph theory is applied by most network analysis software; the algorithm in InFlow spatially orientates nodes in a map based on their relationship with each other. Once the network was mapped, connectivity metrics were calculated, specifically network size, density, and path lengths, along with measures of an individual's network centrality (degree, betweenness, closeness). Degree is defined as the total number of links an individual has with other nodes and is a measure of the activity level of the individual (Scott 2000; Monge and Contractor 2003). Betweenness is a core measure of centrality in a network based on a position of the shortest path between other nodes in a network (Freeman 1977). Someone with higher values of betweenness is most efficiently linking different individuals. For instance, an individual positioned between two clusters of nodes that are not otherwise connected would have a high value of betweenness. Closeness measures how close an individual is to everyone else; individuals with the highest values of closeness have the shortest path to everyone else and are in the best network position to monitor network information flow (Scott 2000; Monge and Contractor 2003).

8.4.4 Findings: Atlantic Herring Management

The communication network map of Atlantic herring fisheries management (Figure 8.2) reflects a snapshot in time (winter/spring 2007) among 146 individuals communicating weekly, consisting of members of state agencies from Maine to North Carolina; U.S. and Canadian federal fisheries agencies;

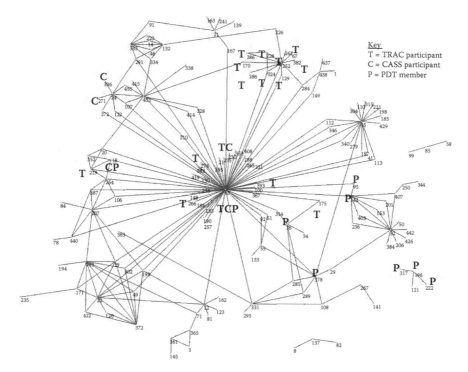

FIGURE 8.2
Atlantic herring management network map (weekly communication) with TRAC, CASS, and PDT participants noted.

several industry sectors, for example, the directed herring fishery, the lobster fishery, and the hook-and-line sector; and four nongovernmental organizations. The map represents individuals (referred to as nodes or actors) as the endpoints of lines and (at least weekly) communication channels between two nodes as lines. The large network (146 individuals) communicates weekly, which illustrates iterative communication.

The network's density was 1%. Density reflects how many links exist between different members; if everyone spoke with everyone else, then the network's density would be 100%. The weighted average path length of the entire network was 2.5. In other words, any two individuals in the network were on average fewer than three links away from each other within the horizon of observability and were likely generally aware of each other. For a network this large and with such a low density, to maintain a weighted average path length below 3.0, important bridging services were being provided by members of the network.

The node located in the center of the map demonstrated on average four times more links to others than the next ranked member of the network. That individual's network position and function illustrated the highest overall activity level in the network. He had the highest betweenness score (0.20), three times higher than the next ranked network member (0.06), and thus was forging strong connections between subgroups. Such bridgers connect otherwise disparate groups and allow the network to share information relatively efficiently, broadly, and quickly (within a week) across the breadth of the network. This bridger enables iterative communication across sectors.

Coleman (1990) showed that the presence of many links in a network (high density) contributes to the sense of belonging and group identity. Bodin et al. (2006) suggested that density is an indication of the strength of trust among individuals. Therefore, the Atlantic herring FMP network was less likely to form a strong sense of group identity or belonging. The stakeholders remain independent, and the network structure illustrated the disparate subgroups. Fisheries management in the United States is often characterized as a competitive public deliberation among disparate interest groups (Orbach 1989; Hilborn 2007), consistent with the network structure findings in this study.

Not all communication is equal; some has more significance in transmitting particular types of information or resources (Granovetter 1983), so pathways to the PDT are important for potential influence and for facilitating the science-to-management process. Both TRAC and CASS members have access to the PDT through their members who participate directly on the PDT. The institutionalized role of TRAC gives it preferential weight in the stock assessment over CASS. However, CASS had twice as many individuals serving bridging roles as TRAC. With greater connections to the PDT, CASS held far more communication pathways to access the PDT to build salience, credibility, and legitimacy in CASS scientific information with its uncertainties among the PDT members. Consequently, CASS likely overcame the institutional disadvantage of being a nonfederal trawl survey.

Figure 8.2 identifies the CASS, the TRAC stock assessment process, and the PDT members in the overall Atlantic herring FMP network. Two individuals participated in both CASS and TRAC. One was from industry (upper central node in Figure 8.2) and one from a state regulatory agency (the central node in Figure 8.2), so they bridged between those subgroups. The state regulatory agency member is the individual discussed above with the highest value of betweenness (0.20) and the most overall links. In fact, that individual also participated on the PDT and so bridged all three subgroups. A second representative from the state regulatory agency participated in the acoustic survey and the PDT. Hence, the acoustic survey subgroup had more representatives on the PDT (two) than TRAC (one), although TRAC had more of its subgroup members (16 out of the total of 28, or 57%) involved in the FMP process at a weekly frequency than the acoustic survey (5 out of the total of 24, or 21%). The network positions of CASS participants within TRAC demonstrated sufficient access to ensure that the information and data along with the uncertainties were available to the stock assessment scientists. A document review confirmed that the TRAC scientists were aware of the availability of the CASS data (Overholtz et al. 2006).

8.5 Case Study 2: Fisheries Habitat Management in the Chesapeake Bay, United States

Chesapeake Bay is North America's largest estuary. The watershed is home to 16.9 million people (more than double the 1950 population), covers 166,000 km², and includes six states (New York, Pennsylvania, West Virginia, Maryland, Delaware, and Virginia) and the District of Columbia (DC). Over half the watershed is forested, and another quarter is in agricultural use. Over 5000 domestic and industrial waste discharges are made into the watershed (Boesch and Goldman 2009; Chesapeake Bay Program 2010).

Excess nutrients, particularly nitrogen and phosphorous from nonpoint source run-off, and atmospheric deposition have led to a significant decline in water quality and accelerated eutrophication. It was estimated that by the mid-1980s, the Bay was receiving seven times more nitrogen and 16 times more phosphorous annually than when the English arrived in 1607 (Boynton et al. 1995). Sediment inputs remain high—two million tons of sediments from nontidal rivers entered the Bay in 2009, and from 1990 to 2009 the average annual sediment load was four million tons (Chesapeake Bay Program 2010). In turn, these factors precipitated a shift in the ecosystem state from a clear, seagrass-based system to a more turbid, phytoplankton-based ecosystem. The predominant habitats and biological communities changed accordingly. Fishing and habitat alterations—for example, dams interrupting alosine fish migration—led to changes in the dominant species and altered stock sizes. The collapse of oyster stocks and the dramatic loss of oyster reefs, coupled with the loss of wetlands and riparian forest, further degraded water quality by removing critical nutrient and sediment sinks (Chesapeake Bay Fisheries Ecosystem Advisory Panel 2006; Boesch and Goldman 2009).

The ecosystem-based management (EBM) efforts in Chesapeake Bay are often cited as stellar examples of EBM that link watershed, estuarine, coastal, and fishery resource management (e.g., Appleton 2008; UNEP/GPA 2006; U.S. Commission on Ocean Policy 2004). However, on-the-ground implementation has fallen short, as consensus on state and federal policy goals and objectives have not always led to local action and successful implementation (Posner 2009; Boesch and Goldman 2009). Boesch (2006) concluded that more effective bridging and integration of science and management were needed to achieve EBM in the Chesapeake. An ecosystem-based fisheries-management (EBFM) pilot project in Chesapeake Bay sought to bridge scientific understanding with management needs.

Starting in 2008, over 80 state and federal agency partners and research institutions (88 individuals), with funding from the U.S. Environmental Protection Agency (USEPA) Chesapeake Bay Program Office and support from the NMFS Chesapeake Bay Office, developed a new operational approach for EBFM in Chesapeake Bay. The EBFM project, coordinated by

Maryland Sea Grant (MDSG), aimed to contribute to the adoption of FMPs that consider the interconnections between five key indicator species in Chesapeake Bay (striped bass, menhaden, American shad, blue crab, and oyster); their physical and living environments; and the human influences on the species and environments. Each species team was charged with defining the essential biological background, stressors, and specific issues each species face, that is, the overall ecosystem context for the species.

8.5.1 Methods: Chesapeake Bay EBFM

We asked each species team to identify critical coastal habitats for their species based on the use of those habitats during significant stages of the species lifecycle, the vulnerability of habitat to degradation, and their professional judgment. Overlaying the critical habitats for each species on a map of Chesapeake Bay, we identified counties throughout the Chesapeake where land-use decisions affecting local habitats were most impactful to fisheries. Accomack County on the eastern shore of Virginia and Somerset County in Maryland were chosen, in part, because they are two adjacent counties that represented comparable populations, multiple jurisdictions, and a relatively manageable universe of fisheries, habitat and land-use planning, and development professionals. We compiled the list of local and state government staff and elected officials, nongovernmental organizations, civic leaders, and other central stakeholders in fisheries and land-use planning in these counties through key informant interviews and a review of public records.

The 88 participants in the EBFM project and 223 individuals from the two counties were surveyed or interviewed. The respondents reviewed a roster of individuals and provided a 1–5, Likert-scaled weighting for communication frequency and usefulness of information in decision-making for each individual listed. Standard data collection procedures and quality control standards were used in the design and administration of the questionnaire (Dillman 1999). Demographic measures were included (employer, level of education, and discipline). As with the Atlantic herring case study, data were imported into the InFlow software for the generation of network maps and the centrality measures of the network structure and function.

8.5.2 Findings: Chesapeake Bay EBFM

Figure 8.3 illustrates the weekly iterative communication network maps for the EBFM project participants (fisheries managers and scientists) and the land-use/habitat planning and development professional networks in Accomack County, Virginia; and Somerset County, Maryland. No individuals participated in both the fisheries and land-use/habitat networks, and only one bridging link existed between the two professional networks. Weekly communication is often at an operational level, illustrating individuals

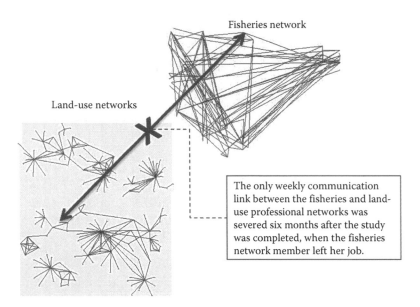

Fisheries network

Land-use networks

The only weekly communication link between the fisheries and land-use professional networks was severed six months after the study was completed, when the fisheries network member left her job.

FIGURE 8.3
Fisheries and land-use weekly communication networks, Chesapeake Bay.

regularly working together on a project. The one bridging individual in the fisheries network was an EBFM project facilitator whose contact node in the land-use/habitat network was in the State of Maryland's Department of Natural Resources. Within six months of completing the network analysis in 2011, this facilitator left her position, thus severing the only weekly communication link between the two professional networks.

The land-use/habitat network is arranged spatially by the InFlow software, reflecting the outcome of the network algorithm, and illustrating the four distinct, weekly communication subnetworks among the land-use professional community. The largest subnetwork in the center contains subclusters comprised predominantly of participants from either Maryland or Virginia but not both. A few bridgers exist, but little iterative communication occurs between the two states' land-use professional network. In contrast, the EBFM network is arranged according to the organizational design chart for the EBFM project. The EBFM network illustrates extensive iterative communication across the manager and science sectors. Thus, greater salience, credibility, legitimacy, and trust in science with its uncertainties were likely among managers than other, nonparticipating stakeholders.

Figure 8.4 contains the usefulness network among land-use/habitat planning and development professionals in Accomack County, Virginia; and Somerset County, Maryland. A usefulness network illustrates information sources that are trusted, salient, and credible enough for an individual to use that information when they make decisions. Information from these sources was identified as "very useful" in making decisions by respondents. The

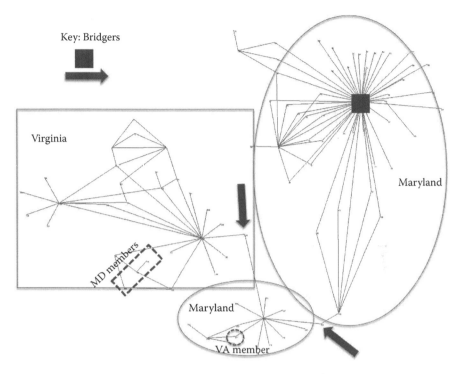

FIGURE 8.4
Usefulness network: information sources very useful for decision-making.

usefulness network consists of a 92-member network as opposed to the four subnetworks composed of 167 individuals in the weekly communication network. There were distinct subclusters in the usefulness network and a low density (1%), which illustrate the critical role specific individuals were playing to maintain connectivity across the entire usefulness network. However, bridgers were effective, and the weighted average path length, 1.29, was also low, illustrating that reaching across the network is relatively easy. No one is so far apart that others may not be aware of their usefulness and actually using others' information in their own decisions. Three distinct bridgers exist in the network. The bridger marked by a black square in Figure 8.4 has the most connections to otherwise isolated individuals; thus, that person's betweenness measure is nearly ten times greater than that for any other individual (0.29 versus 0.03 for the next highest betweenness score). Further, those individuals identified by the black arrows in Figure 8.4 are the only connections between subclusters, including the only connection between Maryland and Virginia usefulness subnetworks. While a small number of individuals from the opposite state were located within each of these state-specific subnetworks, in general Marylanders found other Marylanders most useful in making their decisions, and Virginians saw Virginians as most useful.

8.6 Networks, Boundary Spanners, and Trust in Scientific Information

Trust, salience, and credibility in science come from iterative discussion among stakeholders (Habermas 1984, 1987). Tactics and strategies that seek to expand the review of scientific information and uncertainties or to increase stakeholder involvement in the generation of scientific information have been developed to advance openness and transparency and iterative communication (Dankel et al. 2012). Consideration of the governance networks of fisheries management also provides additional insights into how trust, salience, and credibility are built and maintained in order to advance the use and influence of information. The patterns in the network structure and function of the Atlantic herring FMP and Chesapeake Bay EBFM project illustrated some of those advancements. Table 8.2 summarizes the findings from these two case studies.

Bridgers can synthesize a large pool of knowledge and learn about the organizational dynamics and interests of the subgroups that they connect, aiding the integration of information across subgroups because they can illustrate the salience and credibility of information to both subgroup audiences of which they are a member (Burt 2003). Both the Atlantic herring FMP and Chesapeake Bay EBFM networks had important bridgers who were connecting otherwise disparate members. Examples in both cases illustrated single individuals who were linking large subgroups. Nonetheless, these networks were fragile. If the individuals serving the bridging roles retired, took a new job, or otherwise left the network entirely, the networks would fracture substantially. When bridgers become overwhelmed with their network role or become highly specialized in their expertise, they can begin to inhibit iterative communication rather than advance it. Under these circumstances, a bridger is serving as a gatekeeper, regulating information flow, and hoarding information and resources (Cranefield and Yoong 2007; Long et al. 2013).

Further, individuals simultaneously participate in several networks, serving different roles in each. A critical bridger in one communication network may not be a critical bridger in the usefulness network and thus will not have the same influence in decision-making as he/she did in network communication. An iterative communication network may be fundamental in building mutual understanding, and while that arrangement may also be fundamental in establishing trust, salience, and credibility, the communication network alone is not what makes individuals and their information useful in decision-making. Our understanding of what enables bridgers—what resources and capacities they need—is still emerging. Given that bridging roles and structures may be different in communication versus usefulness networks, a bridger's skills and abilities may need to differ too.

TABLE 8.2

Insights From Network Analysis on Building Trust in Scientific Information

Network Feature	Description
Bridgers and the bridging function	Network-wide bridgers and individuals serving bridging functions between particularly important subgroups, e.g., industry and scientists, connect otherwise disparate groups and allow the network to share information relatively efficiently, broadly, and quickly.
Network density	Density measures indicate the strength of trust and the potential of social control, mutual understanding, group identity, and respect among members of the group(s). However, given the diversity of stakeholder participants and low densities in fisheries management networks, fisheries management and science will start from a point of low trust, salience, and credibility.
Influence	Network structure illustrates whether an individual might have access to decision makers and provides a venue where his/her voice will be heard. Additional network link measures are needed to examine other dimensions of influence, e.g., whether that information is understood and acted upon by the decision makers. Usefulness network measures contribute to filling this gap.
Networks disconnected	Management networks across sectors, e.g., fisheries and land-use/habitat, and jurisdictions, e.g., states and municipalities, are not likely sufficiently connected or nested to advance information integration. Governmental stovepipes are real and inhibit information flow.
Fragile nested networks	While connections between disparate subnetworks of sector or jurisdictional networks are rare, when they do exist they can be fragile, with just a few individuals enabling the connections. If an individual changes jobs, retires, or changes roles, the nested networks fracture and come apart.
Communication pathways	Communication and information flow does not necessarily track the chain of command in the organizational chart; additional pathways of information flow may exist.
Usefulness networks	Usefulness networks share general structural and functional features with communication networks. They have bridgers, and they can be tight and dense or loose and broad networks. However, an individual who is a bridger in a communication network may not serve the same role in a usefulness network.

Bodin et al. (2006) noted that density is an indication of the strength of trust and the potential of group identity and social conformity in a group. The fisheries management communication and usefulness networks here had low densities and highly diverse stakeholder membership. Effective bridgers between network components increased connectivity across the networks and iterative communication across sectors and boundaries, thus facilitating information flow and use. Low weighted average path lengths in the two networks studied here illustrated that reach across the network was well within the horizon of observability—members likely knew of other members and the general usefulness of other information sources. But these networks did not show high levels of group cohesion from high densities, inherent trust, obvious salience, and automatically granted credibility. It is critical to earn trust, to build mutual understanding in order to illustrate salience, and to establish professional relationships that engender credibility.

In spite of the shortcomings identified in the two networks studied here, the network structure was providing the members with access to each other—with a venue for their voices to be heard. The usefulness networks indicate that for some, those messages were heard, accepted, and used. Conducting both communication and usefulness network analyses provided insights into whether individuals and their information were influential, that is, going beyond iterative communication to achieve salience, credibility, and legitimacy of scientific information with uncertainties.

Networks are very dynamic, multifaceted phenomena. As seen in the Chesapeake Bay EBFM communication network, information and resources flow through many pathways, including those beyond the organizational chain of command. We cannot always or fully guide the flow of information and resources through our organizational designs alone. Stovepipes in government and coastal and marine resource management and disconnected subgroups across sectors and jurisdictions are real. There is little connection between regional fisheries management in Chesapeake Bay and local land-use/habitat planning and development decisions that impact critical fish habitat. Further, little connectivity exists across state boundaries on land-use/habitat planning and development between Maryland and Virginia. Without these fundamental communication connections, there can be little hope for informing stakeholders and building trust in the science among each resource management domain and less hope for influencing decisions and outcomes.

8.7 Conclusion

Donald Rumsfeld provided a simple, nontechnical classification of information and uncertainty—knowns, known unknowns, and unknown unknowns. Previous research has shown that different stakeholders view information and these uncertainties differently, with some believing that uncertainties bolster credibility and others feeling that they undermine credibility and trust. Network analysis provides insights into how information can flow between subgroups and how iterative communication can be fostered to build mutual understanding, which in turn, over time, can establish trust, salience, credibility, and legitimacy in science and information.

Some prominent fisheries scientists and managers have voiced concern over how to address uncertainty (e.g., Rosenberg 2007). Does an overemphasis lead to decision paralysis or does underemphasizing uncertainty and not being open and transparent increase the risk of actually undermining the credibility of and trust in science? Since communication and usefulness networks can be structurally and functionally different, and stakeholders perceive uncertainty and strategies for addressing uncertainty differently, it appears that if the public deliberations surrounding uncertainty do not progress beyond generally increasing awareness of scientific outcomes and processes, there is a risk that overemphasizing uncertainty could damage trust and credibility.

To move stakeholders beyond general awareness of information requires more iterative communication than is prevalent in the current stovepiped communication networks. Iterative communication occurs over time, includes multiple communication exchanges between individuals, provides time for reflection and asking clarifying questions, and builds a deeper understanding of each other's perspectives. There were not adequate nested networks across sectors and jurisdictions in our two case studies to establish the salience, credibility, and legitimacy of the scientific information with its uncertainties. More long-term, iterative communication across multiple sectors and stakeholders through participatory modeling and monitoring and collaborative research may be more likely to enhance credibility and trust in science and information in the face of uncertainty than strategies designed only to be more open and transparent, for example, pedigree analysis, uncertainty matrices, and extended peer reviews.

However, our understanding of how governance networks evolve over time and how communication networks may grow into usefulness networks is limited because longitudinal studies of networks are rare. Networks are not static, and it is likely that network structures and function ebb and flow temporally. Thus, networks show great potential to achieve a high level of iterative communication, provide mechanisms for nested networks among sectors and jurisdictions, and develop more open and transparent decision strategies that are more iterative and participatory.

Lastly, network analysis methods and software are becoming easier to use and more frequently applied, although network analysis remains a relatively new, emerging field in the coastal and marine resource management context. Network analysis shows a lot of promise, but the field needs to advance in some critical areas. Networks are far more dynamic and multifaceted than broad social-capital and social-network studies may imply, and thus we need to be cautious in interpreting network structures and functions and the ramifications of network structures and functions for addressing uncertainty, illustrating influence, and guiding management. Nonetheless, the further expansion of connectivity measures to consider salience, credibility, trust, and other dimensions of information flow, influence, and the science-to-management interface presents exciting opportunities for the growth of governance network theory and practice.

References

Agranoff, R. 2007. *Managing within Networks: Adding Value to Public Organizations.* Washington, DC: Georgetown University Press.

Appleton, A. 2008. Perspective: A bird's eye view of the Chesapeake. *Marine Ecosystem and Management* 1 (4): 7–8. http://depts.washington.edu/meam/MEAM4.html.

Atlantic States Marine Fisheries Commission. 2006. Amendment 2 to the interstate fishery management plan for Atlantic herring. Washington, DC: Atlantic States Marine Fisheries Commission.

Bentley N. and T. K. Stokes. 2009. Contrasting paradigms for fisheries management decision making: How well do they serve data-poor fisheries? *Marine and Coastal Fisheries: Dynamics, Management and Ecosystem Science* 1: 391–401.

Betsill, M. M. and E. Corell, eds. 2008. *NGO Diplomacy: The Influence of Nongovernmental Organizations in International Environmental Negotiations.* Cambridge, MA: MIT Press.

Bodin, Ö. and B. I. Crona. 2009. The role of social networks in natural resource governance: What relational patterns make a difference? *Global Environmental Change* 19: 366–374.

Bodin, Ö., B. I. Crona, and H. Ernstson. 2006. Social networks in natural resource management: What is there to learn from a structural perspective? *Ecology and Society* 11: 1–8.

Boesch, D. F. 2006. Scientific requirements for ecosystem-based management in the restoration of Chesapeake Bay and Coastal Louisiana. *Ecological Engineering* 26: 6–26.

Boesch, D. F. and E. B. Goldman. 2009. Chesapeake Bay, USA. In *Ecosystem-Based Management for the Oceans*, edited by K. McLeod and H. Leslie, 268–293. Washington, DC: Island Press.

Boynton, W. R., J. H. Garber, R. Summers, et al. 1995. Inputs, transformations, and transport of nitrogen and phosphorus in Chesapeake Bay and selected tributaries. *Estuaries and Coasts* 18: 285–314.

Burt, R. S. 2003. The social capital of structural holes. In *The New Economic Sociology: Developments in an Emerging Field*, edited by M. F. Guillen, R. Collins, P. England, and M. Meyer, 148–189. New York: Russell Sage Foundation.

Butler, M. J., L. L. Steele, and R. A. Robertson. 2001. Adaptive resource management in the New England groundfish fishery: Implications for public participation and impact assessment. *Society and Natural Resources* 14: 791–801.

Carr, L. M. and W. D. Heyman. 2012. It's about seeing what's actually out there: Quantifying fishers' ecological knowledge and biases in a small-scale commercial fishery as a path towards co-management. *Ocean & Coastal Management* 69: 118–132.

Chesapeake Bay Fisheries Ecosystem Advisory Panel (National Oceanic and Atmospheric Administration Chesapeake Bay Office). 2006. *Fisheries Ecosystem Planning for Chesapeake Bay.* American Fisheries Society, Trends in Fisheries Science and Management 3. Bethesda: American Fisheries Society.

Chesapeake Bay Program. 2010. Bay barometer: A health and restoration assessment of the Chesapeake Bay and watershed 2009. Annapolis: Chesapeake Bay Program. http://www.chesapeakebay.net/content/publications/cbp_50513.pdf.

Coleman, J. S. 1990. *Foundations of Social Theory.* Cambridge, MA: Harvard University Press.

Conway, F. D. L. and C. Pomeroy. 2006. Evaluate the human—as well as the biological—objectives of cooperative fisheries research. *Fisheries* 31: 447–454.

Cook, G. and B. Daley. 2003. Mistrust between scientists, fishers mars key mission. *The Boston Globe.* 27 October. http://www.boston.com/news/local/massachusetts/articles/2003/10/27/scientists_fishermen_in_standoff/?page=full.

Cranefield, J. and P. Yoong. 2007. Interorganisational knowledge transfer: The role of the gatekeeper. *International Journal of Knowledge Learning* 3: 121–138.

Dankel, D. J., R. Aps, G. Padda, et al. 2012. Advice under uncertainty in the marine system. *ICES Journal of Marine Science* 69: 3–7.

Danielson, S., T. Webler, and S. P. Tuler. 2010. Using Q method for formative evaluation of a public participation process. *Society and Natural Resources* 23: 92–96.

DFO (Fisheries and Oceans Canada). 2009. Transboundary Resources Assessment Committee (TRAC): Process. http://www2.mar.dfo-mpo.gc.ca/science/trac/process.html.

Dillman, D. A. 1999. *Mail and Internet Surveys: The Tailored Design Method.* New York: John Wiley.

Freeman, L. C. 1977. A set of measures of centrality based on betweenness. *Sociometry* 40: 35–41.

Friedkin, N. E. 1983. Horizons of observability and limits of informal control in organizations. *Social Forces* 62: 54–77.

Gibbs, M. T. 2008. Network governance in fisheries. *Marine Policy* 32: 113–119.

Granovetter, M. S. 1983. The strength of weak ties: A network theory revisited. *Sociological Theory* 1: 201–233.

Gulf of Maine Cod Working Group. 2012. Gulf of Maine (GOM) Cod Working Group summary. Gloucester: NOAA National Marine Fisheries Service. http://www.nero.noaa.gov/nero/hotnews/gomcod/GOM%20Cod%20Working%20Group%20Meeting_summary_v5.pdf.

Habermas, J. 1984. *The Theory of Communicative Action. Vol. 1, Reason and the Rationalization of Society,* translated by T. McCarthy. Boston: Beacon Press.

Habermas, J. 1987. *The Theory of Communicative Action. Vol. II, The Critique of Functionalist Reason,* translated by T. McCarthy. Boston: Beacon Press.

Hartley, T. W. 2010. Fishery management as a governance network: Examples from the Gulf of Maine and the potential for communication network analysis research in fisheries. *Marine Policy* 34: 1060–1067.

Hartley, T. W. and R. A. Robertson. 2006. Emergence of multi-stakeholder driven cooperative research in the Northwest Atlantic: The case of the northeast consortium. *Marine Policy* 30: 580–592.

Hartley, T. W. and R. A. Robertson. 2008. Cooperative research program goals in New England: Perceptions of active commercial fishermen. *Fisheries* 33: 551–559.

Heinz (The H. John Heinz III Center for Science, Economics and the Environment). 2000. *Fishing Grounds: Defining a New Era for American Fisheries Management.* Washington, DC: Island Press.

Hilborn, R. 2007. Defining success in fisheries and conflicts in objectives. *Marine Policy* 31: 153–158.

Johnson, B. B. and P. Slovic. 1995. Presenting uncertainty in health risk assessment: Initial studies of its effects on risk perception and trust. *Risk Analysis* 15: 485–494.

Johnson, T. R. and W. L. T. van Densen. 2007. Benefits and organization of cooperative research. *ICES Journal of Marine Science* 64: 862–868.

Jones, A., S. J. Slade, A. J. Williams, et al. 2007. Pitfalls and benefits of involving industry in fisheries research: A case study of the live reef fish industry in Queensland, Australia. *Ocean & Coastal Management* 50: 428–442.

Kaplan, I. M. and B. J. McCay. 2004. Cooperative research, co-management and the social dimension of fisheries science and management. *Marine Policy* 28: 257–258.

King, P. 1999. Practitioner's profile: The Fishermen and Scientists Research Society. *Common Property Resource Digest* 49: 9–11.

Keepin, B. and B. Wynne. 1984. Technical analysis of IIASA energy scenarios. *Nature* 312: 691–695.

Kloprogge, P. and J. P. van der Sluijs. 2006. The inclusion of stakeholder knowledge and perspectives in integrated assessment of climate change. *Climatic Change* 75: 359–389.

Legault, C. M. 2009. Report of the retrospective working group, January 14–16, 2008, Woods Hole, MA. Northeast Fisheries Science Center reference document 09-01. Silver Spring: National Oceanic and Atmospheric Administration/ National Marine Fisheries Service. http://www.nefsc.noaa.gov/nefsc/publications/crd/crd0901/crd0901.pdf.

Long, J. C., F. C. Cunningham, and F. Braithwaite. 2013. Bridges, brokers and boundary spanners in collaborative networks: A systematic review. *BMC Health Services Research* 13: 158–171. http://www.biomedcentral.com/1472-6963/13/158.

Manring, S. L. 2007. Creating and maintaining interorganizational learning networks to achieve sustainable ecosystem management. *Organization and Environment* 20: 325–346.

Monge, P. R. and N. S. Contractor. 2003. *Theories of Communication Networks.* New York: Oxford University Press.

New England Fisheries Management Council. 2003. *The Role of Atlantic Herring, Cuplea harengus, in the Northwest Atlantic Ecosystem.* Woods Hole: Northeast Fisheries Science Center.

Newman, L. and A. Dale. 2005. Network structure, diversity, and proactive resilience building: A response to Tompkins and Adger. *Ecology and Society* 10: 1–4. http://www.ecologyandsociety.org/vol10/iss1/resp2/.

Nies, T. 2012. Challenges in rebuilding New England groundfish. Presentation to committee on the evaluating the effectiveness of stock rebuilding plans of the 2006 Fishery Conservation and Management Reauthorization Act, June 8–9, 2012, Boston, MA. Washington, DC: National Research Council.

Northeast Consortium. 2006. Commercial vessel acoustic survey of coastal herring spawning units: Year 5. Durham, NH: Northeast Consortium. http://www.northeastconsortium.org/ProjectView.pm?id=1494.

NRC (National Research Council, National Academies of Science). 2014. *Evaluating the Effectiveness of Fish Stock Rebuilding Plans in the United States*. Washington, DC: The National Academies Press.

O'Boyle, R. and W. Overholtz. 2006. *Proceedings from the Transboundary Resource Assessment Committee (TRAC) Benchmark Review of Stock Assessment Models for Gulf of Maine and Georges Bank Herring*. Dartmouth, Nova Scotia: Fisheries and Oceans Canada.

Okey, T. A. 2003. Membership of the eight regional fishery management councils in the United States: Are special interests over-represented? *Marine Policy* 27: 193–206.

Orbach, M. 1989. Of mackerel and menhaden: A public policy perspective on fishery conflict. *Ocean and Shoreline Management* 12: 1–18.

Overholtz, W. J., L. D. Jacobson, G. D. Melvin, et al. 2004. Stock assessment of the Gulf of Maine–Georges Bank Atlantic complex, 2003. Woods Hole: Northeast Fisheries Science Center.

Overholtz, W. J., J. M. Jech, W. L. Michaels, et al. 2006. Empirical comparisons of survey designs in acoustic surveys of Gulf of Maine–Georges Bank Atlantic herring. *Journal of Northwest Atlantic Fisheries Science* 36: 127–144.

Parma, A. M., J. M. Orensanz, I. Elias, et al. 2003. Diving for shellfish and data: Incentives for the participation of fishers in the monitoring and management of artisanal fisheries around southern South America. In *Towards Sustainability of Data-limited Multi-sector Fisheries: Australian Society for Fish Biology Workshop Proceedings, Bunbury, Australia, 23–24 September 2001*, edited by S. J. Newman, D. J. Gaughan, G. Jackson, M. C. Mackie, B. Molony, J. St. John, and P. Kaiola, 8–29. Perth, Western Australia: Department of Fisheries, Fisheries Occasional Publications.

Posner, P. 2009. Networks in the shadow of government: The Chesapeake Bay program. In *Unlocking the Power of Networks: Keys to High-Performance Government*, edited by S. Goldsmith and D. F. Kettl, 62–94. Washington, DC: Brookings Institution Press.

Rabinovich, A. and T. A. Morton. 2012. Unquestioned answers or unanswered questions: Beliefs about science guide response to uncertainty in climate change risk communication. *Risk Analysis* 32: 992–1002.

Reed, M. S. 2008. Stakeholder participation for environmental management: A literature review. *Biological Conservation* 141: 2417–2431.

Röckmann C., C. Ulrich, M. Dreyer, et al. 2012. The added value of participatory modelling in fisheries management—What has been learnt? *Marine Policy* 36: 1072–1085.

Rosenberg, A. A. 2003. Managing to the margins: The overexploitation of fisheries. *Frontiers in Ecology and the Environment* 1: 102–106.

Rosenberg, A. A. 2007. Fishing for certainty: Science advisors should have confidence in their data, or risk being drowned-out by more dogmatic stakeholders. *Nature* 449: 989.

Rumsfeld, D. H. February 12, 2002. DoD news briefing—Secretary Rumsfeld and General Myers. Washington, DC: U.S. Department of Defense. http://www.defense.gov/transcripts/transcript.aspx?transcriptid=2636n.

Scott, J. 2000. *Social Network Analysis: A Handbook*. 2nd ed. Thousand Oaks, CA: Sage Publications.

Stojanovic, T. A., I. Ball, R. C. Ballinger, et al. 2009. The role of research networks for science–policy collaboration in the coastal areas. *Marine Policy* 33: 901–911.

UNEP/GPA (United Nations Environment Programme/Global Programme for Action for the Protection of the Marine Environment from Land-based Activities). 2006. *Ecosystem-Based Management: Makers for Assessing Progress.* The Hague: UNEP/GPA.

U.S. Commission on Ocean Policy. 2004. *An Ocean Blueprint for the 21st Century.* Washington, DC: U.S. Commission on Ocean Policy.

U.S. Department of Commerce. 2012. Magnuson-Stevens Fishery Conservation and Management Act provisions; Fisheries of the northeastern United States; Northeast multispecies fishery; Interim action; Prepublication. 50 CFR Part 648. *Federal Register* 77: 25623–25630.

U.S. Department of Commerce. National Oceanic and Atmospheric Administration. March 12, 2007. Fisheries of the northeastern United States, Atlantic herring fishery, amendment 1, final rule. *Federal Register* 72: 11252–11281.

van der Sluijs, J. P., A. C. Petersen, P. H. M. Janssen, et al. 2008. Exploring the quality of evidence for complex and contested policy decisions. *Environmental Research Letters* 3: 1–9.

Verschuren, P. and B. Arts. 2004. Quantifying influence in complex decision making by means of paired comparisons. *Quality and Quantity* 38: 495–516.

Walker, W. E., P. Jarremoes, J. P. van der Sluijs, et al. 2003. Defining uncertainty: A conceptual basis for uncertainty management in model-based decision support. *Integrated Assessment* 4: 5–7.

Williams, B. A. and A. R. Matheny. 1995. *Democracy, Dialogue, and Environmental Disputes: The Contested Languages of Social Regulation.* New Haven, CT: Yale University Press.

Wilson, D. C. 2009. *The Paradoxes of Transparency: Science and the Ecosystem Approach to Fisheries Management in Europe.* Amsterdam: Amsterdam University Press.

Witherell, D. 2004. Managing our nation's fisheries: Past, present and future. In *Proceedings of a Conference on Fisheries Management in the United States,* 13–15 November 2003, Washington, DC. Anchorage: North Pacific Fishery Management Council.

Wyborn, C. and B. Bixler. 2013. Collaboration and nested environmental governance: Scale dependency, scale framing, and cross-scale interactions in collaborative conservation. *Journal of Environmental Management* 123: 58–67.

9

Designing Usable Environmental Research

Elizabeth C. McNie, Angela Bednarek, Ryan Meyer, and Adam Parris

CONTENTS

9.1 Introduction

Society funds science to investigate natural phenomena and to inform and solve many discrete problems, including those related to coupled human–environmental systems, marine resources, and more (Lester et al. 2010; America Competes Act 2007; OECD 2002; Bush 1945). This has led scholars to suggest that society values research "not for what it *is* [but] for what it's *for*" (Stokes 1997, p. 98, italics in original). Society can also use relevant scientific information to help inform policy decisions, explore alternatives, manage uncertainties, clarify choices, and develop solutions (Sarewitz and Pielke 2007; Bednarek et al. 2011). A broad body of scholarship has indicated a need to more actively link those producing knowledge to those who will use it (Clark and Dickson 2003). A systemic shift to producing more useful science requires attention to the practice of science, as well as to the approaches for funding and evaluating science (Sarewitz and Pielke 2007). However, the science system—its institutions, incentives, and cultural norms—is largely failing to address this challenge (McNie 2007).

In this chapter, we introduce a typology that can help guide thinking about how science can be more useful in all phases of a research process, from the design of a funding program, to the execution of a research project, and to the evaluation of research outcomes. This perspective is that of four environmental scientists. We discuss the challenge of producing useful information in Section 9.2 and the role of science values and user values in shaping research approaches in Section 9.3. In Section 9.4, we introduce a typology that can be used to help characterize research approaches. In Section 9.5, we introduce two case studies to demonstrate how the typology can be used. Concluding remarks are presented in Section 9.6.

9.2 The Challenge of Producing Useful Information

Producing useful information is a challenging prospect. Useful scientific information has three broad characteristics. First, it must be salient, or relevant to the problem at hand. Second, it must be credible and produced in a manner consistent with expected standards and practices. Third, it must be legitimate, in that users believe the information produced is free from political suasion or bias (Cash et al. 2002). However, the worlds of science and policy are shaped by different values and characteristics, making it challenging to develop a universally accepted definition of valid knowledge. In science, valid information and knowledge usually arises from the testing of hypotheses and through accepted experimental or research methods. In society, validation may also come from other sources and lived experiences, representing a broader array of experts. Even when credible scientific information has been produced, it may still be difficult to integrate it into existing decision-making systems (Eden 2011; NRC 2006). Producing information that is also salient and legitimate to problem solving and trusted by potential users can be even more challenging. Even when done effectively, the timing and content of the linkage may be out of alignment with what is needed by decision makers (McNie 2007).

The dichotomy of basic and applied research presents another challenge to producing useful information. Research is generally lumped into one of these two large and ill-defined conceptual buckets (McNie et al. 2015). Together, basic and applied research form a compelling and pervasive *linear model* of innovation: the idea that knowledge flows in just one direction, from curiosity-driven basic research to applied research focused on a practical problem, and then development (Stokes 1997, p. 84). This model has been shown by many researchers to be an inaccurate conception of how science advances and how science becomes useful. Contrary to what the linear model would suggest, there is no need to divorce basic research from a consideration of use. Likewise, applied research may often be initiated based on a scientist's conception of what society needs, rather than through a process of engaging directly with users. Because they are weighted heavily toward science values, the linear model of innovation can cause scientists and science funders to believe their research to be more relevant to users than may actually be the case (Kropp and Wagner 2010).

Rather than rely solely on the linear model of knowledge transfer, we need a framework and a research approach that actively connect scientists and the intended users of the science (Dilling and Lemos 2011; Campbell et al. 2015). These users constitute a broad array of individuals and groups and can be governmental employees, resource managers, elected officials, members of civil society, other researchers, or anyone who has a vested interest in addressing a given problem. Engagement with these users can start with the research questions and research design itself. As Stokes explains, "the character of scientific knowledge, the intended use of science, and the role of users in the research process will often be directly pertinent to appropriate research design" (Stokes 1997, p. 3). Thus, it follows that engaging with some users even before grants have been awarded could help to clarify the larger research questions in a process often described as *coproduction* (Lemos and Morehouse 2005). Engaging with users throughout the research process may also be necessary to ensure usefulness, making it important to create, maintain, and use adequate social capital to build relationships based on mutual trust and respect (McNie 2013).

However, to date, science-policy decision makers—those responsible for funding research and setting research objectives—have relied largely on the science-heavy, basic versus applied distinction to set priorities. This means that in funding and designing *basic research*, decision makers and scientists rarely ask pertinent questions about the relevance of the funded research, the nature of the problem context at which it is aimed, or the practical ways in which the proposed research will engage that context. Through this approach, the production of scientific information is not actively designed to meet the demands of decision makers, leading to missed opportunities to link research with users (Sarewitz and Pielke 2007; NRC 2006) (see Figure 9.1).

Some have warned that involvement of nonscientists in science can undermine research or, worse, harm the institution of science (Bush 1945), yet research has not borne this out (Logar 2009). Science is a social process, just like many others, and is influenced by societal norms and individual values within its own community, as well as external ones (Jasanoff 2004; Latour 1987; van den Hove 2007). Ideas about what science should be funded are of direct interest to society at large, especially when public funds are being spent (Kitcher 2001). When political stakes related to the problem are high, when the problem is politically charged, or when there is significant power asymmetry between producers and users of knowledge, it may be even more critical to include users early in the process, particularly those who are politically marginalized and most affected by decisions (McNie 2013).

DEMAND : Do users have specific information needs?

		YES	NO
SUPPLY: Is scientific information produced?	**YES**	**Supply and demand reconciled:** Users' information needs reconciled with the production of scientific information.	**Missed opportunity:** Research priorities misaligned or users are unaware of possible utility of information produced.
	NO	**Missed opportunity:** Research priorities need modification in order to respond to users' information needs.	**Supply and demand reconciled:** Information not produced nor needed by users.

FIGURE 9.1
Missed opportunity matrix. (Adapted from Sarewitz, D. and Pielke, R. A. Jr. 2007 *Environmental Science & Policy* 10: 5–16.)

9.3 Science Values and User Values

Despite cultural expectations of experimental objectivity, a variety of values shape and guide decisions about research at many levels. We can think of these values as falling along a spectrum, with science values at one end and user values at the other end of the spectrum. Most research projects and programs fall between the two extremes and have a mix of both science and user values to varying degrees.

9.3.1 Science Values

Science values are those which prioritize generating new knowledge over other activities (Meyer 2011; OECD 2002). Basic research, sometimes called *pure* or *fundamental*, is generally seen as being driven by curiosity. The results are unpredictable, and the findings are meant to be generalizable (Calvert 2006). The National Science Board (NSB) defines basic research as the "systematic study directed toward fuller knowledge or understanding of the fundamental aspects of phenomena and of observable facts without specific applications toward processes or products in mind" (NSB 2010, p. 9).

Applied research is also closely aligned with science values. Applied research is the "systematic study to gain knowledge or understanding necessary to determine the means by which a recognized and specific need may be met" (NSB 2010, p. 9). It also involves "considering the available knowledge and its extension in order to solve particular problems" (OECD 2002, p. 78). But scientists are generally the ones assessing which problems to solve and how to solve them. If users are considered, engagement with them is often one-way, from applied researcher to user. Sometimes what researchers call *applied research* is not that, at least according to the definitions described above. There are cases where applied researchers engage more directly with users and the research activities are aligned more closely with use-inspired research, that is, research that is guided more by user values than by science values (see Chapter 16 of this volume).

Both basic and applied approaches may result in the production of information that is not useful for informing societally relevant problems (NRC 2011; NSB 2010). The research outputs for both approaches are often detached from users' needs so that research can be unfettered by practical demands and be driven primarily by curiosity (Bush 1945). While these approaches usually lead to novel discoveries, they may fail to address the inherent uncertainties in the "problem-solving work itself" (Funtowicz and Ravetz 1993, p. 740).

9.3.2 User Values

Science driven primarily by science values may still produce useful information, but research shows that this is less likely, even with an applied approach, if the user is not more directly engaged (Meyer 2011). Approaches to research that prioritize engagement with users are driven by user values. Such research considers in its design a range of factors related to the utility of information, such as the problem context, effective processes of engagement and collaboration, and outcomes beyond patents and journal articles. Though producing new knowledge is a goal in research driven by user values, this is not viewed as sufficient for success (Meyer 2011). Research that is guided by user values supports *coproduction* in which researchers and users collaborate to produce useful information (Lemos and Morehouse 2005).

Use-inspired research approaches are undertaken when there is a problem that has been identified that can benefit from a tighter integration between information production and use. At times this research may be quite basic in nature, as in Stokes's "use-inspired basic research," in which research is undertaken both by consideration of use and the quest for fundamental understanding of a phenomenon (Stokes 1997). Results of use-inspired research may contribute to a fundamental understanding of nature or may result in the production of context-sensitive information, or both. At other times, the problem may be quite discrete, so salient and legitimate information is needed to help develop solutions to the problem. In these cases, use-inspired research requires direct involvement of users in order to help clarify the scope of the problem, shape research agendas, and link knowledge with action (Clark and Dickson 2003). User-based approaches may have implications for decisions about how resources—for example, financial and human capital—are allocated to research, but the needs for supporting use-inspired approaches are less well-known than for the other approaches (NSB 2010).

Shifting to a more user-focused research approach may require experts beyond scientists (e.g., Guston 2001; Weber and Khademian 2008). Knowledge brokers or boundary organizations can help connect the users and researchers who have relevant knowledge in order to reconcile the users' needs with existing information, create a strategy for an effective process, and mediate that process (Meyer et al. 2015; Bednarek et al. 2015). For example, these experts can identify a community of relevant users, identify which scientists are working on a particular issue, and facilitate interactions among all involved.

Use-inspired research can also involve capacity building, which is undertaken when users' information needs are satisfied, but they do not have the individual, organizational, or institutional capacity to use or integrate the information into their existing knowledge systems. Research (often social science) and knowledge brokers or boundary organizations may be needed to help further clarify the users' understanding of how the information can be deployed and/or integrated into their knowledge systems. Educating and training both researchers and users about how to communicate and interact with each other, and how to integrate the research findings into existing decision systems, may be necessary to ensure that knowledge integration is accomplished.

Of course, the need for basic, applied, or use-inspired research will vary according to the kind of problem that society seeks to inform. In some cases, the need for use-inspired information may be quite low, for example, when our fundamental understanding of phenomena is limited. Basic research and much of applied research may be sufficient. In such conditions, the knowledge produced is likely to be scientifically credible, but the risk is greater that it lacks salience to any particular problem or legitimacy if users are not involved in informing the research design. As the demand for more useful information increases, the need for information to be credible, salient, *and* legitimate increases. Increasing these characteristics, however, comes at a price, as numerous trade-offs have to be made when choosing credibility, for example, over salience (Sarkki et al. 2014).

Whether it is basic, applied, or use-inspired research, we argue that coastal and ocean scientists and science-policy decision makers—those people who conduct research and allocate resources (fiscal, human, etc.) to support research—need better tools to help them design, deliberate, and implement research approaches. These tools should provide guidance and help inform what approaches to research one should undertake, as well as evaluate ongoing research efforts, to make sure that resources support the production of useful information.

9.4 Typology

In this section, we introduce a multidimensional research typology (see Figure 9.2) describing research activities and attributes that characterize research along a value-based continuum from science values to user values (adapted from McNie et al. 2015). An extensive literature review informed the development of the typology. The research typology presented here (Figure 9.2) can help clarify those qualities related to different research approaches. It can also be used to inform the design and implementation of research. The typology divides research into three general activities—knowledge production, learning and engagement, and organizational and institutional processes—each of which is subdivided into more specific attributes. The left side of the spectrum represents research criteria focused on achieving ends internal to science, or science values, while the right side represents research criteria focused on achieving ends external to the research itself, or user values. As one moves from left to right on the spectrum, the value of basic and applied research wanes, while the relative value of use-inspired research increases on the right side of the spectrum.

The typology is not used as a ranking system: high scores are not better than low scores, nor are user values inherently more valuable than science values. The choice of research approach and scoring it on the spectrum is completely

Activity	Attribute	Spectra of research criteria		
		Science values ⟵⟶ User values		
Knowledge production	Expertise	Epistemic		Experiential
	Relevance	General		Contextual
	Disciplinary focus	Singular, narrow		Transdisciplinary, diverse
	Uncertainty	Reduce uncertainty		Manage uncertainty
	Goals for research	Exploratory		Outcome oriented
Learning and engagement	Learning	Theoretical		Social, practical
	Knowledge exchange	Narrow		Iterative, influential
	Network participation	Homogeneous		Heterogeneous
	Social capital	Negligible		Significant
Organizational and institutional processes	Accessibility	Constrained		High
	Outputs	Narrow		Diverse
	Evaluation and effectiveness	Science-centric		Public-value oriented
	Flexibility	Constrained		Responsive
	Human capital	Narrow		Broad
	Boundary management	Limited		Broad

FIGURE 9.2
Typology of research activities and attributes. (From McNie, E. et al. 2015. A typology for assessing the role of users in scientific research: Discussion paper. Consortium of Science, Policy and Outcomes, Arizona State University.)

context sensitive and varies by the research problem. Only under idealized conditions do research projects or programs fall completely at one end of the spectrum or the other and, in reality, most fall somewhere in between.

Use of the typology can help inform the efficient allocation of resources to support the desired research outcomes. The typology can be used descriptively to assess research approaches in existing projects and programs in order to determine how well they support the desired knowledge production outcomes. It can also be used prescriptively and can help inform the development of research projects and programs based on desired knowledge production outcomes. The typology does not, however, suggest some normative bias. One approach to knowledge production is not inherently more valuable than any other: basic research is not better than use-inspired research and vice versa. Such valuation is completely context sensitive and is dependent on the particular expectations of knowledge production and use. For example, should the production of new knowledge be required, perhaps basic research is best suited. Alternatively, if policy makers are calling for information to inform solutions to societally-relevant problems, then use-inspired research may be better suited than basic or applied.

The following activities and attributes are adapted from McNie et al. (2015), while the questions posed for each attribute are taken directly from McNie et al. (2015). The typology is broken into three primary activities of research: knowledge production, learning and engagement, and organizational and institutional processes.

9.4.1 Knowledge Production

Knowledge production asks who is credible, what ways of creating knowledge are credible, and what knowledge is credible (Epstein 1995). Knowledge may be objective and consisting of facts, but may also be subjective and include lived experiences. Producing knowledge that is entirely free from social influence and bias may be difficult to do (Jasanoff 2004; Latour 1987). The five attributes of knowledge production are discussed in the following sections.

9.4.1.1 Expertise

Who has the credibility to produce knowledge? What constitutes expertise varies between academic disciplines and between science and society.

> *Epistemic*: Epistemic experts are typically researchers who have specific training and customs consistent with their research community (Knorr Cetina 1999) and are considered experts in their field.
>
> *Experiential*: Experiential experts have broader areas of expertise and may include economic, lay, and indigenous expertise (Edelenbos et al. 2011). Experiential experts may be needed to solve problems that "academic" experts may not solve alone (Epstein 1995).

9.4.1.2 Relevance

What is the source of relevance to solving the specific problem? What constitutes relevance varies significantly depending on the goals of research.

> *General*: Research that is general aims to test hypotheses and build theories and is most relevant at larger scales (NSB 2010; OECD 2002).
>
> *Contextual*: Contextual research better addresses all relevant scales (Cash et al. 2002).

9.4.1.3 Disciplinary Focus

How discipline-driven are the knowledge production activities?

> *Single/narrow*: A single or narrow disciplinary focus is guided by a reductionist worldview where problems are divided into smaller parts that can be isolated and studied (Funtowicz and Ravetz 1993).

Transdisciplinary/diverse: Transdisciplinary approaches integrate the physical, social, and natural sciences and are oriented toward problem solving (Clark and Dickson 2003; Ziegler and Ott 2011).

9.4.1.4 Uncertainty

How do researchers understand and address the problem of uncertainty in knowledge production? Researchers strive to reduce uncertainty but how, and to what extent, varies by discipline and research project.

Reduce: Reducing uncertainty involves minimizing or eliminating errors while increasing accuracy and precision.

Manage: Complex problems may be irreducible and instead may need to be managed (van den Hove 2007). More information does not necessarily reduce uncertainty and may even increase it (Sarewitz 2004).

9.4.1.5 Goals for Research

Is the knowledge produced to provide insights into science itself, or into questions and problems outside science? Desired goals and expected outcomes vary by research project.

Exploratory: Curiosity, and not specific goals, drives research agendas (NSB 2010; OECD 2002).

Outcomes-oriented: People who will use the research outputs help to shape the research.

9.4.2 Learning and Engagement

Learning requires information and is a process of transformation in which knowledge, skills, and behavior, and so on, are modified, developed, or changed. Social learning is contextual and iterative and requires systems thinking, communication, and negotiation (Pahl-Wostle 2009). Social learning also requires strong communication and negotiation skills as well as systems thinking skills (Keen and Mahatny 2006). As problems become less structured and more complex, learning becomes more difficult (Argyris 1976). The four attributes of learning and engagement are discussed in the following sections.

9.4.2.1 Learning

In what ways do the research outputs change the knowledge or decision-making system? The goals of research projects affect the nature of learning that occurs.

Theoretical: Focuses on the transfer of explicit knowledge related to theories and knowledge that can be easily transferred between people through procedures, documents, and patents, etc. (Nonaka 1994).

Social, practical: Learning is also about developing policies and plans and changing behavior. To do so, tacit knowledge, which is embedded in relationships and is difficult to codify, is exchanged (Nonaka 1994).

9.4.2.2 Knowledge Exchange

To what extent, and how, is knowledge exchanged? Knowledge exchange is about "generating, sharing, and/or using knowledge through various methods appropriate to the context, purpose, and participants involved" (Fazey et al. 2013, p. 19).

Restricted, linear: The exchange of knowledge between researchers and users is typically one way from researchers to users, using peer-reviewed publications, reports, and patents. Users are often other researchers.

Iterative, influential: Communication between researchers and users needs to be bi-directional, occur early and iteratively, and be understandable to both researchers and users alike (Lemos and Morehouse 2005). Achieving social impact requires that productive engagement be incentivized and supported (Spaapen and van Drooge 2011). Knowledge brokers—those people who help to mediate and negotiate between researchers and users—are often required when problems or solutions involve highly complex problems (Clark and Dickson 2003).

9.4.2.3 Network Participation

Who participates in the knowledge network? Information and resources move through structures of relationships between organizations, groups, and individuals. Networks enhance the ability for individuals, groups, or organizations to achieve goals through collective actions that they may not have been able to accomplish unilaterally (Weber and Khademian 2008).

Homogeneous: The network is limited, existing primarily in the researchers' own research community.

Heterogeneous: Networks include researchers, users, and other stakeholders, are distributed more broadly, and involve more complex relationships.

9.4.2.4 Social Capital

How important is the development and deployment of social capital? Social capital is about the relationships and "goodwill that others have toward us," that flow from the "information, influence and solidarity such goodwill makes available" (Adler and Kwon 2002, p. 18). Sharing knowledge, especially tacit knowledge, requires social capital and trust.

Negligible: While social capital is necessary to forge successful research collaborations, strict rules and accepted methods of research limit the amount of social capital needed to produce credible research outputs.

Significant: Sharing knowledge requires the production and deployment of social capital and trust (Levin and Cross 2004). People are more likely to accept and absorb new information when it comes from someone who is trusted.

9.4.3 Organizational and Institutional Processes

The organization of work, research, incentives, and both formal and informal rules, all shape the process of work, knowledge production, and interactions between groups (Trist 1981). Like all forms of work, research processes and organizations are subject to the same factors (Geels 2004). Significant research has been done to identify the best ways to organize research to improve outcomes (Hellström and Jacob 2000). Organizational and institutional processes describe the variables that influence the shaping of organizations and research activities. The six attributes of organizational and institutional processes are discussed in the following sections.

9.4.3.1 Accessibility

How accessible to users are the researchers and their organizations or institutions? Access to organizations is shaped by the organizations' proximity to others as well as institutional characteristics (Boschma 2005).

Constrained: The geographic location of research organizations and their placement within other organizations or institutes (departments, centers, institutes, and colleges) may limit accessibility to nonresearchers.

High: Organizations are located closer to the populations they wish to serve and are designed to support easy access to the researchers.

9.4.3.2 Outputs

How varied are the research outputs? Broadly speaking, outputs include reports, workshops, patents, peer-reviewed publications, proposals, trainings, meetings, and so on. Outcomes constitute a change in knowledge, resources conserved or depleted, policies adopted, and other valuable assets or conditions that are gained or lost (Lasswell 1971).

Narrow: The variety of outputs produced is limited, for example, to peer-reviewed publications, reports, workshops, and patents. New knowledge is the primary outcome.

Diverse: The variety of outputs is expanded and may include training, press releases, new methods, and processes, etc. Outcomes include changes to knowledge, but also improved performance, implementation of new policies, relationships built, or networks expanded, and so on. Outcomes may also include the development of social capital and trust.

9.4.3.3 Evaluation and Effectiveness

What factors shape the evaluation of research? Evaluation methods can use both quantitative and qualitative approaches. The evaluation processes can inform research design and practice, as can the choice of metrics (Mahieu et al. 2014; Molas-Gallart et al. 2014).

Science-centric: Quantitative methods are used most frequently to evaluate outputs, such as the citation index or impact factor for peer-reviewed publications (Best and Holmes 2010). Typically, outcomes are not evaluated, although that is beginning to change in some research fields.

Public-value oriented: Quantitative methods are used; however, qualitative methods are used more widely and may include users and other stakeholders in the evaluation process (Funtowicz and Ravetz 1993). Evaluation of outcomes should include analysis of products and processes to best capture the widest array of outcomes (Fazey et al. 2014).

9.4.3.4 Flexibility

How easy is it to alter research to better respond to users' needs, and changes in those needs? In order to respond to emerging threats or opportunities, organizations need to be nimble and flexible to reallocate resources depending on changing conditions.

Constrained: Organizational rules, funding, and operational conditions limit their flexibility and ability to reallocate resources.

Responsive: Organizational rules and operational conditions support flexibility in order to better respond to users' information needs and to emerging or changing problems.

9.4.3.5 Human Capital

What kinds of skills and training are needed to do the research? Human capital describes the skills and training necessary to perform a given job.

Hard skills: Most researchers earned a PhD as their primary form of training, including rigorous methods and research skills. Hard skills involve scientific and technical activities and are largely rules based.

Soft skills: These are focused more on relationships and behavior and facilitate communication, mediation, and negotiation activities.

9.4.3.6 Boundary Management

To what extent must efforts be made to actively manage the boundary between science and society? The boundary between science and society needs to be managed to satisfy the needs of users while simultaneously ensuring that science remains credible and legitimate. Boundary work involves communicating between science and society, translating information, and mediating and negotiating across the boundary (Guston 2001).

Low: Managing the boundary is less important because the risk that science becomes politicized or its credibility questioned is low.

High: As the involvement of users in shaping research increases, so too does the risk of science becoming politicized, or the credibility of science being questioned. Individuals or organizations who are skilled at managing the boundary may be needed to ensure the relevance, credibility, and legitimacy of science.

9.5 Case Studies

In this section, we present two case studies to illustrate how the typology can be used to assess existing research agendas. Two of the authors applied the typology to marine science projects in which they are directly involved, through work at the California Ocean Science Trust and the Lenfest Ocean Program. Application of the typology was informed by extensive experience with these cases and by informal conversations with colleagues and other stakeholders with knowledge of the projects. The results reflect current operational conditions and the opinions of the authors. Completing the typology is a subjective process and may lead to disagreement among researchers, or other participants, as to the "right" scoring for a particular attribute. This is to be expected and welcomed, as decisions about how to score the typology lead to important discussions and deliberations that enrich one's understanding of the research project in question. Scoring can lead to greater awareness about program priorities, that is, how to balance across attributes, and what adjustments might be needed to meet the project or program goals and expected outcomes. As stated earlier, higher scores are not normatively better than lower scores and vice versa.

9.5.1 MPA Watch

MPA Watch is a network of citizen science programs in California, in which trained volunteers collect data about human activities in and around state marine protected areas (MPAs). Currently, nine programs operate within the MPA Watch network, each run by one or more organizations, most of which are environmental nongovernmental organizations (NGOs). The primary motivation of this effort is to generate useful data about patterns and trends in human use of coastal natural resources, such as fishing, boating, scuba diving, paddle sports, sunbathing, and wildlife watching. Individual MPA Watch programs also pursue goals such as environmental education and stewardship. Over the last few years, MPA Watch programs have been working closely with a boundary organization—the California Ocean Science Trust—to align methods across programs and develop stronger partnerships with potential users beyond the local level. Beyond its role with the MPA Watch network, the Ocean Science Trust leads a public–private partnership to advance innovative, cost-effective MPA monitoring that meets the information needs of natural resource managers.

Ocean Science Trust's work with MPA Watch has focused in part on identifying managers' needs that could be met, at least in part, by analyses of MPA Watch data. Early conversations among MPA Watch programs, scientists, and natural resource managers suggest a variety of potential applications of MPA Watch data related to enforcement of laws and management of coastal natural resources. For example, one goal of the Marine Life Protection Act, which underpins state MPAs, is to enhance recreational opportunities. Currently, little knowledge exists about patterns of recreation along the California coast, and MPA Watch data could help to fill this gap. MPA Watch data could also help natural scientists who are monitoring ecosystems in and around MPAs, to understand better the human impacts on those systems. This understanding in turn could inform adaptive management of California's MPA network. In addition, data about patterns of violations such as poaching could help to shape outreach, education, and enforcement efforts related to MPAs.

Application of the typology to this case study is based on experiential knowledge based on 2 years working with the MPA Watch network, a wealth of organizational knowledge developed from building California's MPA monitoring program, and focused conversations with the MPA Watch Advisory Committee, which is composed of potential users in natural science, social science, and ocean resource management. As can be seen in Table 9.1 and Figure 9.3, this case study is highly focused on user needs,

TABLE 9.1

Activities and Attributes for the MPA Watch

Attributes	Score	Explanation
Knowledge		
Expertise	5	Expertise held by NGOs, citizen volunteers, boundary organization collaborators, and diverse potential users. Also natural and social scientist advisors.
Relevance	4	Research is aimed at place-specific knowledge that, using theory, may be susceptible to extrapolation over time. Extrapolation may be useful, but is not a prerequisite to utility.
Disciplinary focus	2	Currently, the focus is on social science methods that can work with data generated by the volunteers. There is a desire to integrate with other social science data, and with natural science data in order to expand management relevance, that is, the aspiration is toward a higher score here.
Uncertainty	5	Uncertainty is acknowledged and accepted as a reality because of many different scientific and programmatic constraints. Actions have been taken to reduce uncertainty through improvement in methods and analysis, but always in balance with other priorities.
Goals for research	5	The motivation for this program is to support a variety of goals related to coastal resource management.
Learning and Engagement		
Learning	5	Because this is a network of NGO-run citizen science programs informing coastal decision makers at multiple scales, iterative social learning is extremely important. But technical learning is also occurring, related to methods, experimental design, and analysis.
Knowledge exchange	3	Even before involving users, MPA Watch programs must explain technical concepts to volunteer participants. The programs are also translating this technical work into ideas about how MPA Watch creates value related to conservation outcomes and other goals of the primary funder. Finally, MPA Watch programs have begun iteratively engaging with user groups: state managers and natural scientists.
Network participation	4	The MPA Watch network extends throughout most of the California coast, but consists largely of program staff and volunteers. Creation of an advisory group is meant to expand the network into social and natural science communities and into state agency communities.
Social capital	3	Social capital is absolutely needed in this case, because the approach and resulting data are, in the eyes of many, as yet unproven. Social capital will be built through effective communication about program governance, quality assurance/quality control (QA/QC) measures, and relationship building with users. But likely also through traditional means such as peer review of reports, journal articles, or collaboration with established academic experts.

TABLE 9.1   (CONTINUED)

Activities and Attributes for the MPA Watch

Attributes	Score	Explanation
Organizational and Institutional Processes		
Accessibility	3	A new information management system and online data visualization tools are significantly opening up this program to users of any stripe. But the target audiences, for example, state agencies, do not have a deep understanding of the data, let alone the capacity to do serious analysis of them.
Outputs	3	From a scientific perspective, outputs and outcomes have been limited and narrow thus far. However, MPA Watch programs are also pursuing stewardship and environmental education, among other goals.
Evaluation and effectiveness	5	Evaluation has been quite informal, or at least, not very deep to this point. But the program is very much driven by values such as ocean health, education, and community.
Flexibility	3	MPA Watch programs have been refining methods with user needs in mind. But, going forward there is limited leeway for changes due to resource constraints, the need to keep volunteers happy and engaged, and unwillingness to alter protocols for long-term monitoring. Changes also take a very long time, given the distributed nature of program governance.
Human capital	4	Almost anyone who meets a few basic physical requirements can volunteer for MPA Watch. Furthermore, program staff need a range of skills related to volunteer management, program management, fundraising, and data management. Social science expertise is one area where there is a need for more capacity.
Boundary management	5	The need for boundary management is high for two reasons: (1) controversy often surrounds the issues (such as MPA violations) that volunteers are observing, and (2) understanding among potential users of the methods underlying, and potential applications of, the data is limited.

Source: Adapted from McNie, E. et al. 2015. A typology for assessing the role of users in scientific research: Discussion paper. Consortium of Science, Policy and Outcomes, Arizona State University.

as expressed by state partners involved in the day-to-day work of MPA monitoring and management. As the lower numbers on network participation, social capital, accessibility, and outputs indicate, more work is needed to fully realize the goal of meeting user needs. For example, the nine MPA Watch programs themselves have developed a strong network. But while avenues for participation exist, for example, by California State Parks and the California Department of Fish and Wildlife, these interactions are still in their infancy. Similarly, more technical work is needed to improve the credibility of, and demand for, MPA Watch outputs. As a first step, an expert

Activity	Attribute	Spectra of research criteria				
		Science values				User values
		1	2	3	4	5
Knowledge production	Expertise	Epistemic				Experiential ⊘
	Relevance	General			⊘	Contextual
	Disciplinary focus	Singular, narrow ⊘				Transdisciplinary, diverse
	Uncertainty	Reduce uncertainty				Manage uncertainty ⊘
	Goals for research	Exploratory				Outcome oriented ⊘
Learning and engagement	Learning	Theoretical				Social, practical ⊘
	Knowledge exchange	Restricted, linear			⊘ Iterative, influential	
	Network participation	Homogeneous		⊘		Heterogeneous
	Social capital	Negligible		⊘		Significant
Organizational and institutional processes	Accessibility	Constrained		⊘		High
	Outputs	Narrow		⊘		Diverse
	Evaluation and effectiveness	Science-centric				Public-value oriented ⊘
	Flexibility	Constrained		⊘		Responsive
	Human capital	Narrow			⊘	Broad
	Boundary management	Limited				Broad ⊘

FIGURE 9.3
Attributes along spectra of the MPA Watch program. Each attribute is scored on the value spectrum from one, strongly science values to five, strongly user values. (Adapted from McNie, E. 2015. A typology for assessing the role of users in scientific research: Discussion paper. Consortium of Science, Policy and Outcomes, Arizona State University.)

analyst has been contracted to develop recommendations on basic analytical approaches that the program can use to identify patterns and trends in the data. The MPA Watch programs guide this work, as well as the MPA Watch Advisory Committee, mentioned previously.

The fact that MPA Watch is a place-based, citizen science program, built with users in mind, explains the very high scores in many of the attributes. Participation is broad (volunteers range widely in age and background), with low barriers to entry; there is neither cost nor prerequisite to participation, and both training requirements and volunteer commitment expectations are fairly modest. Many kinds of knowledge are important, human capital is diverse, and boundary management is of particular concern. Expertise and evaluation both score fives, representing strong user values, yet a greater emphasis on science values, including the addition of peer-reviewed products and greater involvement of social science experts, are seen as factors that can help to build credibility in the eyes of state partners.

The research undertaken in this case represents predominantly use-inspired research. The need to align research approaches with user values is moderately high to high, because mandates and policy goals call for decision-making based in part on, and understanding of, evolving human use of coastal natural resources. However, research and monitoring capacity in this topic is relatively low, compared with capacity in natural science related to ocean resources. Brokering activities can be seen by MPA Watch's work with citizen scientists across communities and geographic areas, gathering and linking knowledge together to clarify problems and inform solutions. Data related to the phenomena exist, yet researchers themselves do not possess or cannot produce this knowledge themselves, hence the need to access a broader knowledge network to acquire such knowledge. Some capacity-building activities are also being undertaken in terms of MPA Watch's work with stakeholders, for example, scientists and state agencies, to improve the utilization of research and build from contextualized data toward more generalized knowledge about the system.

9.5.2 Operationalizing Fishery Ecosystem Plans

Fisheries management in the United States is organized around fishery management plans (FMPs), traditionally focused on a single species or an associated group of species. Ecosystem-based fisheries management builds on single-species management by accounting for the relationships among ecosystem components—marine organisms, humans, and the environment. To begin implementing this approach, some U.S. regional fishery councils have adopted or are drafting fishery ecosystem plans (FEPs) as frameworks for FMPs.

The Lenfest Ocean Program, a grant-making program focused on supporting policy-relevant research about marine ecosystems and connecting it to users, conducted an extensive scoping exercise regarding FEPs (interviewing over 80 stakeholders).* Program staff found that these plans differ substantially, and that there is no standard for what they should contain. To address these issues, the Lenfest Ocean Program decided to support a task force comprising natural and social scientists to create a practical blueprint that managers can use to make ecosystem-based fisheries management operational. The task force intends its main output to be an outline of the components of effective FEPs and recommendations about how to implement them. To help ensure that its recommendations are realistic and compatible with existing data and management structures, the task force is working collaboratively with an advisory panel made up of fisheries managers, stock assessment scientists, and fishermen. To encourage regional specificity, the task force is holding meetings in four different fishery management jurisdictions within the United States and has invited regional stakeholders to those meetings.

We applied the typology for this case study (see Table 9.2 and Figure 9.4) using focused conversations about the design and intended outcomes of the

---

* The Lenfest Ocean Program has funds from the Lenfest Foundation and is managed by The Pew Charitable Trusts. The Program supports research on policy-relevant topics concerning the world's oceans and fisheries and engages with managers to connect its supported research results to decision-making about ocean ecosystems.

TABLE 9.2

Activities and Attributes of Operationalizing Fisheries Ecosystem Plans

Attributes	Score	Explanation
Knowledge		
Expertise	3	The task force includes scientists with resource management experience, and the advisory body includes fisheries managers and fishermen. In addition, the task force invites other stakeholders to its meetings to provide additional perspectives.
Relevance	5	The task force is aimed at ensuring that its findings reflect specific management contexts, ecosystem dynamics, and socioeconomic circumstances.
Disciplinary focus	4	Both natural and social scientist members of the task force can assess essential ecosystem and socioeconomic components of an effective fishery ecosystem plan.
Uncertainty	4	The task force plans to incorporate considerations of uncertainty in ecosystem-based management within its recommendations.
Goals for research	5	The task force aim is explicitly outcome-oriented: to provide feasible options for implementing ecosystem-based fisheries management.
Learning and Engagement		
Learning	4	The task force is developing tools and guidance to make it easier for fisheries managers to implement ecosystem-based management so the focus is on practical learning.
Knowledge exchange	5	This project is designed to include user feedback early in, and throughout, its life cycle. The task force engages regularly with an advisory body made up of fisheries managers and includes various user groups in each meeting. Lenfest Ocean Program staff will be involved throughout the project to support other engagement opportunities.
Network participation	3	The target audience is at the regional level, rather than at the state or local community levels, and stakeholders are represented by a constituency, for example, fishermen, rather than by individual stakeholders.
Social capital	3	This project requires significant social capital because of the challenges inherent in shifting to a more holistic vision of resource management. The project was designed to build a certain level of trust by engaging with a variety of users from the beginning of the project. Development of social capital has yet to be fully realized.
Organizational and Institutional Processes		
Accessibility	3	The task force is accessible to some users during regional meetings and outreach activities. So far, no comment sessions have been held or calls for public engagement made, so not all stakeholders may find the task force accessible.
Outputs	4	The project is still in its early stages, so few products exist. However, the task force plans to release products throughout the life cycle of the project, rather than just at the end, to facilitate additional feedback and engagement opportunities.

TABLE 9.2 (CONTINUED)

Activities and Attributes of Operationalizing Fisheries Ecosystem Plans

Attributes	Score	Explanation
Evaluation and effectiveness	4	Plans exist to evaluate project outcomes based on the uptake of the information in fisheries management plans.
Flexibility	4	As the project progresses, the task force intends to be responsive to changing circumstances, such as incorporating lessons learned from ecosystem-based planning that occurs while the task force is in process.
Human capital	3	The task force itself consists of experts with doctoral degrees. The advisory body includes a wider variety of skill sets, including fishermen.
Boundary management	5	The Lenfest Ocean Program operates as a boundary organization for the task force. Program staff conducted the scoping exercise to assess what fisheries managers might need to operationalize ecosystem-based management and developed the framework of a scientific task force coupled with a resource managers' advisory panel. Program staff also help to find ways to engage with additional users and translate key findings for specific stakeholder groups.

Source: Adapted from McNie, E. 2015. A typology for assessing the role of users in scientific research: Discussion paper. Consortium of Science, Policy and Outcomes, Arizona State University.

project with Lenfest Ocean Program staff and other experts involved in the project. The project is still in its early stages (circa September 2015) so some of the project's scores in the typology may change over time. Because the overarching goal of the project is for managers to be able to implement ecosystem-based management using existing data and in a way that is useful for their specific management contexts, ecosystem dynamics, and socioeconomic circumstances, the project receives high scores for several attributes in the typology, including relevance, knowledge exchange, and boundary management. Knowledge exchange is particularly important given the large volume of data that need to be shared and synthesized. Some attributes are expected to become more user-value oriented as the project matures, for example, social capital and network participation.

The approach undertaken here represents predominantly use-inspired research. Brokering activities can be seen by the task force's work in synthesizing large volumes of existing data in order to develop a set of guidelines. Researchers possess a fundamental understanding of the phenomena and are capable of working with that information to produce the relevant knowledge. Capacity-building activities are also used to ensure that the information produced by the task force is relevant to problem solving and is capable of being used for the development of future fishery plans. The need for useful information is moderately high to high in the program, justifying research approaches that align with user values.

Activity	Attribute	Spectra of research criteria				
		Science values			User values	
		1	2	3	4	5
Knowledge production	Expertise	Epistemic		●		Experiential
	Relevance	General				Contextual ●
	Disciplinary focus	Singular, narrow			● Transdisciplinary, diverse	
	Uncertainty	Reduce uncertainty			● Manage uncertainty	
	Goals for research	Exploratory				Outcome oriented ●
Learning and engagement	Learning	Theoretical			●	Social, practical
	Knowledge exchange	Restricted, linear			● Iterative, influential	
	Network participation	Homogeneous		●		Heterogeneous
	Social capital	Negligible		●		Significant
Organizational and institutional processes	Accessibility	Constrained		●		High
	Outputs	Narrow			●	Diverse
	Evaluation and effectiveness	Science-centric			● Public-value oriented	
	Flexibility	Constrained			●	Responsive
	Human capital	Narrow		●		Broad
	Boundary management	Limited				Broad ●

FIGURE 9.4
Attributes along spectra of operationalizing fishery ecosystem plans. (Adapted from McNie, E. et al., A typology for assessing the role of users in scientific research: Discussion paper. Consortium of Science, Policy and Outcomes, Arizona State University, 2015.)

Each of the two cases presented here is unique and involves very different approaches to research and investigation of phenomena. Together they show that two cases that are predominantly use-inspired research can look very different along the value spectrum. This outcome is to be expected based on the context-sensitive characteristics of each case. Each case can be characterized by its own set of attributes along the value spectrum, which provides an informative snapshot of the current state of each project (see Figures 9.3 and 9.4). Using the typology can help decision makers decide where they may need to alter research trajectories and reallocate resources. For example, in the MPA Watch program, the California Ocean Science Trust has identified the need to continue to expand its network participation. With the operationalization of fisheries management plans, more work may be needed over time to improve the stakeholders' access to the research. As stated earlier, high scores are not normatively better than lower scores, yet higher scores may represent a better fit given the need, in both cases, to produce highly relevant information aimed at satisfying users' information needs. Nevertheless, some low scores are also expected given, for example, the need for more peer-reviewed publications in the first case and the broad array of expertise in the second case. Both cases scored boundary management as five, given that they navigate complex political and social systems with a broad array of stakeholders and given the need to produce information that is seen as salient, credible, *and* legitimate.

9.6 Conclusion

Linking science with decision-making is difficult for many reasons. One hypothesis is that science's dominant research approaches—basic and applied—have traditionally isolated knowledge production from users. These approaches work well if the goal of research is to produce new knowledge, but are problematic if the goal of research is to inform decision-making. Even in the case of applied research, in which scientists may define a problem according to what they think users might need, research is largely undertaken without directly engaging users. Research oriented toward user values may be better suited to reconciling the supply of scientific information with users' demands.

We used two case studies to demonstrate the use of a typology that can help identify patterns in the attributes of projects aimed at producing useful information. Both cases were explicitly aimed at producing useful information, so, not surprisingly, attributes for each project or program tended toward the user-value end of the spectrum. However, despite their shared intentions to produce usable science, the cases showed that multiple kinds of use-inspired approaches may be required depending on the specific characteristics of the problem(s). The cases also show that the typology can point out how and where projects might need to be adjusted to better align with project goals and expectations. In some cases, this may require moving to the left on the spectrum, at other times to the right. Understanding how research is aligned, or should be aligned, with a particular approach can help science-policy decision makers identify the necessary resources to allocate to optimize the research based on project and programmatic goals.

The effective use of the typology depends on two factors. First, adequate time and resources need to be deployed to clarify the problem at the heart of the research question, project, or program. This may require a robust examination of current trends and conditioning factors that characterize the problem and an estimation of the expected outputs and outcomes that may result from the research project. Users who are vested in the outcome of the research should be consulted during the problem-clarification process, even in the preliminary stages of developing the call for proposals to undertake the research. Careful attention needs to be paid to identify all relevant knowledge and ensure that it is incorporated into the research design.

Second, adequate resources need to be deployed not only to facilitate knowledge exchange and learning, but also to aid in the implementation of each approach. Personnel in the research team or organization must have the appropriate depth and breadth of training to conduct research under each approach, especially when engaging users is a necessary function of research. Researchers need greater flexibility to spend funds and allocate resources after a grant or project begins. This flexibility enables them to be more nimble, adapt to evolving research questions, and better manage knowledge exchange between science and society. User-oriented projects may take longer to complete than expected, especially if needs change rapidly. Appropriate incentives must be implemented to promote and reward the appropriate level of engagement with users. Finally, those undertaking these kinds of projects must have the skill sets that allow them to facilitate engagement with users, develop networks of users, and recognize changes in the policy or decision landscape over time.

Ultimately, the value of a typology that can characterize use-inspired research depends on the willingness on the part of program and project managers to think beyond the dominant basic and applied research dichotomy. They may need to recognize that research is more like an ecosystem, with different needs and processes relevant to different stages of knowledge creation and dissemination. The typology is not intended to facilitate the eradication of basic and applied research—these approaches are essential to knowledge generation—but rather, to help to identify those conditions in which other approaches may result in a greater likelihood that useful information will be produced and used. The scale and complexity of environmental problems that we face today, especially related to the coasts and oceans, need science to serve them to the best of its ability. This means that we, as program and project managers and researchers, need to consider more novel ways to investigate phenomena underlying the problems.

References

Adler, P. S. and S. W. Kwon. 2002. Social capital: Prospects for a new concept. *The Academy of Management Review* 27 (1): 17–40.

America Competes Act. 2007. H.R. 2272, 100th Congress, 2007–2009.

Argyris, C. 1976. Single-loop and double-loop models in research on decision making. *Administrative Science Quarterly* 21 (3): 363–375.

Bednarek, A. T., A. B. Cooper, K. A. Cresswell, et al. 2011. The certainty of uncertainty in marine conservation and what to do about it. *Bulletin of Marine Science* 87 (2): 177–195.

Bednarek, A. T., B. Shouse, C. G. Hudson, et al. 2015. Science–policy intermediaries from a practitioner's perspective: The Lenfest ocean program experience. *Science and Public Policy* Mar. 17, 1–10.

Best, A. and B. Holmes. 2010. Systems thinking, knowledge and action: Towards better models and methods. *Evidence & Policy: A Journal of Research, Debate and Practice* 6 (2): 145–159.

Boschma, R. 2005. Proximity and innovation: A critical assessment. *Regional Studies* 39: 61–74.

Bush, V. 1945. *Science, the Endless Frontier*. Washington, DC: United States Government Printing Office.

Calvert, J. 2006. What's special about basic research? *Science, Technology & Human Values* 31: 199–220.

Campbell, C. A., E. C. Lefroy, S. Caddy-Retalic, et al. 2015. Designing environmental research for impact. *Science of the Total Environment* 534 (15): 4–13.

Cash, D., W. Clark, F. Alcock, et al. 2002. Salience, credibility, legitimacy and boundaries: Linking research, assessment, and decision making. Working Paper RWP02-046. Cambridge, MA: Kennedy School of Government, Harvard University.

Clark, W. C. and N. M. Dickson. 2003. Sustainability science: The emerging research program. *Proceedings of the National Academy of Sciences* 100 (14): 8059–8061.

Dilling, L. and M. C. Lemos. 2011. Creating usable science: Opportunities and constraints for climate knowledge use and their implications for science policy. *Global Environmental Change* 21: 680–689.

Edelenbos, J., A. van Buuren, and N. van Schie. 2011. Co-producing knowledge: Joint knowledge production between experts, bureaucrats, and stakeholders in Dutch water management projects. *Environmental Science & Policy* 14: 675–684.

Eden, S. 2011. Lessons on the generation of usable science from an assessment of decision support practices. *Environmental Science & Policy* 14: 11–19.

Epstein, S. 1995. The construction of lay expertise: AIDS activism and the forging of credibility in the reform of clinical trials. *Science, Technology and Human Values* 20 (4): 408–437.

Fazey, I., L. Bunse, J. Msika, et al. 2014. Evaluating knowledge exchange in interdisciplinary and multi-stakeholder research. *Global Environmental Change* 25: 204–220.

Fazey, I., A. C. Evely, M. S. Reed, et al. 2013. Knowledge exchange: A review and research agenda for environmental management. *Environmental Conservation* 40: 19–36.

Funtowicz, S. O. and J. R. Ravetz. 1993. Science for the post-normal age. *Futures* 25: 739–755.

Geels, F. W. 2004. From sectoral systems of innovation to socio-technical systems. *Research Policy* 33 (6–7): 897–920.

Guston, D. H. 2001. Boundary organizations in environmental policy and science: An introduction. *Science, Technology & Human Values* 26 (4): 399–408.

Hellström, T. and M. Jacob. 2000. Scientification of politics or politicization of science? Traditionalist science-policy discourse and its quarrels with mode 2 epistemology. *Social Epistemology* 14 (1): 69–77.

Jasanoff, S., ed. 2004. The idiom of co-production. In *States of Knowledge: The Co-Production of Science and Social Order*, edited by S. Jasanoff, 1–12. London: Routledge.

Keen, M. and S. Mahatny. 2006. Learning in sustainable natural resource management: Challenges and opportunities in the Pacific. *Society and Natural Resources* 19 (6): 497–513.

Kitcher, P. 2001. *Science, Truth and Democracy.* Oxford: Oxford University Press.

Knorr Cetina, K. 1999. *Epistemic Cultures: How Sciences Make Knowledge.* Cambridge, MA: Harvard University Press.

Kropp, C. and J. Wagner. 2010. Knowledge on stage: Scientific policy advice. *Science, Technology & Human Values* 35: 812–838.

Lasswell, H. D. 1971. *A Pre-View of Policy Sciences.* New York: Elsevier.

Latour, B. 1987. *Science in Action: How to Follow Scientists and Engineers through Society.* Cambridge, MA: Harvard University Press.

Lemos, M. C. and B. J. Morehouse. 2005. The co-production of science and policy in integrated climate assessments. *Global Environmental Change* 15: 57–68.

Lester, S. E., K. M. McLeod, H. M. Tallis, et al. 2010. Science in support of ecosystem-based management for the U.S. west coast and beyond. *Biological Conservation* 143: 576–587.

Levin, D. Z. and R. Cross. 2004. The strength of weak ties you can trust: The mediating role of trust in effective knowledge transfer. *Management Science* 50 (11): 1477–1490.

Logar, N. 2009. Towards a culture of application: Science and decision making at the National Institute of Standards & Technology. *Minerva* 47: 345–366.

Mahieu, B., E. Arnold, and P. Kolarz. 2014. Measuring scientific performance of improved policy making: Summary of a study. Science and technology options assessment Working Paper, European Parliamentary Research Service, PE 527.383. Brussels: European Union.

McNie, E. C. 2007. Reconciling the supply of scientific information with user demands: An analysis of the problem and review of the literature. *Environmental Science & Policy* 10: 17–38.

McNie, E. C. 2013. Delivering climate services: Organizational strategies and approaches for producing useful climate-science information. *Weather, Climate, and Society* 5: 14–26.

McNie, E. C., A. Parris, and D. Sarewitz. 2015. A typology for assessing the role of users in scientific research: Discussion paper. Consortium of Science, Policy and Outcomes, Arizona State University.

Meyer, R. 2011. The public values failures of climate science in the US. *Minerva* 49: 47–70.

Meyer, R., S. McAfee, and E. Whiteman. 2015. How California is mobilizing boundary chains to integrate science, policy and management for changing ocean chemistry. *Climate Risk Management* 9: 50–61.

Molas-Gallart, J., P. D'Este, Ó. Llopis, et al. 2014. Towards an alternative framework for the evaluation of translational research initiatives. INGENIO Working Paper Series No. 2014–03. Valencia: INGENIO (CSIC-UPV). Accessed 15 September 2015. http://www.ingenio.upv.es/sites/default/files/working-paper/2014-03.pdf.

Nonaka, I. 1994. A dynamic theory of organizational knowledge creation. *Organization Science* 5 (1): 14–37.

NRC (National Research Council). 2006. *Linking Knowledge with Action for Sustainable Development: The Role of Program Management. Summary of a Workshop.* Washington, DC: National Academies Press.

NRC (National Research Council). 2011. *Measuring the Impact of Federal Investments in Research. Workshop Summary.* Washington, DC: National Academies Press.

NSB (National Science Board). 2010. *Key Science and Engineering Indicators: 2010 Digest*, edited by C. Roesel. NSB No. 10-02. Washington, DC: National Science Foundation.

OECD (Organisation for Economic Co-operation and Development). 2002. *Frascati Manual: Proposed Standard Practice for Surveys on Research and Experimental Development*. Paris: OECD Publication Services.

Pahl-Wostle, C. 2009. A conceptual framework for analysing adaptive capacity and multi-level learning processes in resource governance regimes. *Global Environmental Change* 19 (3): 354–365.

Sarewitz, D. 2004. How science makes environmental controversies worse. *Environmental Science & Policy* 7 (5): 385–403

Sarewitz, D. and R. A. Pielke, Jr. 2007. The neglected heart of science policy: Reconciling supply of and demand for science. *Environmental Science & Policy* 10: 5–16.

Sarkki, S., J. Niemelä, R. Tinch, et al. 2014. Balancing credibility, relevance, and legitimacy: A critical assessment of trade-offs in science-policy interfaces. *Science and Public Policy* 41: 194–206.

Spaapen, J. and L. van Drooge. 2011. Introducing "productive interactions" in social impact assessment. *Research Evaluation* 20: 211–218.

Stokes, D. 1997. *Pasteur's Quadrant: Basic Science and Technological Innovation*. Washington, DC: Brookings Institution.

Trist, E. 1981. The evolution of socio-technical systems: A conceptual framework and an action research program. Ontario Quality of Working Life Centre occasional paper No. 2. Toronto: York University.

van den Hove, S. 2007. A rationale for science–policy interfaces. *Futures* 39: 807–826.

Weber, E. P. and A. M. Khademian. 2008. Wicked problems, knowledge challenges, and collaborative capacity builders in network settings. *Public Administration Review* 68: 334–349.

Ziegler, R. and K. Ott. 2011. The quality of sustainability science: A philosophical perspective. *Sustainability: Science, Practice, & Policy* 7: 31–44.

10

The Balancing Act of Science in Public Policy

Peter Gluckman and Kristiann Allen

CONTENTS

10.1 Introduction

This chapter starts from the premise that much public policy-making in complex areas such as environmental and conservation policy depends on both scientific* evidence, which is almost always incomplete, and public values, which are almost always in dispute. This means that scientific input into the policy process will almost always be contested territory. It is unlike other forms of scientific information, such as an academic paper shared among peers or the provision of health information to the public. This is not because such scientific input is qualitatively different from any other insight derived from peer-reviewed scientific methods, but because of its position within the policy-making processes of modern democracies, which by their nature are a normative exercise in balancing public values.

For this reason, those charged with bringing science into the policy arena are themselves thrown into a challenging position. Today's scientists are increasingly called upon to take on a public-facing role by producing and conveying knowledge that is relevant to society's most pressing social, environmental, and economic concerns. Yet this *impact imperative* can also be a difficult balancing act for scientists. In highly disputed and values-charged environments, knowledge and knowledge production will always be subject to scrutiny, with interest groups across the political spectrum looking for footholds in policy debates. What appears to be objective advice to some will inevitably seem like advocacy to others. Scientists must tread very carefully along this divide, especially when it is often a personal passion or interest in a topic area that draws them to professionally study it in the first instance.

At the same time, it would be naïve to assume that scientific evidence that informs a public policy problem should be the singular basis on which a solution is formulated. Policy decisions are necessarily a balance of fiscal considerations, the electoral contract, competing priorities, and evidential input. Science provides for one among many types of input into public policy decisions, but its internationally accepted protocols and alertness to mitigating bias and influence can offer it a privileged place in the framing of public policy.

To parse these issues, this chapter proceeds in four parts. We first describe some of the unique challenges for science posed by the public

---

* In this chapter, references to "science," "scientific," and "scientist" are intended to be inclusive and encompass science from across the spectrum of disciplines, including social sciences and engineering, as relevant. Similarly, we do not attempt to distinguish use-inspired or "applied" research from curiosity-driven or "basic" research because each of these will have their own pathway to the policy context, which is beyond the scope of this discussion. Rather, what is relevant to the present discussion is that, regardless of type or discipline, the key feature is scientific rigor according to international standards such as formalized skepticism, replicability, and peer review.

policy context. Against this background, we then discuss how these challenges play out in new and changing expectations of scientists. This leads into a discussion of the principles, structures, and tools that can help balance the inherent tensions for scientists taking up an explicitly public-facing role, regardless of whether they are situated within the academy, public service, or another type of research organization. Finally, we draw on the example of the development of the Research and Scientific Information Standard for the management of New Zealand fisheries to illustrate the importance of robust tools and processes that can balance scientific requirements, inclusive policy imperatives, and the realpolitik of public decision-making, particularly in areas of high public concern and contention.

10.2 Three Main Challenges for Policy-Relevant Knowledge Production

10.2.1 The Unlikelihood of Simple Technical Responses to Policy Questions (Post-Normal Science)

From the perspective of anyone outside of the policy-making machinery, the process seems straightforward enough. Indeed, the very existence of the term "policy cycle" gives the distinct impression of a tight and textbook process, albeit one that is iterative. The reality of policy formation is far messier: those charged with working up the options must contend with multiple inputs and influences including public opinion, fiscal commitments, legislative obligations, reigning government ideology, and, hopefully, scientific and technical knowledge of the issue.

Yet the greatest challenge, perhaps, is the fact that the policy questions for which scientific input is most often sought are not the straightforward technical matters, but instead issues that have all the hallmarks of what has been called "post-normal science" (Funtowicz and Ravetz 1993). These are the questions that defy Thomas Kuhn's "normal" approach to structuring science (Kuhn 1962). They are the urgent issues of high public and political concern, where the people involved hold strong positions based on their values, and where the science is complex, incomplete, and uncertain. Diverse meanings and understandings of risks and trade-offs inevitably dominate, making the politics itself equally complex. Examples might include protecting ecosystem biodiversity and controlling invasive species; legalization of recreational psychotropic drugs; addressing the multifactor antecedents of obesity, mental health, and suicide; attending to the social impacts of an aging population; investment in early-childhood education; sustainable city planning; fisheries management; and balancing economic growth and environmental sustainability, particularly where the application of new technologies is under consideration.

Viewed within the landscape of "post-normality," it is easy to see how the policy context can amplify the public effect and impression of scientific uncertainty. Thresholds of evidence that may be subject to long held scientific consensus suddenly can be thrown into question when interrogated from a public policy or media perspective. Similarly, levels of uncertainty common to the science community can be exploited for political gain (Oreskes and Conway 2010; Michaels 2008). Even the most robust available evidence—whether it is shedding light on a problem or pointing to a potential technological solution—may be mercilessly challenged in the *court of public opinion*, a court that matters to decision makers in a democracy. This tension is normal practice in government and it is the reason why the production and application of scientific information *for public policy* often requires bespoke skills, tools, and approaches, and the presence of boundary structures to interpret and navigate across the two cultures of science and policy-making. In an increasing number of democratic jurisdictions, such boundary structures are emerging through the establishment of offices or panels of science advisors to governments. In addition, such boundary approaches are also drawing increasingly from the social sciences to help serve their translational roles.

10.2.2 The Complex Societal Context for Public Policy Questions That Are Most in Need of Scientific Input (Post-Trust Society)

Together with the challenge of providing post-normal knowledge production for the policy environment is a related challenge of what has been called our "post-trust society" (Löfstedt 2005). Access to 24-hour global news, the proliferation of science and pseudoscience information across the Internet, and changes to most Western political landscapes that increasingly encourage corporate interactions with academia for greater economic development have had a combined effect on public perceptions of science and how it is produced. In addition, public crises such as bovine spongiform encephalopathy (BSE)* and, more recently, the British Petroleum Deepwater Horizon leak in the Gulf of Mexico, or the unanticipated consequences of the L'Aquila and Fukushima earthquakes, have contributed to an erosion of public confidence in science (Alexander 2014; Yeo 2014). As scientists take on more public roles, it is sometimes unclear to the observer whether they are simply conveying what is known (and not known) about an issue or whether they are acting as an advocate of a specific course of action and, in the latter case, whether such advocacy is legitimately supported by the evidence, or whether it is

---

* An epizootic illness traced to the feeding of beef cattle with meat and bone meal. The crisis sparked mass culling of herds in the United Kingdom and a widespread ban on British beef exports from the mid-1990s to mid-2000s.

conjecture that reaches beyond available information. Sorting claims from counterclaims can be increasingly difficult for lay observers, particularly as they will also bring their own frames of reference to the information.

Yet at the same time, there has never been a more urgent public call and need for an active role for scientific expertise in the work of governing, lawmaking, and achieving public consensus. The challenges every society faces make science an essential tool of democracy, but as such it is now a critical public and *political* resource across the ideological spectrum. And this is what makes scientific knowledge production and translation processes so vulnerable to criticism, and risks undermining public trust in the scientific enterprise in a highly charged public policy context (Gluckman 2015).

Thus, we rely on the hallmark processes that define science to protect scientific integrity and foster trust. The use of recognized scientific methods, peer review, and the publication of methodology and results are the pillars of science and the basis on which peer and public trust is built. These pillars are founded on the important principle of scientific skepticism that demands and enables the testing and retesting of hypotheses. Importantly, the processes of science have been developed to minimize influence and bias in the collection and analysis of data. In this way, science provides the only processes by which we can gather relatively reliable information about our world (Marks 2009). Protecting and promulgating the integrity of these processes is the key feature that legitimates scientific expertise over other forms of knowledge (e.g., doctrinal, traditional, anecdotal, etc.).

Yet, in the context of policies built on post-normal science and for a post-trust public, "expertise" itself is rarely immune to critical questioning. Sociologist Jürgen Habermas was the first to problematize the concept of expertise a generation ago, critiquing it as elitist and even counterproductive to democracy (Habermas 1970). While opening up a new area of critical theory, his work also served as an important forerunner to influential empirical studies that followed, which began to view the legitimacy of expert advice as a combination of authority, built on access to specialized knowledge, and— importantly—trust (Callon et al. 2009). It is noteworthy that the legitimating features of expertise are necessarily a matter of both technical and social considerations.

The wide-ranging body of social and behavioral science literature on the concept of trust, while a testament to its importance, is beyond the scope of the present discussion. But it is self-evident that trust is a quality that must be actively earned and maintained, and it is not easily regained once it is lost. For all scientists, but especially those working in boundary roles at the science–policy interface, trust in their expertise is paramount. Section 10.4 of this chapter describes the tools, mechanisms, and strategies that can be built into boundary-spanning roles and processes with a view to building and maintaining trust.

10.2.3 Differing Culture and Practices of Science and Public Policy

Whereas short timeframes, simplicity of message, and an inherently normative decision-making framework characterize the policy-making context, science is by nature a long-term and iterative undertaking that strives to be objective. Moreover, scientific uncertainty can be anathema to the policy (or political) imperative for decisiveness. While many scientists would rather avoid reaching a conclusion until "more research is done," policy makers often do not have the luxury of time. On the surface, the cultures and practices of these separate worlds seem almost irreconcilable such that both the scientist and the policy maker will be frustrated by any attempt to cast an appropriately scientific lens onto public policy issues.

Key to bridging this divide is to clearly delineate the role that science can play in the policy context and to acknowledge the limits of that role. These limits are not necessarily due to scientific uncertainty, but also arise in the face of legitimate values-based considerations for policy. Scientific input must be clearly distinguished from other policy considerations, but should not supplant them. Indeed, recognizing and distinguishing the legitimate place of public values in policy-making can help to protect science from being strategically co-opted as a proxy for normative debates.

For instance, calling into question the rigor of climate science or casting doubt on compelling evidence of the benefits of harm reduction approaches to addictions are by now recognized tactics to subvert a normative debate by disguising it as a supposed scientific one, even (and perhaps especially) where the science is largely settled. The deliberately confusing public messaging undermines trust in scientific input, and renders impossible any meaningful public debate on the normative values-based aspects of a policy issue. Only when these aspects are clearly identified and not obfuscated is science best able to inform the public and decision makers on the associated trade-offs.

It is not only the values-rich contexts of policy debates that can prompt the questioning of the legitimacy of science. The sheer operational realities of policy-making can conflict with the very processes that are designed to protect the integrity of scientific knowledge production. Ideal methodologies may be deemed too expensive, full peer review may be too time-consuming, and the required testing to eliminate uncertainties may be untenable from a policy perspective.

Both the degree of operational compromise and the essential role that values play in policy-making can be a source of discomfort for many scientists who may prefer to adhere to an idealized Mertonian view of knowledge production (Merton 1942). But this is an impossible and indeed unacceptable framing for modern science. If science is to be of practical value in meaningfully informing public policy decisions, it must acknowledge and embrace its own coproduction with other societal processes (Jasanoff 2004; van Kerkhoff and Lebel 2015). However, this requires a clearer and more mature understanding of the (real and potential) mutual influences between societal and scientific processes. Making such influences visible allows them to be carefully managed, which ultimately protects the integrity of science, while also making it more inclusive.

Moreover, scientific knowledge production itself cannot be assumed to be value-free (Douglas 2009); it is important for scientists to identify and manage the appropriate ways in which value judgments enter the scientific process, from framing questions and choosing methodologies, to making an inferential leap as to the sufficiency of evidence for a given course of action. Recognizing that these are normative decisions common across all professional scientific practice allows them to be defended using collective standards of scientific consensus. This in turn renders them less vulnerable to becoming targets of criticism by advocates seeking to undermine evidence for the policy context.

Beyond knowledge production, human values will and must enter into the question of the *application* of knowledge. Indeed, the issue of *social license* for scientific and technological innovation is expanding rapidly. This is understandable because science is a public tool and a common good, but the application of new knowledge is rarely without some risk. Arguably, the scientific community has been insufficiently attentive to the necessary societal conversation in technology-driven, yet values-based, public debates. Inherent in such discussions is the need to maintain integrity and public confidence in the scientific endeavor because the way in which the public discourse unfolds has the potential to cast a long shadow by influencing public perceptions of science in general and for the long term. Arguably, democracy is hurt when science becomes a proxy for debates that should legitimately occur under a different label.

10.3 Changing Culture of Science Systems and New Public Roles of Scientists

It is clear that the application of scientific information to policy development in an era characterized by post-normal science and a post-trust society is both essential and vulnerable. These pressures and tensions are particularly apparent for those working at the interface of science and public policy, who must simultaneously maintain the trust of the science community, policy makers, politicians, the media, and the public in order to perform their increasingly essential translational role. In this section, we widen the analytical frame to consider how the very existence of an "interface" between science and public policy is part of a larger and more explicit commitment to ensuring that science has "impact." Ironically, however, fulfilling this commitment can also challenge some of the long held tenets of scientific practice.

What Michael Gibbons famously called "science's social contract with society" is now undergoing a major upheaval with a move to a much more relational and iterative "social compact" between multiple stakeholders within and outside the science system (Gibbons 1999). This compact increasingly prioritizes societally-relevant research (de Jong et al. 2015). Although many consider such relevance primarily through the lens of economic development and technology commercialization, it is critical also to the mobilization of new knowledge to inform public policy and address health, societal, and environmental challenges (HEFCE n.d.; Harland and O'Connor 2014).

Though broad in scope, this increasing focus on the relevance and potential impact of science carries the risk of adversely affecting what are currently considered the fundamental pillars of the scientific endeavor. For instance, whereas scientists and the peer review processes that assess their work have traditionally focused on scientific *excellence* and not the normative questions of *relevance*, there is now an increasing expectation within public science systems to move beyond excellence alone and give explicit priority to research that meets end-user needs.

There is an ironic tension here that must be confronted: The increased focus on the relevance and impact of research also increases the risk of real or perceived external influence on the pillars of the research process, which can undermine public trust in science. And *trust* in the science is precisely what facilitates its application to public policy and thus its relevance. Indeed, this inherent tension is echoed more broadly in the fact that a closer relationship between science and society is welcome and necessary, but the mechanisms to achieve this can threaten the very feature that makes science most useful to informed societal debate; namely, its objectivity. Scientists are thus left to balance these competing imperatives and carefully manage the consequences.

To better understand the complexity of this situation for science and for scientists, a brief look at history may be instructive. The professional sciences and public science systems as we know them today are relatively recent phenomena and products of global events and societal preconditions. The earliest Western scientists of the seventeenth and early eighteenth centuries were rare and fortunate hobbyists who could survey, experiment, and invent at their leisure. Universities existed, but far from the model of research and scholarship we know today; rather, these institutions were primarily conservative bastions of tradition, faith, and received wisdom, most often intended to perpetuate a scholarly religious lineage. It was not until the late eighteenth and early nineteenth centuries that the concept of professional scientist emerged. This was a period of considerable global activity in refining many of the norms and operational standards of science as a self-regulating professional activity. And it is on the basis of these enduring norms and standards—for example, hypothesis testing, replicability, publication of results, and assessment by peer experts—that society could begin building its trust in science.

In 1918, the Haldane report in the United Kingdom ushered in the era of public science that is more recognizable to today's professional science practitioner (Haldane 1918). Haldane's principles were central to establishing the way in which science has evolved free of political interference, but they also had the effect of reinforcing a separation between science and society by establishing a science-centric view of how to manage the public research system. The hallmarks of scientific autonomy and self-governance were also championed by the influential U.S. science policy advisor Vannevar Bush (1945) and academics such as Robert Merton (1942) and Michael Polanyi (1962). By mid-century, their concept of an autonomous culture of science standing apart from the rest of society, while also instructing it, was the dominant perspective, and some may consider this still to be the case (Gluckman 2015).

Yet, the public science system has long been a highly contested space, where the need for objectivity through autonomy wrestles with more utilitarian imperatives, which were perhaps first invoked by Bernal (1939) and which are commonplace today. Indeed, the contemporary turn toward a more responsive mission-driven science system has its roots in the relationship between science and society that was starting to emerge as a result of geo-political needs through hot and cold wars and, later, as a response to the opportunities and challenges of globalization (e.g., pandemics, climate change, and the global marketplace).

But, whereas the structure of science systems may be changing globally to prioritize societal relevance and impact of research, the corresponding behavior by scientists thrust into a more publicly responsive role is less easily scripted. With some exceptions, conventional disciplinary science training in most universities has not fully embraced the potential of a "civics of science" approach to integrated training that could also impart the skills and tools increasingly needed by scientists. Some of these would include, for instance, critical awareness to identify and mitigate personal biases and the potential risks of external influences; responsible public communication of technical information, scientific uncertainty, and risk; and providing science advice.

A new critical awareness and reflexive practice within the sciences is needed as policy makers and scientists alike struggle to balance the multiple objectives and changing expectations of the public science system while protecting the integrity of the knowledge it produces. Thus, the key challenge becomes reconciling the inherent tensions in the evolving role of the public scientist to ensure an appropriate place for science and scientists in societal decision-making and public policy-making. This is an especially urgent challenge as bias and conflicts will become more common as public science budgets globally require more funding partnerships, whether with the private sector or civil society interest groups.

10.4 What Works: Key Structures, Tools, and Approaches to Reconcile Tensions, Balance Demands, and Maintain Trust

The contested landscape of "relevant" science—including science for public policy—is reshaping the science system and is bestowing more public-facing roles and responsibilities onto scientists, often without commensurate training or preparation. Here we consider some of the structures, tools, and approaches that may be deployed to better enable scientists to contribute their expertise meaningfully to the public policy process. These observations are largely based on the principles of science advice as described from the perspective of the individual government science advisor (Gluckman 2014). However, we argue that they can be usefully extrapolated to apply to the multiple contexts and types of science advice for public policy— whether the information needs are acute or chronic, and whether the means are panels, committees, expert witnesses, or individual government science advisors.

Across these multiple contexts and varying delivery models, trust in the integrity of the scientific evidence remains the key, if elusive, requirement. Here we regroup the main considerations within the three high-level domains that can have the most potential impact on the real and perceived trustworthiness of science as it is applied to public policy. They are structural, operational, and cultural considerations.

10.4.1 Structural: Governance Arrangements for Science Advice

- Quality science: The assured quality of scientific input is the singular criterion on which the public and knowledge users can trust evidence. We have already shown not just how scientific information applied to the public policy context is particularly vulnerable to criticism, but also how that criticism may be legitimate given the many ways in which values and lack of objectivity can (intentionally or unintentionally) enter into scientific knowledge production and translation at any step—from defining the research to communicating the results. Robust methods of peer review are the pillar of scientific quality assurance. Where science for public policy is produced in-house or through commissioned (non-academic) processes, attention to the governance of peer review and quality control structures is particularly important.

- Independent advice: As an input to the policy process, even the highest quality scientific information will pass through multiple levels and filters as it makes its way across the desks of decision makers. It is a challenge to ensure the integrity of the core scientific conclusions through this process. The translational mechanism—whether individual or collective—needs to be independent from political or other influence. Where scientists are appointed to provide expertise and advice, carefully structured terms of reference can ensure independence by, for instance, making use of part-time secondments from within the academic community, or timing longer-term appointments to extend beyond electoral cycles.

- Diversity of mechanisms: We have argued elsewhere for the merits of a mixed science information and advisory system that can combine the access to decision makers, flexibility, and quick response time of an individual advisor providing informal advice with the deliberative processes often grounded in scientific collectivities, such as national academies or expert committees attached to specific government departments (OPMCSA 2014). Structuring a system that offers access to *both* rapid response, discrete and informal advice, backed by longer-term formal and deliberative studies, can ensure both practicality and public trust in the face of the iterative messiness of real policy-making.

- Responsiveness to context: The variety of situations for which policy-relevant scientific information and advice is required must be clearly distinguished. For instance, in-depth studies of long-term and chronic issues such as the effects of climate change, ecosystem health, or the assessment of new technologies are often best undertaken by standing or task-oriented committees of academics, whether through national academies or under the aegis of a commission or other body. More precise and circumscribed problems for which policy analysts and managers need direct data inputs, such as monitoring fish stocks, may best be undertaken by dedicated research units or expert groups within or contracting to government departments. By contrast, acute situations of crisis such as pandemic, industrial, natural, and other disasters will require rapid advice and action. In such time-sensitive situations, science advisors often become *de facto* decision makers and focus should be on clear and singular communication of what is known (and not known) about a rapidly evolving situation. Systems such as the UK's Scientific Advisory Group for Emergencies (SAGE) mechanism, within the Cabinet Office Briefing (COBRA) executive emergency response mechanism, include pre-assigned roles and responsibilities of scientific and technical experts in the event of an emergency (Government Office for Science 2014). Key structural features of this mechanism are (1) maintaining an up-to-date national risk register, (2) preidentification and briefing of rosters of technical experts to be deployed as needed, and (3) the national Chief Science Advisor as the central hub and outward face of SAGE to ensure efficiency of operations and clarity of message.

10.4.2 Operational: Science Advice in Practice

- Operating as an honest broker of knowledge: While appropriate structural elements can create the conditions for robust and broadly credible scientific input to reach the policy process, operationalizing any of these structures in practice requires a particular approach from scientists. Roger Pielke Jr. (2007) has famously contrasted—within a larger typology—the scientists acting as "issue-advocates" from those adopting the role of "honest broker." The role of the honest broker is to transmit and interpret knowledge by clarifying and contextualizing what is known and not known about an issue. Honest brokers elucidate the conclusions that can justifiably be inferred from the available data and parse the implications of a range of policy options without advocating for any. While the honest broker concept has been critiqued for its tendency to essentialize both the scientific and policy processes into idealized models, it nonetheless provides a useful heuristic for the required attributes in operationalizing information for decision-making. By contrast, the issue-advocate may intentionally (and often unintentionally) be seen to support a particular course of action by lending to it the weight of (real or assumed) expertise. Particularly complex is the situation of national academies, which may provide technical input into policy but are also in a legitimate position to play an overt advocacy role owing to imperatives of academic freedom. Globally, many representative academic bodies are considering the implications of this tension as they assume a greater role in public policy dialogues. For a recent example, see the Science Council of Japan's renewed code of practice developed in the wake of the Fukushima Daiichi disaster (SCJ 2013) and the OECD (2015) report on the public roles and responsibilities of scientists. Similar sentiments are expressed in the Singapore Statement on Research Integrity (WCRI 2010).

- Inclusiveness and appropriate coproduction: We have suggested that scientific information applied to the public policy context will inevitably be subject to a different type of scrutiny to that of scientific peer review alone. In highly contentious cases, both the structural conditions that protect the integrity of science advice and the role of honest broker may still not satisfy the trust of stakeholders. Bremer and Funtowicz (2015) have suggested that the area of "sustainability science" and the science of environmental stewardship are such examples. They suggest that, to the extent that such issues are most often place-based and values-rich, they may benefit from a deliberative and inclusive approach. The authors suggest the practice of "extended peer review committees" that can include a diversity of stakeholders to operationalize transparency and coproduce the science-based advice given to decision makers. It is clear that such an approach may be seen to compromise the essential processes of science that allow it to remain relatively objective, but here the distinction remains between how science is produced and how it is interpreted and applied. The latter is not a decision of the scientific community alone, and broadening the scope of the review committee may add this dimension to the analysis.

Thus, the key challenge in coproduction will always be to protect the integrity of the science on one hand, while on the other hand enhancing the knowledge base with locally resonant inputs. Sheila Jasanoff's (2004) pioneering theories of coproduction first drew attention to how the structures of science and society are unavoidably mutually influencing, but her analytical frame also serves to promote an awareness of the type and extent of mutual influences. This information in turn can be used to inform and set thresholds of data quality and acceptability of evidence. Applying the descriptive concept of coproduction to promote deliberate (and arguably normative) operational practice can help to build mutual understanding and reach agreement on those thresholds in such a way that both scientific and societal conditions can be met (van Kerkhoff and Lebel 2015; Corburn 2009).

10.4.3 Cultural: How We Think about Science for Public Policy

- Cultures of public reason: Every policy-making jurisdiction will have a particular position regarding what Jasanoff has called "public reason" (Jansanoff 2012). This is the culturally and historically inscribed framework against which we reach societal decisions. Thus public laws, regulations, and the parameters of acceptable evidence to inform decision-making are ultimately cultural and not scientific tools. Yet, in the majority of today's democratic economies, science-derived evidence plays an increasingly significant, even privileged, role in structuring public reason. To be sure, we can rely on internationally accepted standards of science to structure the process to produce fairly reliable information. However, assigning a relative weight to this information against other considerations cannot be prescribed or universalized, nor can methods that deliver it into the decision-making processes. Events in the European Union that saw the disestablishment of the Chief Science Advisor position, amid both vocal support and opposition from across Europe, illustrate the diversity of deeply held convictions about how best to structure methods of public reason. It is useful to be mindful of this diversity and tension in establishing the pathways by which scientific information makes its way into public policy.
- The role and limits of science: Understanding the historical and cultural context of public reason is an apt reminder that science

can only inform policy, it cannot make it. Public policy formation is fundamentally a normative exercise of weighing values, balancing trade-offs, and making choices that are ideally informed by, but not necessarily directed by science. There is some evidence to suggest that the research community is gaining a better understanding of the uncertain place (and indeed the normative limits) of policy-relevant scientific information. Whereas many scientists might still assume it to be self-evident that science should drive policy (policy-based evidence), a cursory bibliometric analysis* of the peer-reviewed literature shows a perceptible rise in the use of the subtler and nuanced term "evidence-informed" and a recent decline of the more confident label "evidence-based." Although this rapid appraisal of the research vernacular does not show how many of these instances apply directly to the policy context, the rise in use of the label "evidence-informed" by researchers who are overwhelmingly from the health and social sciences may be a useful proxy for the general understanding of the realities of how evidence operates in the public (and in this case "social") policy context. Perhaps significantly, there is less evidence of the term's popularity among researchers in the policy-relevant environmental sciences.

10.5 Illustration: The Research and Scientific Information Standard for New Zealand Fisheries

To conclude this chapter, we consider how the key principles, structures, and operational practices of applying scientific information to the public policy context can be given effect. Here, we focus on the quality assurance of scientific information intended to inform public policy decisions. Assuring the quality of scientific information is a central structural pillar, but as we have shown, standards of quality and evidential thresholds can be vulnerable to criticism where issues are contentious. Here, it is important to have robust tools and processes that can withstand scrutiny from the science community and issue advocates alike.

---

* The database Scopus shows that the term *evidence-informed* first appeared in 1997 with a total of 996 instances of its use since that time, almost exclusively in health and human sciences. Its usage is rising. By contrast, the usage of the term *evidence-based* totals 171,938 instances up to 2014, with early sporadic usages (from as early as 1918) mostly in the physical sciences, where it did not relate to the policy context. The term began to make a mark on the academic vernacular in 1995. In 1996, medicine and allied health professions accounted for over 75% of 493 instances of its use that year. Usage of the term seems to have peaked in 2013 (15,735 instances) and it has been in decline since then.

One such tool is the *Research and Scientific Information Standard for New Zealand Fisheries*, which came into force in 2011. This science quality assurance standard is not unique globally, but it does contain exemplary features that provide an apt illustration of the preceding discussion.

10.5.1 New Zealand Fisheries

New Zealand is a small, geographically isolated, democratic country in the southern Pacific Ocean. Its multicultural population of just over 4.5 million lives primarily in the urban centers of Auckland, Wellington, and Christchurch, and many people have significant rural ties. The country's economy is largely driven by land- and water-based industries.

New Zealand is proud of its unique environmental ecosystems, rare flora and fauna, and record of publically supported conservation and environmental stewardship. The country has more than 17,000 protected areas, covering more than 8.6 million hectares (some 32%) of the total land area. There are 44 marine reserves and, since 2005, a growing network of other marine protected areas addressing the uniqueness of its marine ecosystems. New Zealand has the world's fourth largest marine estate. These issues are high in the public and political consciousness as the balance between the use and protection of land and water resources are necessarily central features of public discourse.

New Zealand's commercial fisheries are subject to a quota management system whereby the total allowable catch of any commercial species is apportioned through an individual transferrable quota mechanism. New Zealand was the first country to adopt such a system as policy nationwide in 1986. Shortly after the introduction of the system, quota allocations were suspended while New Zealand's indigenous people (Maori) entered into negotiations with government on the provisions for Maori fisheries in light of the Treaty of Waitangi, which is New Zealand's founding document dating from 1840 and details the relationship between the Crown and Maori. The Treaty and its implications and interpretation are the subject of ongoing reconciliation and redress.

The Treaty claims regarding fisheries were effectively resolved through a commensurate transfer of quota holdings to Maori. The Crown also enacted legislation to protect Maori customary fishing rights outside of the quota management system. New Zealand applies a 200 nautical mile economic exclusion zone, which includes both inshore and deepwater fisheries, for which only citizens and New Zealand-owned companies can own quota (Lock and Leslie 2007). New Zealand's fisheries management has been widely regarded as one of the world's leading systems, not least for the extent to which the quota system is applied and for close collaboration between the regulatory authority and the Department of Conservation (Pitcher et al. 2009; Lock and Leslie 2007; Worm et al. 2009).

10.5.2 Science in Fisheries Management and Ecosystem Sustainability Decisions

New Zealand's quota management system is administered by the Fisheries Management directorate within the amalgamated Ministry for Primary Industries (MPI). Both MPI and its predecessor responsible for fisheries (the Ministry of Fisheries, MFish) have relied on expert stock and ecosystem health assessments to make management and marine stewardship decisions. In a small country such as New Zealand, both the scientific and financial capacity to carry out such research is inherently limited. In response to this challenge, MFish actively sought to promote price competition among expert knowledge providers by outsourcing the necessary scientific work. It also put in place fiscal measures through the quota system, which were aimed at industry bearing a share of the cost of the research necessary to ensure stock sustainability. Commercial fishers thus either help support research or can themselves be direct purchasers of scientific knowledge with a view to influencing a government's fisheries and marine environmental management decisions (Mace et al. 2014; P. Mace, personal communication, May 2015).

With this opening up of the scientific market for both research purchasers and providers, MFish recognized the need for a method to ensure the quality of any scientific information intended for use in informing policy and management decisions. The inherent difficulty in conducting marine research in offshore ecosystems (Craig et al. 2000) and the challenges for monitoring inshore systems at sufficiently local scales (Chuenpagdee 2012) made the need for such a quality assurance standard particularly acute. So, too, would the perennial shortage of expert human resources, not to mention an alertness to the possibility of biased science provided by industry or other interested parties (Bremer and Glavovic 2013).

To meet these challenges, MFish, together with relevant agencies across government, undertook a consultative process to develop a definitive tool to assure the quality of science used to inform policy decisions related to the marine environment. The *Research and Scientific Information Standard for New Zealand Fisheries* was thus launched in April 2011 as a succinct yet comprehensive guide to both purchasers and providers of scientific knowledge (Ministry of Fisheries 2011). This policy statement covers the expected roles and responsibilities of government and other such knowledge purchasers in procuring sound science for decision-making, but it goes into greatest depth on the expectations of knowledge providers and the various forms of peer review that can be applied according to context.

To operationalize the standard, MPI oversees a number of standing theme-based "scientific peer review working groups," which undertake approximately 100 meetings each year. These are supported by "participatory workshops" that provide a more consultative and localized process of review. These workshops are held less frequently and are employed "where issues have broad geographic or disciplinary scope, or where the science is addressing new methods or information, or where the questions attract considerable attention from stakeholders and advocates" (Ministry of Fisheries 2011).

Three key features of this science quality assurance mechanism stand out. Firstly, stakeholder trust is prioritized through the peer review principles of independence, balanced expertise, and inclusiveness. On the surface, these principles may appear to be contradictory, emphasizing the inherent tensions discussed earlier. After all, how can a review be conducted independently if it is also open to interested stakeholders? But such inclusiveness has the potential to help bolster the independence of the review by engaging multiple points of view, demonstrating transparency and balanced expertise.

When an issue is highly contentious, the science that informs it will inevitably be scrutinized and often denounced by aligned interests, but a "broad tent" approach to scientific review that is open to observers and engages diverse expert opinions can better achieve consensus on the policy recommendations to be drawn from the science (Ministry of Fisheries 2011). Contested spatial allocation for marine protected areas, changes to total allowable catch limits, specific arrangements that affect customary indigenous fisheries, or the use of new and controversial technologies in fisheries development are all examples where post-normal practices such as "extended peer communities" could apply without jeopardizing the rigor of standard scientific review (Bess and Rallapudi 2007; Bremer and Funtowicz 2015).

Secondly, the peer review working groups do not limit their work to assessing the end products of the research process, but are also able to provide staged technical guidance to aim for maximum effectiveness of research and efficiency of resources. They are able to stop or redirect research that shows no early relevance or lacks quality, thus ensuring appropriate stewardship of research funds and raising the likelihood of a higher quality end product to inform a management or policy decision.

Finally, in terms of scope of application—regardless of whether the information is generated by government, industry, or an advocacy organization—if it is intended to be used to inform relevant public policy, then the quality assurance standard and procedures for peer review will automatically apply to its development and final knowledge product. In this way, officials can be sure that any scientific input that reaches the desks of decision makers will have been appropriately assessed, with little scope for industry or other interested parties to circumvent the vetting with science that is "more favorable" to their cause.

Despite the level of detail, impressive scope, procedural rigor, and inclusiveness of this quality assurance tool, there are limits to how and where it can be used in the policy process. Clearly, it is designed to lift the level of science generated or taken in by Ministry staff, and for this it appears to be exceeding expectations. However, the contexts to which it applies best are most often the predictable areas of long-term monitoring and stock/ecosystem assessment that can be planned for. Indeed, the review process is time- and labor-intensive to coordinate and manage on an ongoing basis. It is unlikely that such an administration-heavy process could be deployed to respond nimbly to scientific knowledge needs in emergencies for instance. Were an ecological disaster to occur, there would not be the luxury of intensive review in responding to threatened species. In such cases, appealing to a SAGE-type model within a mixed system would allow for scientific information to flow directly and quickly through a single reliable channel and from pre-identified reliable sources.

10.6 Concluding Comments

There is an increasing acknowledgment globally that science is truly a societal endeavor, yet the implications of putting it into practice are far-reaching and still poorly understood. Conceptually, it means working more closely with knowledge end-users to make science more useful and impactful, democratizing the research agenda, and listening carefully to public discourse about technology and social license, among other things. Practically, each of these actions requires a rebalancing of the conventional structures and approaches to knowledge production and application, which will inevitably create tensions.

In this chapter we have tried to extract and render visible these tensions that exist both at the interface of science and policy and within the science sector as it strives to become more immediately relevant to policy makers and society generally. Acknowledging and understanding these tensions allows them to be confronted. To this end, we have tried to offer a number of tools and approaches that may assist, in particular by helping to build and maintain stakeholder trust in the science system and in the preparation and delivery of science advice.

While stakeholder trust may be the product of greater inclusion and of deliberate marshaling of the inherent coproductive capacities of science in the policy context (van Kerkhoff 2014), care must be taken to ensure that this is done without jeopardizing the standards and practices that make science trustworthy in the first place. Similarly, the scientist who is willing to be the public face of a complex issue within an area of their expertise can also be a pillar around which public trust is built, provided they can balance the demands of the hungry media while conveying complexity and uncertainty, and not step beyond the data. This not an easy task; unfortunately, the qualities most often sought by the media, such as the ability to provide catchy punchlines and colorful content on demand or to take sides in a supposed debate, do not lend themselves easily to the maintenance of expert integrity. Scientists stepping into the public eye must manage these risks constantly, particularly with a fast-developing story.

When the integrity of science is threatened by external influence and bias or when scientific input extends beyond the reasonable boundaries of inference that the data allow, the explanatory power—and ultimately the relevance—of science will be lost. There is an important distinction between advocating for the application of science in public policy contexts and science advocating for a particular course of action. To the extent that this is advocacy, it is advocating for the better use of science.

References

Alexander, D. 2014. Communicating earthquake risk to the public: The trial of the L'Aquila seven. *Natural Hazards* 72: 1159–1173.

Bernal, J. D. 1939. The social function of science. *The Social Function of Science*. London: G. Routledge.

Bess, R. and R. Rallapudi. 2007. Spatial conflicts in New Zealand fisheries: The rights of fishers and protection of the marine environment. *Marine Policy* 31 (6): 719–729.

Bremer, S. and S. Funtowicz. 2015. Negotiating a place for sustainability science: Narratives from the Waikaraka Estuary in New Zealand. *Environmental Science & Policy* 53 (Part A): 47–59.

Bremer, S. and B. Glavovic. 2013. Exploring the science–policy interface for integrated coastal management in New Zealand. *Ocean & Coastal Management* 84: 107–118.

Bush, V. 1945. Science: The endless frontier. *Transactions of the Kansas Academy of Science* 48 (3): 231–264.

Callon, M., P. Lascoumes, and Y. Barthe. 2009. *Acting in an Uncertain World: An Essay on Technical Democracy*. Cambridge, MA: MIT Press. (Originally published in French in 2001.)

Chuenpagdee, R. 2012. Global partnership for small-scale fisheries research: Too big to ignore. *Traditional Marine Resource Management and Knowledge Information Bulletin* 29: 22–25.

Corburn, J. 2009. *Toward the Healthy City: People, Places, and the Politics of Urban Planning*. Cambridge, MA: MIT Press.

Craig J., S. Anderson, M. Clout, et al. 2000. Conservation issues in New Zealand. *Annual Review of Ecology and Systematics* 31: 61–78.

de Jong, S. P. L., J. Smit, and L. van Drooge. 2015. Scientists' response to societal impact policies: A policy paradox. *Science and Public Policy*, p. 1–13.

Douglas, H. 2009. *Science, Policy, and the Value-Free Ideal*. Pittsburgh: University of Pittsburgh Press.

Funtowicz, S. O. and J. R. Ravetz. 1993. The emergence of post-normal science. In *Science, Politics and Morality: Scientific Uncertainty and Decision-Making*, edited by R. von Schomberg, 85–123. Dordrecht: Springer.

Gibbons, M. 1999. Science's new social contract with society. *Nature* 403 (suppl.): 81–84.

Gluckman, P. 2014. Policy: The art of science advice to government. *Nature* 507 (7491): 163–165.

Gluckman, P. 2015. Trusting the scientist. Keynote speech presented to the New Zealand Association of Scientists. http://www.pmcsa.org.nz/wp-content/uploads/NZAS-Speech_Trusting-the-scientist.pdf.

Government Office for Science. 2014. Annual report of the Government Chief Scientific Advisor 2014. Innovation: managing risk, not avoiding it. London: The Government Office for Science. https://www.gov.uk/government/publications/innovation-managing-risk-not-avoiding-it.

Habermas, J. 1970. Technology and science as ideology. In *Toward a Rational Society*. Boston: Beacon.

Haldane, R. B. 1918. Report of the Machinery of Government Committee under the chairmanship of Viscount Haldane of Cloan. London: HMSO.

Harland, K. and H. O'Connor. 2015. Broadening the scope of impact: Defining, assessing and measuring impact of major public research programmes, with lessons from 6 small advanced economies. Small Advanced Economies Initiative, Ireland. http://www.smalladvancedeconomies.org/wp-content/uploads/SAEI_Impact-Framework_Feb_2015_Issue2.pdf.

HEFCE (Higher Education Funding Council for England). n.d. Research impact case studies database. http://impact.ref.ac.uk/CaseStudies/.

Jasanoff, S., ed. 2004. *States of Knowledge: The Co-production of Science and the Social Order*. London: Routledge.

Jasanoff, S. 2012. *Science and Public Reason*. New York: Routledge.

Kuhn, T. S. 1962. *The Structure of Scientific Revolutions*. Chicago: University of Chicago Press.

Lock, K. and S. Leslie. 2007. *New Zealand's Quota Management System: A History of the First 20 Years*. Auckland: Ministry of Fisheries, Motu Economic and Public Policy Research.

Löfstedt, R. E. 2005. *Risk Management in Post-Trust Societies*. London: Earthscan Press.

Mace, P. M., K. J. Sullivan, and M. Cryer. 2014. The evolution of New Zealand's fisheries science and management systems under ITQs. *ICES Journal of Marine Science* 71 (2): 204–215.

Marks, J. 2009. *Why I Am Not a Scientist. Anthropology and Modern Knowledge*. Berkeley: University of California Press.

Merton, R. 1942. Science and technology in a democratic order. *Journal of Legal and Political Sociology* 1: 115–126.

Michaels, D. 2008. *Doubt Is Their Product: How Industry's Assault on Science Threatens Your Health*. Oxford: Oxford University Press.

Ministry of Fisheries. 2011. Research and science information standard for New Zealand fisheries. http://www.fish.govt.nz/NR/rdonlyres/D1158D67-505F-4B9D-9A87-13E5DE0A3ABC/0/ResearchandScienceInformationStandard2011.pdf.

OECD (Organisation for Economic Co-operation). 2015. Scientific advice for policy making: The role and responsibilities of expert bodies and individual scientists. OECD Science, Technology and Industry Policy Papers, No. 21. Paris: OECD Publishing.

OPMCSA (Office of the Prime Minister's Chief Science Advisor). 2014. Synthesis report: Science advice to governments conference, 28–29 August 2014, Auckland, New Zealand. Auckland: Office of the Prime Minister's Science Advisory Committee. http://www.pmcsa.org.nz/wp-content/uploads/Synthesis-Report_Science-Advice-to-Governments_August-2014.pdf.

Oreskes, N. and E. M. Conway. 2010. *Merchants of Doubt. How a Handful of Scientists Obscured the Truth on Issues from Tobacco Smoke to Global Warming*. New York: Bloomsbury Press.

Pielke, R. A. 2007. *The Honest Broker: Making Sense of Science in Policy and Politics*. Cambridge: Cambridge University Press.

Pitcher, T. J., D. Kalikoski, K. Short, et al. 2009. An evaluation of progress in implementing ecosystem-based management of fisheries in 33 countries. *Marine Policy* 33 (2): 223–232.

Polanyi, M. 1962. The republic of science: Its political and economic theory. *Minerva* 1 (1): 54–73.

SCJ (Science Council of Japan). 2013. Code of conduct for scientists—Revised version. http://www.scj.go.jp/en/report/Code%20of%20Conduct%20for%20Scientists-Revised%20version.pdf.

van Kerkhoff, L. 2014. Developing integrative research for sustainability science through a complexity principles-based approach. *Sustainability Science* 9 (2): 143–155.

van Kerkhoff, L. E. and L. Lebel. 2015. Coproductive capacities: Rethinking science-governance relations in a diverse world. *Ecology and Society* 20 (1): 14.

WCRI (World Conference on Research Integrity). 2010. Singapore statement on research integrity. http://www.singaporestatement.org/.

Worm, B., R. Hilborn, J. K. Baum, et al. 2009. Rebuilding global fisheries. *Science* 325 (5940): 578–585.

Yeo, M. 2014. Fault lines at the interface of science and policy: Interpretative responses to the trial of scientists in L'Aquila. *Earth-Science Reviews* 139: 406–419.

11

Measuring Awareness, Use, and Influence of Information: Where Theory Meets Practice

Suzuette S. Soomai, Peter G. Wells, Bertrum H. MacDonald, Elizabeth M. De Santo, and Anatoliy Gruzd

CONTENTS

11.1 Introduction

It is a truism that human activities cannot be managed in an ecosystem that is poorly understood. Information and knowledge about aquatic and marine ecosystems and the role of information in the various guiding frameworks for integrated coastal and ocean management (ICOM) are well recognized and important components of successful ocean management (GESAMP 1996; Doody et al. 1998; Ehler 2003; DFO 2002; Pickaver et al. 2004; Levin et al. 2009; Nova Scotia Government 2009; Wilson 2009; ICES 2013; among others). This tenet of ICOM is especially important for understanding major global issues such as climate change (Tribbia and Moser 2008) and the loss of marine biodiversity, and recognizing the linkage between such issues and other key challenges facing our societies (Watson 2005). To work with the extensive literature about coastal and ocean issues can be daunting due to the complexities related to the diverse types of knowledge systems that have been emphasized as important in decision-making (e.g., Cash et al. 2003; Cicin-Sain and Knecht 1998). Recently, a coastal knowledge system has been described, with a novel analytical framework covering areas of focus, forms of knowledge, interaction between forms of knowledge, and influencing barriers or filters (Coffey and O'Toole 2012; see also Chapter 3 in this volume). Clearly, marine information and its management and use play a vital role in every phase of ICOM (MacDonald et al. 2011, 2012; Soomai et al. 2011a, 2011b, 2013).

Throughout this book, we strive to show the linkages between the various guiding frameworks underlying ICOM, for example, the information cycle, the policy cycle, and the ICOM process cycle, in order to illustrate the pivotal role of information in the ocean management process, namely, information awareness, access, use, and influence. Information's role can be examined using suitable indicators for "measuring the progress and outcomes of ICOM" (UNESCO 2006), reflecting the various pressures on the coasts and open oceans (Tribbia and Moser 2008).

In this chapter, we describe the important concepts and classical methods for measuring awareness, use, and influence (sometimes referred to as impact or effectiveness) of information, largely from the perspective of the information management specialist. However, it is recognized that the role of information in ICOM can also be examined from the point of view of the ICOM practitioner, identifying what the operational managers and senior policy advisors require as timely information and determining how they can acquire it. This perspective is similar to the two-directional flow of information at the science–policy interface(s), the so-called *science push, policy pull* interaction so essential for effective management. The case studies (Chapters 12–18 in this volume) illustrate these perspectives and approaches, and show how the information production, that is, science, information management, and ICOM worlds merge in an essential symbiosis.

Knowledge of currently available methods in information management, ranging from advanced searching, citation analysis, and webometrics, to the use of social media and network analysis, is fundamental to understanding how information supports ICOM. This knowledge should be in the tool kit of ICOM researchers and practitioners, as some methods can be used directly to account for and understand information behavior and information use. Knowing the needs of ICOM specialists should also encourage information management professionals to interface with ICOM's many disciplines and practitioners to optimize the use of available information, wherever it is needed.

Given recent advances in information science, computer science, and technology, the methods themselves must continue to be evaluated, improved, or newly developed. The incentives for optimal use of information are numerous. From the point of view of the senior decision maker, manager, or policy maker, accountability for use of public funds is critical. Justifying the essential need for but high cost of research, state of marine environment reporting, mitigation measures, and so on, is reason enough to demand evidence that important existing information is used effectively. Studies using the methods described below also contribute to a better understanding of processes at the science–policy interface and of the continuous need for evidence-based and evidence-informed policy-making. Applying these methods and knowing how information is used will support more effective ICOM.

11.2 Theoretical and Practical Perspectives

11.2.1 Use of Scientific Information in the Policy-Making Process

Since the 1970s, several models about information use in policy-making have been described (e.g., Caplan 1979; Glasziou and Haynes 2005; Knott and Wildavsky 1980; Landry et al. 2001a, 2001b; Oh 1996; Weiss 1979). The models have varied from fixed typologies to more fluid descriptions. Walt (1994) and Weiss (1977) outlined an *ideal model* of research as a linear series of events that lead to the dissemination of various published outputs with the end stage occurring when those outputs, that is, scientific advice, reach the policy makers and decision makers. Weiss (1977) also described an *enlightenment* model in which the links between research and policy are often indirect or less obvious, as a single piece of research is unlikely to influence policy change directly. Rather, over time the cumulative weight of research information permeates gradually into the policy process via a number of information channels, for example, through the involvement of expert groups, such as various national academies; interest groups, such as environmental nongovernmental organizations (see Chapters 15 and 18 in this volume); the news media and increasingly social media; and a gradual change in the viewpoints of policy makers occurs. Weiss (1979) later described seven typologies of information use in policy-making where use can be viewed as *knowledge-driven*, that is, the information serves an educational purpose for an intended audience, or use can be seen as *problem solving*, namely, the information provides advice to policy makers and is intended to guide the selection of management solutions. Three broad categorizations of use were identified: direct or instrumental use, indirect or conceptual use, and selective use. In direct or instrumental use, scientific evidence is incorporated into decision-making to reach a specific solution. In indirect or conceptual use, evidence influences or informs how policy makers and practitioners think about issues, problems, or potential solutions. Indirect use of evidence occurs when information is used to establish new goals and benchmarks and it helps to enhance understanding of the complexity of problems and the consequences of action or inaction (Caplan 1979). Selective use is typically strategic, where, for example, information is used to legitimize and sustain predetermined positions, and falls within the *tactical* typology described by Weiss (1979). Selective use can also be viewed as a subset of direct or instrumental use. The degree to which evidence is used directly, indirectly, or selectively may vary in relation to several factors, for example, the management level at which the decision makers operate in the policy process within an organization, that is, upper, middle, or lower levels; how evidence is framed, that is, vaguely versus complexly, or focused versus simple; and the issue itself, which is related to the urgency of decision-making or action (Hallsworth et al. 2011).

Within a typology of use, research use by policy makers can be fluid and iterative or it can proceed in stages after the information is produced (e.g., Glasziou and Haynes 2005; Knott and Wildavsky 1980; Landry et al. 2001b). Knott and Wildavsky (1980) described a seven-stage *chain of utilization* involving reception of the research, cognition, reference, effort, adoption, implementation, and impact. In an assessment of the use of research produced by social scientists in Canada, Landry et al. (2001a) defined a modified *ladder* of research use with six steps similar to Knott and Wildavsky's *chain*: transmission, cognition, reference, effort, influence, and application. While approximately 50% of researchers stated that they transmitted key research findings to relevant policy makers, as much as 30% of findings did not even reach this first stage (Landry et al. 2001a). Glasziou and Haynes (2005) developed a *pipeline model* of seven stages of research use from a healthcare practitioner's perspective: awareness of research findings, acceptance of the findings, the knowledge is perceived to be applicable, the practitioner has the capability of applying the research findings, the practitioner acts on the research, the findings are adopted in practice, and the research findings are adhered to over time. This pipeline model suggests, however, that some elements of the research findings can be ignored at each stage.

The above models generally show research use as a linear activity and imply that it must proceed in a logical order from one stage to the next. The models also do not specify how the stages in information use are structured within the different typologies of information, for example, political and tactical (Glasziou and Haynes 2005; Knott and Wildavsky 1980; Landry et al. 2001b). After reviewing numerous models, Nutley et al. (2007, p. 45) concluded that research information use is "a dynamic process that may not be readily 'boxed' into static typologies." To reconcile the limitations of the typologies of information use and the stage models, they defined information use as a continuum of options for research use in policy-making. In this model, conceptual and instrumental uses of information are placed at opposite ends of a continuum that ranges from raising awareness about research, to enhanced knowledge and understanding of the research causing a shift in policy-making attitudes, to direct use of research information and change in policy (Figure 11.1).

Concurrent with the description of models about the use of research information articulated over the past four decades, attention has also been put on

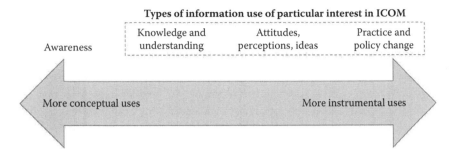

FIGURE 11.1
Continuum of research use. (Nutley, S. et al. 2008. *Using Evidence: How Research Can Inform Public Services*. Bristol: The Policy Press.)

knowledge mobilization or knowledge transfer and exchange (KTE), and knowledge translation, especially in healthcare settings (Contandriopoulois 2012; Contandriopoulois et al. 2010; Van Eerd et al. 2011; Chappells et al. 2015; Azimi et al. 2015). These terms account for "efforts to modify practices or decisions to make them coherent with current scientific evidence" (Contandriopoulois 2012, p. 30) and are closely associated with the concept of research information use depicted in the models about information use in policy-making and decision-making processes discussed above. Despite considerable investment in developing KTE applications, reviews on this subject conclude that evaluation of KTE practices, that is, measurement of the outcome of practices in terms of the use of research information and its influence, continues to be a challenge (Van Eerd et al. 2011; Straus et al. 2011; Hudon et al. 2015). In the view of one KTE scholar, "much of the available practice-oriented advice on knowledge transfer promotes either one particular technique as a solution to the challenges of knowledge exchange and utilization or else very linear, knowledge driven processes" (Contandriopoulois 2012, p. 468). Often, the evidence from actual cases does not match the scenarios in the linear models.

11.2.2 Where Theory and Science Meet Practice

A common theme running through all of the models of the use of research and knowledge transfer is the importance of context. Context matters because of the interplay of many and specific stakeholders, ranging from information creators to decision makers; the availability of a wide variety of communication channels with attending enablers and barriers to information flow; and the variability of institutional and organizational cultures and politics in which decisions are made (see Chapter 2 in this volume). This is especially true for ICOM, which is a multifaceted process, encompassing many players and the consideration of different environmental factors that vary on spatial and temporal scales. Many of the chapters in this book highlight these points, as well as emphasize the importance of understanding the contexts of information use. The many variables contribute to the difficulty in measuring use of information, and in determining the influence of research. Nonetheless, ultimately, the role of information in ICOM has to be shown by example, as the literature on the subject and many of the chapters in this book illustrate.

Several examples of attempts to enhance marine information use, in the context of what is needed for effective ICOM, are available (e.g., Doody et al. 1998; New Zealand Ministry of Fisheries 2011). One of the best international examples is the assessment of assessments (AoA) initiative of the United Nations Environment Programme (UNEP) and the Intergovernmental Oceanographic Commission (IOC) of UNESCO, which is "the most comprehensive initiative undertaken to date by the UN system to better coordinate ocean governance" (UNEP and IOC-UNESCO 2009, p. 12). Assessments, or state of the environment (SOE) reports, aim to assemble "monitoring and research knowledge in a form useful for decision-making" and are prepared on the principle that "regular assessment is an integral part of adaptive management that can respond to changing conditions" (UNEP and IOC-UNESCO 2009, p. 16). In the AoA, the assessments were considered as both process and product, as context matters as noted above. In addition, every report was examined for relevance, legitimacy, and credibility "as these attributes have been identified as central to an assessment's influence and used in identifying best practices" (UNEP and IOC-UNESCO 2009, p. 24). A clear but challenging objective was to make the link "between assessment and policy and management processes." The central recommendation of the AoA "calls for a mechanism that builds on existing global, regional and national institutions and processes while integrating all available information, including socioeconomic data, on how our seas and oceans are actually being used" (UNEP and IOC-UNESCO 2009, p. 12).

The preparation of many SOE reports, or assessments, now follow the DPSIR framework (Driving Forces, Pressures, State Changes, Impacts, Responses) which link the state of the environment (the science component) with societal responses (the use of the data and information, that is, the decision-making and management responses) (see discussion of one such SOE report in Chapter 12 in this volume). Elements of this framework highlight information communication and use that can assist in bridging the science–policy gap (Bell 2012; Gregory et al. 2013; Ness et al. 2010; Tscherning et al. 2012), and in turn offer a means of measuring information use and ultimately its influence, including indicators of use and influence which help to define these often ambiguous concepts.

Other examples of initiatives to enhance marine information use in ICOM include the work of the Atlantic Coastal Zone Information Steering Committee (ACZISC) (ACZISC 2009; see also Chapter 15 in this volume) and Mitchell et al. (2006), whose book was the first rigorous analysis of global environmental assessments and their influence. A further example is the Environmental Information: Use and Influence (EIUI—www.eiui.ca) research program at Dalhousie University, initiated in 2002 because it was recognized that an evaluation of the numerous SOE reports was needed to determine their *value*, which was understood to be both the process of producing a particular report and the use made of the information once assembled (Wells 2003; Cordes 2004). The oceans were in continuing decline, despite the large number of research papers and synthesis reports describing the threats and the options for solutions. Frequently, a disconnect existed between the presence of information indicating a problem and the policy and management efforts to publicly recognize and solve it. Often, the time period for action was very long, for example, climate change and invasive species impacts. In the EIUI program, several projects have examined the use of marine reports produced by governments or intergovernmental groups, for example, a study of the *State of Nova Scotia's Coast Report* (Soomai et al. 2011a, 2013). One seminal conclusion was that the specific influence of information in policy-making and decision-making with marine environmental issues was usually difficult to track and measure. Many factors are at play at the science–policy interface, and at the point of making decisions in evidence-based policies, research-based information is only one element, and sometimes a minor one, in a complex process. Despite good intentions, such as keeping people well-informed and material readily accessible (e.g., the ACZISC program, see Chapter 15 in this volume), it is overly simplistic to assume that the availability of relevant, salient, and legitimate information will automatically be noted and lead to good decisions and effective ICOM.

Numerous other examples of organizations, local to global, that emphasize the role of science and information in ICOM can be mentioned. Among the more notable are the following:

1. The Intergovernmental Panel on Climate Change (IPCC), which has produced five massive synthesis reports as well as summaries for policy makers (IPCC 2013–2014; see also Watson 2005).

2. The World Meteorological Organization (WMO) and UNEP, which have recently published a scientific assessment of ozone depletion and assessment for decision makers (WMO and UNEP 2014), and 11 earlier reports.

3. The International Council for the Exploration of the Sea (ICES), a highly networked scientific organization involved in research on fisheries and marine ecosystems. ICES's strategic plan states that it "will produce the information and advice that decision makers need" and "continue to deliver evidence-based scientific advice on environmental issues and fisheries management" (ICES 2013; see also Wilson 2009).

4. The Millennium Ecosystem Assessment (MEA), which focused on "the impacts of ecosystem changes for human well-being, and synthesizing ... research and making it available in a form relevant to current policy questions, and improved decision-making

concerning ecosystem management and human well-being" (MEA 2005; Watson 2005).

5. The World Health Organization's Health and Environment Initiative, which has produced Global Environmental Outlook Assessments. The objectives of these assessments are "to facilitate the production of accessible but scientifically relevant information to policy makers, and to increase the capacity of governments to use environmental information for decision-making and action planning for sustainable development" (WHO 2015).

6. The Joint Group of Experts on the Scientific Aspects of Marine Environmental Protection (GESAMP), which has produced numerous assessment reports since the 1970s and an evaluation of science's role in integrated coastal management or ICM (GESAMP 1996).

7. The Ecosystem Indicators Partnership of the Gulf of Maine Council on the Marine Environment, which produces comprehensive fact sheets, supported by a mapping tool, on key environmental issues in the Gulf of Maine and Bay of Fundy (www.gulfofmaine.org).

Collectively, the activities of these and other organizations show a growing effort to package and bring the primary messages of ocean and environmental science to wider audiences, including the public, key policy makers and decision makers, and politicians in coastal countries (Levin et al. 2009).

11.3 Measurement Approaches

As shown in this volume, the production, dissemination, and use of information for decision-making regarding coastal and ocean issues is an amalgam of many variables and a myriad of contexts. Developing an understanding of the use and influence of information is, as a result, not a straightforward exercise or application of a single method of measurement. Researchers and ICOM practitioners have approached the subject from numerous points of view and with various measurement methods (see Table 11.1).

11.3.1 Challenges in Selecting Methods to Measure Information Use

Scientific information may or may not be used for its intended purpose, as use is affected by many factors (discussed throughout this book). For example, use may be difficult to measure due to challenges in obtaining evidence; the information may have been produced for one purpose, which could direct such measurement, resulting in the unexpected uses being overlooked. Use of information contained in a publication can be indirect, for

TABLE 11.1

Methods to Measure Awareness and Use of Information

Methods	What Will Be Measured (Information Pathways and Use in Policy Contexts)
Bibliometrics, for example, citation analysis	Citations to publications, for example, who cites, the location and role of citing authors, and direct use of information
Webometrics and altmetrics, for example, web statistics based on link searches and web content	Awareness of publications by diverse audiences; evidence of networking within and outside the organizations, for example, information read, cited, and/or linked; direct use of information
Semi-structured interviews; online surveys	Information activities and collaboration of stakeholders in the *chain of utilization* of research, for example, read, understood, and decisions based on information
Content analysis of print and digital sources, news media and social media, for example, blogs, Twitter records	Awareness and use of information by diverse audiences; interrelationships of online network members, for example, sources that influence ICOM decisions and legislation
Direct observation of meetings; content analysis; discourse analysis	Observation of actors at various stages of research utilization to determine how information sources are used in decisions, for example, research changed an actor's frame of reference (indirect use); or why relevant sources are not used
Network analysis; social network analysis, for example, study of Twitter records	Inventory of networked individuals and institutions at various stages of research use; visualization of network structures including online social networks, showing indirect and direct influence of information, for example, who (person or organization) was involved and which information was used in decisions

example, information may increase a stakeholder's knowledge of an issue, or use can be direct—with a more tangible output—for example, policies are developed. Historically, measuring information use was focused on the direct policy outputs, that is, evidence that information had a direct relationship with the development or implementation of policies, regulations, and management plans (see Chapter 16 in this volume as an example). Definitions of *use* and *influence* of scientific information have often been ambiguous, and measurements of both may be similar. For instance, evidence in policies, peer-reviewed papers, and reports produced by an organization (gray literature) can be considered as indicators of both use and influence. Evidence of attitude changes or paradigm shifts in policy-making can also be interpreted as indicators of use and influence; however, such changes are usually not tangibly obvious and are often more difficult to measure. Indirect or more conceptual influence generally occurs in the early stages after the release of scientific reports when stakeholders' awareness of the reports indicates indirect use of the information (see, e.g., Soomai, MacDonald and Wells 2011a,b). Evidence of awareness of information and publications may be more common than evidence of other forms of use and influence.

Measurement of the use of information can take a limited perspective and focus on selected components of information life cycles in ICOM or attempt to develop a holistic understanding of the flow of information within an organization, within communities of practice (CoP) (see Chapter 15 of this volume), or in decision-making contexts. For example, Environment Canada measured its research and development performance (including indicators of information use) using several methods (Environment Canada 2009). Since holistic measurement is typically complex, some studies account for information use through proxy techniques. For example, the production of systematic reviews of the current state of knowledge in relation to policy issues has been used to measure information use (Holmes and Clark 2008). Such reviews, if expert in perspective and constructively critical, test whether or not a policy has worked, hence showing whether the information behind the policy, assuming that it is evidence-based, has had any *influence*. Examples of this approach also include reviews by the National Academy of Sciences in Washington, DC (e.g., NRC 2015). Proxy measures, such as a systematic review of a topic, rely on an indirect assessment of use and do not reveal the full picture. Researchers can turn, however, to a suite of other measures to uncover more obvious evidence of how research-based information is used.

11.3.2 Qualitative Research Design and Methods

A range of qualitative methods is available to assess how information is communicated in policy- and decision-making processes. These include interviews, based on semi-structured questionnaires; direct observations, of meetings, for example; content analysis; and discourse analysis (Anastas and MacDonald 1994; Jackson 2002). Such methods have been commonly used to collect data to answer questions of awareness and whether or not scientific publications are being used in policy-making, to highlight the pathways of (and barriers to) the flow of scientific information within organizations and among stakeholders, and to reveal the ways in which information is being used in policy-making and decision-making for environmental and resource management (Cano Chacón 2013; Cossarini et al. 2014; Ross 2015; Soomai, MacDonald and Wells 2011a,b, 2013; Soomai, Wells and MacDonald 2011c). Surveys, direct observations, and content analysis of documents are often used in case studies designed to develop breadth and depth of understanding of information use. The data analysis typically involves categorization and interpretation based on common themes emerging from the data.

Surveys or interviews, for example, have yielded data on the contexts in which scientific reports are prepared, disseminated, and used, as well as the personal and organizational networks involved in these activities (e.g., Clark and Holmes 2010; Ouimet et al. 2010; Soomai et al., 2011c). Using largely open-ended questionnaires and semi-structured interviews of the key actors in the policy-making process within organizations, that is, scientists, managers, and policy makers, has provided evidence of the direct use of scientific reports at the policy- and decision-making level, as well as enablers and barriers to communication (Soomai, MacDonald and Wells 2011a). Interviews can reveal stakeholder groups' reasons for acting or not acting on scientific advice. Interviews can also be used to determine their ideas about how to increase communication at the science–policy interface. Interviews can clarify how an individual's thinking about policy problems changed and which information types or sources of information were most catalytic in changing perceptions. Questionnaires can be fielded in person, by mail, or now frequently through online software, such as Opinio (ObjectPlanet 2015). Questionnaires can also be used to reach a wide range of external actors in policy-making, including NGOs and industry, to obtain a fuller picture of awareness, use, and influence of scientific reports (e.g., Soomai et al. 2011b).

Content analysis is often applied to the textual data collected through such qualitative methods, that is, surveys and interviews (Krippendorff 2013). This involves developing labels or codes based on identified themes in the text for the analysis (Coffey and Atkinson 1996). Content analysis can also be used to analyze the text of publications. In one example, Sandström and Rova (2010) analyzed legal documents and written policies on fish stocks to determine whether institutional uncertainties had complicated the establishment of adaptive policy-making systems in Sweden. The approach identified the sources of information that influenced the production of fisheries policy, for example, UN Convention on the Law of the Sea (UNCLOS) documents, Food and Agriculture Organization (FAO) voluntary codes, and ICES reports. The analysis also revealed how information use is influenced by the way issues are framed, a theme that is illustrated in Chapter 4 in this volume.

In other examples of the application of content analysis, Tosun (2011) examined the manifestos of German political parties for federal elections published in a 30-year period (1980–2009) to determine the attention political parties gave, if any, to the issue of marine pollution on a national and global level and how this attention changed when a pollution incident occurred. It was assumed that underlying the election manifestos was the parties' intent to use words deliberately to send *ideological signals* to the electorate by mentioning some words more frequently than others. The texts of the manifestos were coded for the theme of sea pollution, by noting the occurrence of words such as *ocean*, *sea*, and *waters*, and the analysis showed changes in the parties' stances on marine pollution over time and the overall salience of environmental issues for the different party systems. Likewise, Soomai (2015) applied content analysis to data collected through interviews and direct observations of science and policy-making groups in the Department of Fisheries and Oceans Canada, the Northwest Atlantic Fisheries Organization, and FAO to ascertain drivers, enablers, and barriers in the information pathways—production, communication, and use—in these fisheries management organizations.

Content analysis draws on the intellectual capacity of researchers to ascertain meaning largely unaided by the original creators. The analysis is typically rigorous, but even so a researcher may not capture the exact intent of meaning of the creators. A variation on such analysis allows the authors to self-index the textual data they create. David Snowden and his collaborators have developed SenseMaker, a set of data collection and analytical tools (Cognitive Edge 2015) that captures narrative fragments of individuals, for example, conversations, to which individuals add layers of meaning to their experiences. Assembling this data, sometimes in large quantities, may provide much richer and more detailed evidence of how information is used in decision-making contexts. For example, Lynam and Fletcher (2015) applied this method in a study of climate change adaptation.

Discourse analysis is commonly used in social science research to analyze the arguments about an issue to develop an understanding of a decision-making process (Gee 2011; Bering Keiding 2010; Hajer and Versteeg 2005). Through discourse analysis, links and interactions between stakeholders can be mapped to indicate who is involved, the level of discussion or debate, attention given to particular issues, and the level of support surrounding the decision-making activity. For instance, evidence of the ways that policy makers regard an issue can be observed through their consistent reference to specific scientific recommendations and reports in regulations, agency documents, advisory reports, among others. For example, Delaney et al. (2007) examined scientific advice and management decisions from Denmark, the United Kingdom, France, the Netherlands, and Norway from 2001 to 2004 to understand the public debate on the North Sea cod fishery. In this study, the sources included national newspapers, publications of the fishing industry press, newsletters, web discussions, minutes from meetings, and interviews with persons involved in the public debate. Discourse analysis was combined with qualitative interviews to determine the influence of the public debate on the advisory and decision-making system for managing this particular fishery. In another example of discourse analysis, Hajer (1995) examined documents, including the science advisory and economic committee reports of ICES and the Commission of the European Community, to develop a narrative or *story line* to describe the claims of stakeholder groups about the state of North Sea cod (Hajer 1995; Wilson 2001). For instance, a fisher tended to select facts that supported personal economic efforts while an environmentalist chose facts that emphasized the seriousness of the environmental problem. Over time, the participants, momentum, and themes in the discourse changed. Then after the decision-making concluded, scientists claimed that their advice was not followed and fishers stated that they were ignored. This discourse analysis showed the inherent biases of stakeholders, a not uncommon problem where self-interest may trump objectivity and compromise.

In direct observations, individual participants and the interaction among them are observed by the researcher to assemble data that are often compelling and hard to refute (Anastas and MacDonald 1994; Leedy and Ormrod 2012). The researcher does not participate in the events while they are being observed. For instance, in a study of the use and influence of a Nova Scotia coastal report, observations at public consultations showed that mostly members of the *interested public* participated in discussions with government staff members (Soomai, MacDonald and Wells 2011a). In another example, ethnographic studies occur mainly through direct observations and interviews and are ideal for describing organizational cultures (Spradley 1979, see also Chapter 14 in this volume). Carballo-Cárdenas et al. (2013) examined the cultural norms, beliefs, and organizational structures of a fisheries organization to gain an understanding of how the actors in marine protected area (MPA) management, that is, managers, academics, government officials, and environmentalists, constructed meaning differently. Different interpretation of information often led to ambiguous communication and the inability of participants in MPA governance and management to reach mutual understanding of the supporting data.

11.3.3 Quantitative Research Design and Methods

Quantitative methods for tracing the movement of information from production to use have been dominated over the past half-century by extensive application of bibliometric analysis. This method of analysis has extended into webometrics to account for the recent massive movement to digital communication technologies, principally web-based. In addition, quantitative methods also include growing interest in network analysis and social media analysis, as developments over the past decade have demonstrated that information pathways have moved heavily into digital media. Information can reach stakeholders, policy makers, and decision makers through a variety of channels, all of which merit study to advance understanding of information behavior and information use at the science–policy interface.

Bibliometric analyses have been used extensively in research performance evaluation to document and explain citation patterns of research published in peer-reviewed journals. Citations refer to data included in footnotes and bibliographies of publications. Since the 1960s, this field of study has resulted in sophisticated application of citation analyses at the level of individual researchers, research groups or university departments, government departments, academic disciplines, and countries (for an overview see, e.g., Adam 2002; Cronin 1984; De Bellis 2009, 2014; Eom 2009; Meho 2007; van Raan 2005).

The EIUI research team at Dalhousie University has used citation analysis in studies of the publications of international intergovernmental organizations. Data have been drawn from the long-established citation databases of Web of Science, and more recent sources, Scopus, Google, and Google Scholar, as well as directly from monographs, which until recently have not been well represented in any citation source (Avdić 2013; Cordes 2004; Hutton 2009, 2010; MacDonald et al. 2004). Published by Thomson Reuters, Web of Science currently contains over 90 million records extracted from several sources, for example, papers published annually in about 12,000 peer-reviewed journals and 160,000 conference proceedings (Web of Science 2015). Launched by Eugene Garfield in 1960, Web of Science (known previously by other names) dominated the bibliometric field until Elsevier introduced Scopus in 2004, the rise of Google began at the end of the 1990s, and rapid growth occurred in web-based publications (Meho and Sugimoto 2009). Similar to Web of Science, Scopus covers research fields in "science, mathematics, engineering, technology, health and medicine, social sciences, and arts and humanities" through a very large abstract and citation database (Elsevier 2015). While Web of Science and Scopus are very large sources of citation data, their coverage focuses mostly on scholarly literature and generally overlooks information published in other formats, such as gray literature, that is, reports and other publications not released by commercial publishers. For the latter, citation searches in Google Scholar and Google provide more complete coverage (Hutton 2010; Avdić 2013). Citations located through searches in the open Web, conducted in Google and Google Scholar, can extend understanding of the use and influence of an organization's reports related to subjects relevant to ICOM, since such citation data are not restricted to scholarly literature as is the case in Web of Science and Scopus. For example, in an extensive search for citations to biennial editions of FAO's flagship publication, *State of World Fisheries and Aquaculture*, Avdić (2013) found many citations in searches conducted in Google Scholar that were not found in Web of Science or Scopus.

Citation data provide details that can characterize information use such as the geographical range and the time scale as determined by the author(s) of a citing document, the geographic location and institutional affiliation of the citing author(s), the types of citing source, for example, other research journals, and details of the timing of a citation after the release of the original publication, for example, month and year of publication. The level of citation analysis can also be extended to include a measure of *influence* by reading and coding the text surrounding citations in order to identify the reason(s) for which a publication was cited (Zhang et al. 2013). Studies on the use of citation analysis acknowledge that while citations are informative indicators of use of information, authors often do not cite all of the publications that influenced their research (MacRoberts and MacRoberts1989). In research literature, especially in the sciences, well-established facts that have been verified many times, are stated without source attribution, for example, genes are composed of DNA except in some viruses. As well, *text–bibliography discrepancies*, that is, incorrectly referenced facts or sources, and excessive self-citing, are factors that need to be recognized in using citations as an indicator of awareness, use, and influence of information.

With rise of the web and social media since 2000, alternatives to traditional citation analysis have been developed. These include webometrics (Thelwall et al. 2005; Thelwall 2009; Thelwall et al. 2010) and altmetrics (Priem 2013; Cronin and Sugimoto 2014; Mohammadi et al. 2015). Webometric and web-based bibliometric studies rely on open access web data or web usage statistics, including web tracking (e.g., the number of visits to a site, page views, and downloads of publications), web link analysis, and content analysis of web pages of organizations (Bar-Ilan 2001; Thelwall 2009, 2011). Analysis of traffic at a website is now quite easy to undertake with analytical software such as Google Analytics, but this type of analysis is dependent on the collection of such statistics by an organization. Webometrics can provide evidence of awareness and use of publications produced by organizations, for example, link searches can be conducted for individual web pages, whole web domains, or subdomains by typing the URL of the web unit containing the designated information into the web browser (Soomai, MacDonald and Wells 2011a). As with citation analysis, web links provide evidence of use by identifying characteristics of the links to an organization's website. This feature of web searching can identify intermediary organizations, or *boundary organizations*, and highlight their effect on facilitating awareness, use, and potential influence of publications. Web link searching provides the basis for developing network diagrams through an analysis of these links (Björneborn and Ingwersen 2004; Thelwall et al. 2005; Thelwall 2011, 2012).

Altmetrics, a concept first named in 2010, aims to provide "a more expansive view" of the impact of research and other publications (Lin and Fenner 2013). The popularity of recent digital communication channels, such as Facebook and Twitter, means that understanding the pathways of information from production through awareness, use, and influence must consider these channels. Altmetrics embraces the "Web's power to disseminate and filter scholarship more broadly and meaningfully" than previously feasible (Priem 2013, p. 440). Altmetrics attempts to identify and analyze indicators of impact in terms of views (web page views), discussions (via blogs, Twitter, etc.), saves (to databases such as Mendeley and Zotero), citations (as are found in Web of Science and Scopus), and recommendations (reported in various social media sites). While still evolving, altmetric methods are being used to measure the impact of research (Bornmann 2015), and to investigate who is reading research articles (Mohammadi et al. 2015). In the latter study, for example, the researchers examined data about people who registered in Mendeley as readers of publications about clinical medicine, engineering and technology, social sciences, physics, and chemistry.

Social network analysis is increasingly being used to understand the complexities of information pathways within communities of practice and systems. The findings have been used to improve cooperation among researchers, policy makers, and public groups (Berry et al. 2002; Durland and Fredericks 2005; Midgley and Richardson 2007; Crona and Parker 2011). Social network analysis focuses on patterns of relations and information flows among individuals, organizations, and states (see Chapter 8 in this volume). Analysis of data about the links or connections among these entities can describe who is interacting with whom and how they are interacting in the policy process within an organization or external to it. The data for network analysis are collected by both qualitative and quantitative methods. For instance, identifying the institutions and persons for network analysis can be undertaken by webometrics; surveys, for example, interviews of key informants; content analysis; direct observation at research and policy-related meetings; and tracking social media usage (Weiss et al. 2011; White et al. 2015). The output of a network analysis is often a map or network configuration showing each actor's position in a management network, for example, with regard to their roles related to knowledge facilitation and policy influence (Bodin and Crona 2009; Hanneman and Riddle 2005; Sandström and Rova 2010). Network maps are useful for organizations that may wish to evaluate or modify their communication and policy development practices to increase the legitimacy of information to diverse or dispersed stakeholder groups (Carlsson and Sandström 2008; Hanneman and Riddle 2005; Otte and Rousseau 2002; Sandström and Rova 2010; Weiss et al. 2011). The network measures provide a quantitative indication of how much information an actor receives, how much information he/she produces, and how much influence he/she has over policy decisions, respectively. Generally, network analysis may not provide a direct measure of the influence of information since it cannot fully address whether the technical information contained in reports is understood, for example, or acted upon (Hartley 2010; Hartley and Glass 2010). Communication network analysis is a subfield of social network analysis and focuses on the characteristics of specific communication pathways and the patterns of connections that communication produces, for instance, e-mail, face-to-face contact, telephone, and ad hoc meetings, among the actors (Monge and Contractor 2003). Recently, social network analysis has extended to include social media usage patterns (e.g., White et al. 2015; Gruzd and Roy 2016). Network analysis may prove to be invaluable to the ICOM practitioner of the future as it could test or demonstrate how well the various players and programs are actually integrated and coordinated, and how information flows (or not) within networks.

11.3.4 Mixed Methods

Each of the previously described methods contributes to understanding of awareness, use, and influence of scientific information in policy-making. However, no method alone can fully answer questions about information use and influence of publications produced for a wide range of target audiences. As a consequence, it becomes necessary to employ multiple methods in a determination of information use. A mixed methods research design synthesizes concepts from both quantitative and qualitative approaches to research. In a mixed method approach a method can be chosen specifically to fill in the gaps or shortcomings of another. Mixed methods research provides the means for triangulation or data verification to increase the reliability, that is, reduce the uncertainty, of the results.

In mixed methods research, utilizing qualitative methods alone, direct observations, surveys, discourse analysis, ethnographic techniques, and content analysis can be used to determine whether scientific information is considered, what pathways were used to communicate the information, and what methods (e.g., e-mail or face-to-face) or formats (e.g., technical report or briefing memo) were used to communicate the information. These qualitative methods can also determine who the key actors are, what their roles are, what opportunities and constraints to using scientific information exist, and what information networks are involved. Often there is interest in addressing all of these questions and no single method will be adequate. Surveys, for example, can be useful for describing information pathways from the release of information to its use in policy-making. Discourse analysis may be particularly informative in determining how scientific research is embedded in policy arguments. Ethnographic techniques can be used to study the cultural norms, beliefs, and organizational structures of organizations related to information use. Beem (2005, 2007) illustrated the use of such a suite of qualitative methods to examine the interplay of institutions, information, and interests in the development of fisheries policy for management of the blue crab fisheries of Chesapeake Bay, United States. The objective of the research was to identify opportunities for learning about the policy process by determining contacts within a fisheries management system that could facilitate this learning. Beem (2005, 2007) used systematic comparison through content analysis, interview data, and personal observations at public hearings and various relevant workgroups and task forces to determine the role of institutions, information, and interests in affecting policy learning.

The mixed method approach is particularly effective in providing the different data sets needed to develop an overall picture of a network in which different actors or stakeholders operate, for example, government, research institutions, and industry. For instance, while bibliometric analysis produces indicators of awareness and use in the form of citations related to an organization's publications, webometrics identifies links that indicate collaboration and partnerships between organizations during the production and dissemination phases of the organization's reports. Interviews complement quantitative research methods, for example, webometrics, by verifying and describing collaboration between organizations. While webometrics identifies the occurrence of collaboration, interviews describe the nature of the collaboration in the production of information and in policy-making (Thelwall 2009; Thelwall et al. 2010; Vaughan and Shaw 2003). Interviews also compensate for the lack of robustness of webometrics as a network tool (Otte and Rousseau 2002; Thelwall et al. 2010). Furthermore, network analysis utilizes data from surveys, observations, content analysis, and discourse analysis to highlight the actors who provide or receive the most information, those who have the most or least policy influence, the most prominent *brokers* who create links between other actors, actors with overlapping roles and functions, and, overall, how well connected actors are, based on knowledge exchange and policy influence.

Assessing the impact of research on user communities, for example, policy makers, often requires a case-based approach with a diversity of research methods. A multiple case study approach is preferred over a single case study as comparisons and generalizations on particular aspects of information use can be made. Still, single case studies may be particularly informative due to the depth of analysis that mixed methods offers. Common case-based approaches used in the study of information pathways involve forward-tracking or backward-looking studies (Nutley et al. 2007). In forward-tracking studies, qualitative methods can highlight the nonlinear interactions within policy-making, while quantitative methods can highlight the linear pathways between research products and policy output (Nutley et al. 2007). Approaches that highlight the linearity of movement from research output to assessment of impacts tend to simplify the complexity of the processes at work in the uptake of research into policy-making. Assessing the impact of a particular research source on policy choices may be problematic as the research that feeds into policy-making is synthesized with other types of knowledge and expert opinions and may not be easily isolated (Nutley et al. 2007).

11.3.5 Issues in Data Collection

Since the information pathways and decision-making processes in ICOM are extensive and varied (see Chapter 9 in this volume), undertaking studies of information awareness, use, and influence presents many questions about where to begin and what to measure. As noted above, mixed research methods can be used to address some of these questions more effectively than reliance on a single method. Deciding what to measure is particularly important. One may be interested in tracking the movement of information products of various formats, examining the channels and methods of communication, determining the actors involved, considering the decision-making processes at local to international levels, or examining attitudinal, cultural, and political factors influencing the selection of information and decisions. As well, there are other elements of information life cycles and ICOM activities to consider. In other words, a sizeable number of variables warrant investigation and they may co-vary, complicating which methods to use. Deciding on what *use* and *influence* mean and which indicators should or could be employed is generally not a simple matter.

Another decision is when to measure, since use and influence can change over time after a particular scientific publication is released. The researchers who contributed to a report, or the policy makers who received it, may have relocated or been reassigned to other portfolios. Identifying the key policy makers and decision makers who can recall how research was used may be problematic for the same reasons. Establishing and achieving a sample size appropriate for a method of measurement may not be straightforward. Samples sizes for qualitative studies are often smaller than numbers required for quantitative methods, but not necessarily easier to obtain, particularly if interviews are required. A lengthy period of trust-building may be needed to gain access to decision makers. Some actors may simply not be accessible until after they have left their positions, as was the case in the research conducted by Lalor and Hick, who interviewed "former Environment Ministers (senior politicians) and Department Secretaries/Deputy Ministers (senior public servants) to better understand the role of science-based knowledge in the Executive decision-making processes of Westminster-based governments" in Australia and Canada (Lalor and Hick 2013, p. 767).

Sampling techniques may depend on the assistance of participants. For example, network-based sampling, such as *snowball sampling*, in which respondents refer the researcher to other respondents, is quite common in qualitative research. In such cases, the researcher has to begin a study with the assumption that a snowball technique will work. Research completed by Soomai (2015) demonstrated the effectiveness of internships, which allow a researcher to be embedded as an observer within case study organizations. A physical presence within the organization facilitates gaining the trust of the staff and thereby increases their willingness to be interviewed and provide access to meetings for direct observations. Gaining the trust of the staff in the respective offices can support the opening up of opportunities for additional data collection within the organizations.

The ethical implications of research have to be considered when people are the focus of investigation. In interviews and observations of scientists and policy makers, for example, opinions may reflect professional views on awareness and use of scientific information and not include any personal opinions, which may be quite important. Potential respondents must be given the opportunity to accept an invitation to participate in the research through informed consent, while being ensured that their responses will be used only within the context of the research. Survey methods, such as web-based surveys, can be designed to guarantee the anonymity of individuals or groups who participate, but such methods limit the opportunity to probe responses that may be vague or difficult to interpret. While ethical issues are less likely to arise when publicly accessible data are used, other issues may present difficulties with regard to cost or time required to analyze the data. In the case of citation analysis based on Google searches, for example, *cleaning* the data can be exceptionally time-consuming since every hit, which may number in the thousands, needs to be examined to verify and extract relevant data. Meho and Yang described this process as "grueling" (Meho and Yang 2007, p. 2105).

11.4 Conclusion

Understanding of the role(s) of information in ICOM and the influences of new information technologies on ICOM practitioners and practice is not yet well advanced. This is possibly because the ICOM processes play out in many different contexts, involving many factors, and information behavior activities are quite complex (see Chapters 7 and 9 in this volume). Thus, research investigating this field may be limited to date by the sheer dimensions of the phenomena.

It is clear that no single method of measuring information use and influence will result in comprehensive understanding of information-related activities at the science–policy interface. Each method discussed above can contribute to that understanding and, in fact, can be employed in particular studies since it may be the best approach for addressing a particular question and types of data selected as indicators of information use and influence. In general, though, a mixed methods approach is called for, which may be achieved by pursuing numerous case studies employing different measurement methods as appropriate and then conducting meta-analysis of the results of all of the individual cases. Mixed methods are necessary because ICOM encompasses many different actors. For example, there are scientists, program managers, and policy makers, as well as traditional groups and other stakeholders in policy-making, for example, NGOs and industry groups; different levels and jurisdictions of decision-making, for example, operational and policy levels, in local, national, and international contexts; different types of decision-making, for example, operational or strategic; and different formats of information developed for purposes and introduced at different stages in decision-making.

Since rapidly changing technologies are affecting information behavior and communication patterns, sometimes dramatically, the methods described in this chapter may require modification to attend to new developments. Nonetheless, this suite of methods could be incorporated into a tool kit that could be employed in case studies and larger research initiatives. The research strength of the suite draws on a combination of qualitative and quantitative methods. Advancing understanding of the use and influence of information in ICOM will help to resolve problems encountered at science–policy interfaces, and ultimately contribute to informed decisions that address the significant environmental challenges in the world's coasts and oceans.

References

ACZISC (Atlantic Coastal Zone Information Steering Committee). 2009. Discussion paper. The role of the ACZISC in ICOM policy development and implementation in Atlantic Canada. http://coinatlantic.ca/images/stories/documents/role%20of%20aczisc-icom.pdf.

Adam, D. 2002. The counting house. *Nature* 415: 726–729.

Anastas, J. W. and M. L. MacDonald. 1994. *Research Design for Social Work and the Human Services.* New York: Lexington Books.

Avdić, V. 2013. Measuring use and influence: An assessment of the FAO's flagship report *The State of World Fisheries and Aquaculture.* Master's project report, Dalhousie University.

Azimi, A., R. Fattachi, and M. Asadi-Lari. 2015. Knowledge translation status and barriers. *Journal of the Medical Library Association* 103 (2): 96–99.

Bar-Ilan, J. 2001. Data collection methods on the web for informetric purposes—A review and analysis. *Scientometrics* 50 (1): 7–32.

Beem, B. E. 2005. Fisheries policy learning: The interplay of institutions, interests, and information. PhD diss., University of Washington.

Beem, B. E. 2007. Co-management from the top? The roles of policy entrepreneurs and distributive conflict in developing co-management arrangements. *Marine Policy* 31 (4): 540–549.

Bell, S. 2012. DPSIR = A problem structuring method? An explanation from the "imagine" approach. *European Journal of Operational Research* 222: 350–360.

Bering Keiding, T. 2010. Observing participating observation: A re-description based on systems theory. *Forum: Qualitative Social Research* 11 (3): Art. 11.

Berry, B. J. L., L. D. Kiel, and E. Elliott. 2002. Adaptive agents, intelligence, and emergent human organization: Capturing complexity through agent-based modeling. *Proceedings of the National Academy of Sciences* 99 (7): 187–188.

Björneborn, L. and P. Ingwersen. 2004. Toward a basic framework for webometrics. *Journal of the American Society for Information Science and Technology* 55 (14): 1216–1227.

Bodin, Ö. and B. I. Crona. 2009. The role of social networks in natural resource governance: What relational patterns make a difference? *Global Environmental Change* 19: 366–374.

Bornmann, L. 2015. Usefulness of altmetrics for measuring the broader impact of research: A case study using data from PLOS and F100Prime. *Aslib: Journal of Information Management* 67 (3): 305–319.

Cano Chacón, M. 2013. The role of the information of the Marine Stewardship Council certification process in developing countries: A case study of two MSC certified fisheries in Mexico. Master's project report, Dalhousie University.

Caplan, N. 1979. The two-communities theory and knowledge utilization. *American Behavioral Scientist* 22 (3): 459–470.

Carballo-Cárdenas, E. C., A. P. J. Mol, and H. Tobi. 2013. Information systems for marine protected areas: How do users interpret desirable data attributes? *Environmental Modelling and Software* 41: 185–198.

Carlsson, L. and A. Sandström. 2008. Network governance of the commons. *International Journal of the Commons* 2 (1): 33–54.

Cash, D. W., W. C. Clark, F. Alcock, et al. 2003. Science and technology for sustainable development special feature: Knowledge systems for sustainable development. *Proceedings of the National Academy of Science* 100 (14): 8086–8091.

Chappells, H., N. Campbell, J. Drage, et al. 2015. Understanding the translation of scientific knowledge about arsenic risk exposure among private well water users in Nova Scotia. *Science of the Total Environment* 505: 1259–1273.

Cicin-Sain, B. and R. W. Knecht. 1998. *Integrated Coastal and Ocean Management: Concepts and Practices.* Washington, DC: Island Press.

Clark, R. and J. Holmes. 2010. Improving input from research to environmental policy: Challenges of structure and culture. *Science and Public Policy* 37 (10): 751–764.

Coffey, A. and P. Atkinson. 1996. *Making Sense of Qualitative Data: Complementary Research Strategies*. Thousand Oaks, CA: Sage.

Coffey, B. and K. O'Toole. 2012. Towards an improved understanding of knowledge dynamics in integrated coastal zone management: A knowledge systems framework. *Conservation and Society* 10 (4): 318–329.

Cognitive Edge. 2015. About SenseMaker. http://cognitive-edge.com/sensemaker/#sensemaker-about.

Contandriopoulois, D. 2012. Some thoughts on the field of KTE/Réflexions sur l'échange et le transfert de connaissances. *Healthcare Policy* 7 (2): 29–37.

Contandriopoulois, D., M. Lemire, J.-L. Denis, et al. 2010. Knowledge exchange processes in organizations and policy arenas: A narrative systematic review of the literature. *The Millbank Quarterly* 88 (4): 444–483.

Cordes, R. E. 2004. Is grey literature ever used? Using citation analysis to measure the impact of GESAMP, an international marine scientific advisory body. *Canadian Journal of Information and Library Science* 28: 45–65.

Cossarini, D. M., B. H. MacDonald, and P. G. Wells. 2014. Communicating marine environmental information to decision makers: Enablers and barriers to use of publications (grey literature) of the Gulf of Maine Council on the Marine Environment. *Ocean & Coastal Management* 96: 163–172.

Crona, B. I. and J. N. Parker. 2011. Network determinants of knowledge utilization: Preliminary lessons from a boundary organization. *Science Communication* 33 (4): 448–471.

Cronin, B. 1984. *The Citation Process: The Role and Significance of Citations in Scientific Communication*. London: Taylor Graham.

Cronin, B. and C. R. Sugimoto, eds. 2014. *Beyond Bibliometrics: Harnessing Multidimensional Indicators of Scholarly Impact*. Cambridge, MA: MIT Press.

De Bellis, N. 2009. *Bibliometrics and Citation Analysis. From the Science Citation Index to Cybermetrics*. Lanham, MD: The Scarecrow Press.

De Bellis, N. 2014. History and evolution of (biblio)metrics. In B. Cronin and C. R. Sugimoto (eds) *Beyond Bibliometrics: Harnessing Multidimensional Indicators of Scholarly Impact*, 23–44. Cambridge, MA: MIT Press.

Delaney, A. E., H. A. McLay, and W. L. T. van Densen. 2007. Influences of discourse on decision-making in EU fisheries management: The case of North Sea cod (*Gadus morhua*). *ICES Journal of Marine Science* 64 (4): 804–810.

DFO (Department of Fisheries and Oceans). 2002. Canada's oceans strategy: Our oceans, our future. Policy and operational framework for integrated management of estuarine, coastal and marine environments in Canada. Ottawa, ON: Fisheries and Oceans Canada. http://www.dfo-mpo.gc.ca/oceans/publications/cosframework-cadresoc/pdf/im-gi-eng.pdf.

Doody, J. P., C. F. Pamplin, C. Gilbert, et al. 1998. Information required for integrated coastal zone management. Executive summary. http://www.ec.europa.eu/environment/iczm/pdf/themf_ex.pdf.

Durland, M. and K. A. Fredericks. 2005. An introduction to social network analysis. *New Directions for Evaluation* 107: 5–13.

Ehler, C. N. 2003. Indicators to measure governance performance in integrated coastal management. *Ocean & Coastal Management* 46 (3/4): 335–345.

Elsevier. 2015. About Scopus. http://www.elsevier.com/solutions/scopus.

Environment Canada. 2009. *Measuring Environment Canada's Research & Development Performance*. Gatineau, QC: Environment Canada.

Eom, S. 2009. *Author Cocitation Analysis: Quantitative Methods for Mapping the Intellectual Structure of an Academic Discipline*. Hershey, NY: Information Science Reference.

Gee, J. P. 2011. *An Introduction to Discourse Analysis Theory and Method*. 3rd ed. New York: Routledge.

GESAMP (Joint Group of Experts on the Scientific Aspects of Marine Environmental Protection). 1996. *The Contributions of Science to Integrated Coastal Management*. GESAMP Reports and Studies No. 61. Rome, Italy: Food and Agriculture Organization.

Glasziou, P. and B. Haynes. 2005. The paths from research to improved health outcomes. *Evidence Based Nursing* 8: 36–38.

Gregory, A. J., J. P. Atkins, D. Burton, et al. 2013. A problem structuring method for ecosystem-based management: The DPSIR modelling process. *European Journal of Operational Research* 227: 558–569.

Gruzd, A., & J. Roy. 2016. Social media and local government in Canada: An examination of presence and purpose. In M. Z. Sobaci (ed.) *Social Media and Local Governments*, 79–94. Heidelberg: Springer.

Hajer, M. A. 1995. *The Politics of Environmental Discourse: Ecological Modernization and the Policy Process*. Oxford: Clarenden Press.

Hajer, M. and W. Versteeg. 2005. A decade of discourse analysis of environmental politics: Achievements, challenges, perspectives. *Journal of Environmental Policy and Planning* 7 (3): 175–184.

Hallsworth, M., S. Parker, and J. Rutter. 2011. *Policy-Making in the Real World: Evidence and Analysis*. London: Institute for Government.

Hanneman, R. A. and M. Riddle. 2005. *Introduction to Social Network Methods*. Riverside, CA: University of California Press. http://faculty.ucr.edu/hanneman.

Hartley, T. W. 2010. Fishery management as a governance network: Examples from the Gulf of Maine and the potential for communication network analysis research in fisheries. *Marine Policy* 34 (5): 1060–1067.

Hartley, T. W. and C. Glass. 2010. Science-to-management pathways in US Atlantic herring management: Using governance network structure and function to track information flow and potential influence. *ICES Journal of Marine Science* 67 (6): 1154–1163.

Holmes, J. and R. Clark. 2008. Enhancing the use of science in environmental policy-making and regulation. *Environmental Science and Policy* 11 (8): 702–711.

Hudon, A., M.-J. Gervais, and M. Hunt. 2015. The contributions of conceptual frameworks to knowledge translation intervention in physical therapy. *Physical Therapy* 95 (4): 630–639.

Hutton, G. R. G. 2009. Developing an inclusive measure of influence for marine environmental grey literature. Master's thesis, Dalhousie University.

Hutton, G. R. G. 2010. Understanding influence of scientific information in the digital age: A study of the grey literature of a United Nations advisory group. *Proceedings of the Nova Scotian Institute of Science* 45 (2): 91–101.

ICES (International Council for the Exploration of the Sea). 2013. *ICES Strategic Plan 2014–2018*. Copenhagen: ICES.

IPCC (Intergovernmental Panel on Climate Change). 2013–2014. Fifth Assessment Report. http://www.ipcc.ch.

Jackson, W. 2002. *Methods: Doing Social Research*. Toronto: Prentice-Hall.

Knott, J. and A. Wildavsky. 1980. If dissemination is the solution, what's the problem? *Knowledge: Creation, Diffusion, Utilization* 1 (4): 537–578.

Krippendorff, K. 2013. *Content Analysis. An Introduction to Its Methodology*. 3rd ed. Thousand Oaks, CA: Sage.

Lalor, B. M. and G. M. Hick. 2013. Environmental science and public policy in executive government: Insights from Australia and Canada. *Science and Public Policy* 40: 767–778.

Landry, R., N. Amara, and M. Lamari. 2001a. Climbing the ladder of research utilization: Evidence from social science research. *Science Communication* 22: 396–422.

Landry, R., N. Amara, and M. Lamari. 2001b. Utilization of social science research knowledge in Canada. *Research Policy* 30: 333–349.

Leedy, P. D. and J. E. Ormrod. 2012. *Practical Research: Planning and Design*. 10th ed. New Jersey: Pearson Educational.

Levin, P. S., M. J. Fogarty, S. A. Murawski, et al. 2009. Integrated ecosystem assessments: Developing the scientific basis for ecosystem-based management of the ocean. *PLOS Biology* 7 (1): 23–28.

Lin, J. and M. Fenner. 2013. Altmetrics in evolution: Defining and redefining the ontology of article-level metrics. *Information Standards Quarterly* 25 (2): 20–26.

Lynam, T. and C. Fletcher. 2015. Sensemaking: A complexity perspective. *Ecology and Society* 20 (1): 65.

MacDonald, B. H., R. E. Cordes, and P. G. Wells. 2004. Grey literature in the life of GESAMP, an international marine scientific advisory body. *Publishing Research Quarterly* 20: 25–41.

MacDonald, B. H., S. S. Soomai, E. M. De Santo, et al. 2012. Advancing effective integrated coastal and ocean management—Recognizing the critical role of marine information use and influence. Paper presented at the 10th International Conference, Coastal Zone Canada, June 2012, Rimouski, Quebec, Canada.

MacDonald, B. H., P. G. Wells, S. S. Soomai, et al. 2011. Awareness, use and influence of coastal and marine environmental information: Case studies of governmental and intergovernmental organizations. Poster presented at 9th Bay of Fundy Science Workshop, 27–30 September 2011, Saint John, New Brunswick, Canada.

MacRoberts, M. H. and B. R. MacRoberts. 1989. Problems of citation analysis. *Scientometrics* 36: 435–444.

MEA (Millennium Ecosystem Assessment). 2005. Overview of the millennium ecosystem assessment. http://www.millenniumassessment.org/en/About.html.

Meho, L. I. 2007. The rise and rise of citation analysis. *Physics World* 20 (1): 32–36.

Meho, L. I. and C. R. Sugimoto. 2009. Assessing the scholarly impact of information studies: A tale of two citation databases—Scopus and Web of Science. *Journal of the American Society for Information Science and Technology* 60 (12): 2499–2508.

Meho, L. I. and K. Yang. 2007. Impact of data sources on citation counts and rankings of LIS faculty: Web of Science versus Scopus and Google Scholar. *Journal of the American Society for Information Science and Technology* 58 (13): 2105–2125.

Midgley, G. and K. A. Richardson. 2007. Systems thinking for community involvement in policy analysis. *Emergence: Complexity and Organization* 9 (1–2): 167–183.

Mitchell, R. B., W. C. Clark, D. W. Cash, et al. 2006. *Global Environmental Assessments: Information and Influence*. Cambridge, MA: MIT Press.

Mohammadi, E., M. Thelwall, S. Haustein, et al. 2015. Who reads research articles? An altmetrics analysis of Mendeley user categories. *Journal of the Association for Information Science and Technology* 66 (9): 1832–1846.

Monge, P. R. and N. S. Contractor. 2003. *Theories of Communication Networks*. New York: Oxford University Press.

Ness, B., S. Anderberg, and L. Olsson. 2010. Structuring problems in sustainability science: The multi-level DPSIR framework. *Geoforum* 41: 479–488.

New Zealand Ministry of Fisheries. 2011. *Research and Science Information Standard for New Zealand Fisheries.* Wellington: New Zealand Ministry of Fisheries.

Nova Scotia Government. 2009. Our Coast. The 2009 State of Nova Scotia's Coast Summary Report. http://www.gov.ns.ca/coast.

NRC (National Research Council). 2015. *Climate Intervention: Carbon Dioxide Removal and Reliable Sequestration.* Washington, DC: The National Academies Press.

Nutley, S., I. Walter, and H. T. O. Davies. 2007. *Using Evidence: How Research Can Inform Public Services.* Bristol: The Policy Press.

ObjectPlanet. 2015. Opinio (Version 6.6.3) [Software]. Oslo: ObjectPlanet. http://www.objectplanet.com/opinio/.

Oh, C. H. 1996. Information seeking in governmental bureaucracies: An integrated model. *American Review of Public Administration* 26 (1): 41–70.

Otte, E. and R. Rousseau. 2002. Social network analysis: A powerful strategy, also for the information sciences. *Journal of Information Science* 28 (6): 441–453.

Ouimet, M., P. Bédard, J. Turgeon, et al. 2010. Correlates of consulting research evidence among policy analysts in government ministries: A cross-sectional survey. *Evidence and Policy* 6 (4): 433–460.

Pickaver, A. H., C. Gilbert, and F. Breton. 2004. An indicator set to measure the progress in the implementation of integrated coastal zone management in Europe. *Ocean & Coastal Management* 47 (9): 449–462.

Priem, J. 2013. Scholarship beyond the paper. *Nature* 495: 437–440.

Ross, J. D. 2015. What do users want from a state of the environment report? A case study from *The State of the Scotian Shelf Report.* Master's thesis, Dalhousie University.

Sandström, A. and C. Rova. 2010. The network structure of adaptive governance: A single case study of a fish management area. *International Journal of the Commons* 4: 528–551.

Soomai, S. S. 2015. Elucidating the role of fisheries scientific information in policy-making for fisheries management. PhD diss., Dalhousie University.

Soomai, S. S., B. H. MacDonald, and P. G. Wells. 2011a. The 2009 State of Nova Scotia's Coast Report: An Initial Study of Its Use and Influence. Halifax: Dalhousie University.

Soomai, S. S., B. H. MacDonald, and P. G. Wells. 2011b. The State of the Gulf of Maine Report: An Initial Study of Awareness, Use and Influence. Halifax: Dalhousie University.

Soomai, S. S., P. G. Wells, and B. H. MacDonald. 2011c. Multi-stakeholder perspectives on the use and influence of "grey" scientific information in fisheries management. *Marine Policy* 35 (1): 50–62.

Soomai, S. S., B. H. MacDonald, and P. G. Wells. 2013. Communicating environmental information to the stakeholders in coastal and marine policy-making: Case studies from Nova Scotia and the Gulf of Maine/Bay of Fundy region. *Marine Policy* 40: 176–186.

Spradley, J. 1979. *The Ethnographic Interview.* New York: Holt, Rinehart, and Winston.

Straus, S. E., J. M. Tetroe, and I. D. Graham. 2011. Knowledge translation is the use of knowledge in health care decision making. *Journal of Clinical Epidemiology* 64: 6–10.

Thelwall, M. 2009. *Introduction to Webometrics: Quantitative Web Research for the Social Sciences.* San Rafael, CA: Morgan and Claypool.

Thelwall, M. 2011. A comparison of link and URL citation counting. *Aslib Proceedings: New Information Perspectives* 63 (4): 419–425.

Thelwall, M. 2012. Webometric analyst (Version 2.0) [Software]. University of Wolverhampton: Statistical Cybermetrics Research Group. http://lexiurl.wlv.ac.uk/index.html.

Thelwall, M., A. Klitkou, A. Verbeek, et al. 2010. Policy-relevant webometrics for individual scientific fields. *Journal of the American Society for Information Science and Technology* 61 (7): 1464–1475.

Thelwall, M., L. Vaughan, and L. Björneborn. 2005. Webometrics. *Annual Review of Information Science and Technology* 39: 81–135.

Tosun, J. 2011. Political parties and marine pollution policy: Exploring the case of Germany. *Marine Policy* 35 (4): 536–541.

Tribbia, J. and S. C. Moser. 2008. More than information: What coastal managers need to plan for climate change. *Environmental Science and Policy* 11: 315–328.

Tscherning, K., K. Helming, B. Knippe, et al. 2012. Does research applying the DPSIR framework support decision making? *Land Use Policy* 29: 102–110.

UNEP (United Nations Environment Programme) and IOC-UNESCO (Intergovernmental Oceanographic Commission of UNESCO). 2009. An assessment of assessments. Summary for decision makers. Findings of the Group of Experts. Pursuant to UNGA resolution 60/30. Paris, France and Nairobi, Kenya: UNEP and ICO-UNESCO.

UNESCO (United Nations Educational, Scientific and Cultural Organization). 2006. *A Handbook for Measuring the Progress and Outcomes of ICOM*. IOC Manuals and Guides, 46, ICAM Dossier 2. Paris, France: UNESCO.

Van Eerd, D., D. Cole, K. Keown, et al. 2011. Report on knowledge transfer and exchange practices: A systematic review of the quality and types of instruments used to assess KTE implementation and impact. Toronto, Canada: Institute for Work and Health.

van Raan, A. 2005. Measuring science. In H. Moed et al. (eds) *Handbook of Quantitative Science and Technology Research*, 19–50. Dordrecht, the Netherlands: Springer.

Vaughan, L. and D. Shaw. 2003. Bibliographic and web citations: What is the difference? *Journal of the American Society for Information Science and Technology* 54 (14): 1313–1322.

Walt, G. 1994. How far does research influence policy? *European Journal of Public Health* 4: 233–235.

Watson, R. T. 2005. Turning science into policy: Challenges and experiences from the science–policy interface. *Philosophical Transactions of the Royal Society B: Biological Sciences* 360 (1454): 471–477.

Web of Science. 2015. Web of Science. The complete collection. http://wokinfo.com/citationconnection/realfacts/#regional.

Weiss, C. H. 1977. Research for policy's sake: The enlightenment function of social science. *Policy Analysis* 3 (4): 531–545.

Weiss, C. H. 1979. Many meanings of research utilization. *Public Administration Review* 39 (5): 426–431.

Weiss, K., M. Hamann, M., Kinney, et al. 2011. Knowledge exchange and policy influence in a marine resource governance network. *Global Environmental Change* 22 (1): 178–188.

Wells, P. G. 2003. State of the marine environment reports—A need to evaluate their role in marine environmental protection and conservation. *Marine Pollution Bulletin* 46 (10): 1219–1223.

White, B., H. Castleden, and A. Gruzd. 2015. Talking to Twitter users: Motivations behind Twitter use on the Alberta oil sands and the northern gateway pipeline. *First Monday* 20 (10). doi:10.5210/fm.v20i1.5404.

WHO (World Health Organization). 2015. Global Environment Outlook (GEO) assessments. http://www.who.int/heli/tools/geoassess/en/.

Wilson, D. C. 2001. Three complementary approaches to understanding the use of scientific claims in environmental debates. Paper presented to the Nordic Workshop on Conflicts between Protected Species and Fisheries, Finnish Environmental Institute, Helsinki, Finland, 12–13 August 2001. Institute for Fisheries Management Research Publication Series No. 57.

Wilson, D. C. 2009. *The Paradoxes of Transparency: Science and the Ecosystem Approach to Fisheries Management in Europe.* Amsterdam, The Netherlands: Amsterdam University Press.

WMO (World Meteorological Organization) and United Nations Environment Programme (UNEP). 2014. *Assessment for Decision-Makers. Scientific Assessment of Ozone Depletion: 2014.* WMO Global Ozone Research and Monitoring Project, Report No. 56. Geneva, Switzerland: WMO. https://2a9e94bc607930c3d739bec c3293b562f744406b.googledrive.com/host/0BwdvoC9AeWjUazhkNTdXRXUz OEU/gormp_56_en.pdf.

Zhang, G., Y. Ding, and S. Milojević. 2013. Citation Content Analysis (CCA): A framework for systematic and semantic analysis of citation content. *Journal of the American Society for Information Science and Technology* 64 (7): 1490–1503.

Section III

Case Studies

12

What Do Users Want from a State of the Environment Report? A Case Study of Awareness and Use of Canada's State of the Scotian Shelf Report

James D. Ross and Heather Breeze

CONTENTS

12.1 Introduction

The 1972 United Nations (UN) Conference on the Human Environment (United Nations 1972) recognized that effective environmental management needed to be conducted on a global scale and supported by the best available scientific information. Since the conference's declaration, various governments, nongovernmental organizations, and intergovernmental partnerships have sought to address the latter need by producing State of the Environment (SOE) reports, beginning with Ward and Dubos's *Only One Earth*, the unofficial report commissioned for that conference (Ward and Dubos 1972). Though SOE reports vary widely in scope, they generally share the purpose of providing a comprehensive aggregation of available scientific information for a particular ecosystem, written clearly to facilitate understanding and focusing on information relevant to the management and policy issues affecting that ecosystem.

Over 40 years later, in the face of worsening global environmental conditions driven by advancing anthropogenic climate change, the need for effective environmental management informed by high-quality scientific information is greater than ever. But are decision makers aware of the available scientific information that could support solutions to environmental problems? Are SOE reports actually being used for their intended purpose? If not, what steps can be taken to raise awareness and encourage use of such reports in the relevant communities of practice? Answering these questions has become a growing concern (Wells 2003; Mitchell et al. 2006) and, with environmental conditions projected to worsen in the near- and medium-future, time is of the essence for addressing this concern.

12.2 The Eastern Scotian Shelf Integrated Management Initiative and the State of the Scotian Shelf Report

This case study sought to investigate the above questions by focusing on one particular SOE: the State of the Scotian Shelf Report, a product of the Canadian Department of Fisheries and Oceans' (DFO) Eastern Scotian Shelf Integrated Management (ESSIM) Initiative.

Canada's Oceans Act (1996, s. 30) called for a management strategy based on the principle of "integrated management of activities in estuaries, coastal waters and marine waters that form part of Canada or in which Canada has sovereign rights under international law." Following the Act's passage, various integrated coastal and ocean management (ICOM) initiatives were developed across Canada. Founded in 1998, the ESSIM Initiative was the first of these ICOM initiatives to take an offshore focus (McCuaig and Herbert 2013), as well as the first to be established in the Atlantic region.

The ESSIM Initiative's jurisdiction encompassed the Eastern Scotian Shelf and Slope, an oceanic region of approximately 325,000 km^2 off the eastern coast of Nova Scotia. This region was selected for an ICOM initiative for a variety of reasons, including its high levels of living and non-living natural resources and diversifying human use in recent years. The initiative sought to bring regulators and policy makers together with regional industry sectors and other stakeholders with a vested interest in the management of the region's resources to develop a management plan that would provide "long-term direction and commitment for integrated, ecosystem-based and adaptive management of all marine activities in or affecting the Eastern Scotian Shelf" (McCuaig and Herbert 2013, p. viii). The development of this plan was the primary activity for the first eight years of the initiative.

Following the completion of the ESSIM Plan, the Initiative reoriented its efforts during 2006–2011 toward the implementation of the strategies and management actions identified in the Plan, as well as developing a strategy for evaluation of the Initiative's progress toward its planned objectives (McCuaig and Herbert 2013). It was during this period that the SOE reporting that would form the basis of the State of the Scotian Shelf (SoSS) Report was initiated. The intention was that the information produced would provide knowledge support to the ESSIM participants and, eventually, contribute to the evaluation of the Initiative's progress.

The SoSS Report had the benefit of drawing on the development process, format, and even personnel of the State of the Gulf of Maine Report, a project that provided a strong example of how to produce a large-scale regional SOE with limited resources (Gulf of Maine Council on the Marine Environment 2010). Like the State of the Gulf of Maine Report, the SoSS Report was developed in consultation with stakeholders in order to enhance its credibility, legitimacy, and salience in the eyes of likely users (Mitchell et al. 2006). Likewise, the SoSS Report was developed for public release in a modular format, with a context document and a series of theme papers focusing on relevant topics (e.g., ocean acidification and marine waste and debris), while a large-scale technical report was produced for internal purposes (McLean et al. 2013). To reduce financial outlays and enhance the convenience of distribution, the report's producers chose to publish the report in an online-only, digital format via a partner organization, the Atlantic Coastal Zone Information Steering Committee (ACZISC), and its COINAtlantic website. The context document and theme papers were developed and published over a period of 2 and a half years.

Like the ESSIM Plan itself, the SoSS Report was developed with input from ESSIM's Stakeholder Advisory Committee (SAC). This committee included representatives of the shipping, fishing, petroleum, and telecommunications industries; officials from the municipal, provincial, and federal levels of government; representatives from First Nations (aboriginal) organizations; as well as members of community, environmental, and historical preservation groups. SAC was involved in defining the scope of the SoSS Report by identifying appropriate theme paper topics, while a steering committee drawn from the organizations represented by SAC played a more active role in approving drafts and finalizing papers. Table 12.1 provides a detailed overview of the editorial process for the theme papers.

The period of the report's development coincided with the completion of ESSIM's implementation planning (McCuaig and Herbert 2013). The ESSIM Plan had been submitted for the federal minister's approval two years earlier.

TABLE 12.1

Drafting Process for the State of the Scotian Shelf Report Theme Papers

Task	Participants
Review of scope of papers and Driving forces, Pressures, States, Impacts, and Responses (DPSIR) framework	Coordinator and authors
Draft table of contents	Authors
Review of draft table of contents, DPSIR	Steering committee
Draft theme papers	Authors
Steering committee and peer review	Peer reviewers, steering committee
Authors incorporate review comments	Authors
Approval by coordinator and/or steering committee (if needed)	Coordinator, steering committee
Layout	Contracted graphic design firm

However, while the development of the SoSS Report was still underway, it became clear that the federal Minister of Fisheries and Oceans would not provide final approval for the plan. As a result, the Initiative concluded with a formal review and evaluation in 2012. Despite this termination, interest in proactive governance in the Scotian Shelf has continued, most recently with the release of a Regional Oceans Plan for the region and with a section of the region being considered for the development of a network of marine protected areas (Department of Fisheries and Oceans 2015).

Though its original intended primary audience and purpose were lost with the conclusion of the ESSIM Initiative, the SoSS Report was completed and published to provide decision-making support to regional stakeholders, even in the absence of a multistakeholder management program. As originally planned, this publishing effort was a digital-only strategy, carried out by the co-publisher, ACZISC, via its COINAtlantic website.

During the publication of the report, DFO approached the Environmental Information: Use and Influence (EIUI) research program at Dalhousie University (www.eiui.ca) to conduct a study assessing the awareness and use of the report in the relevant communities of practice. The context document and most of the theme papers had been released before the study began; by the study's close, all the theme papers had been released. This study also investigated how users from different potential audiences make use of the report and what qualities these users look for in a SOE report.

12.3 Identifying Useful Scientific Information

One of the most significant challenges to assessing the use of any information product is defining exactly what it means to "use" a piece of information. One example of research use is a policy maker directly citing a piece of research while drafting a policy. However, Nutley et al. (2007, p. 34) propose that information is "often used in much more indirect, diverse and subtle ways," which are commonly referred to as conceptual, rather than instrumental, uses.

A piece of research may have an impact by shaping a policy maker's understanding of an issue, even if that research does not play an instrumental role in the writing of the final policy. Research may be used *ex post facto* to support a decision that has already been made, or may be used by opposing politicians and policy makers to critique a decision they disagree with (Nutley et al. 2007). For this reason, Nutley et al. (2007) present a continuum of research uses (see Figure 11.1), ranging from generating awareness and improving knowledge and understanding at the more conceptual end to shaping attitudes and perceptions and directly impacting policy at the more instrumental end. According to Nutley et al. (2007), this spectrum is further complicated by the facts that policy makers are not the only audience that may make use of research—they offer teachers as a prominent example—and that the other users may ultimately have indirect effects on policy through their own use of research.

Drawing on work by Greenberg and Mandell (1991), Nutley et al. (2007) present a spectrum of information use that subdivides conceptual and concrete usages into three further subcategories: substantive, elaborative, and strategic. Concrete-substantive use involves the use of information to shape the essence of a decision, while conceptual-substantive use implies that a particular piece of information is essential to forming the user's understanding or orientation toward an issue (Nutley et al. 2007). Elaborative use of information is more peripheral: on the concrete end of the spectrum, it involves using information to further refine an established position and, on the conceptual end, it involves using information to enhance an established understanding of an issue. Strategic use does not involve shaping either positions or understanding. Rather, concrete-strategic use involves enlisting a piece of information as argumentative support to justify or defend a position that already has been developed.

The question of whether SOE reports are read or referred to by environmental managers and other policy makers and whether they ultimately have an impact on decision-making raises an obvious corollary: what characteristics define a SOE report as an effective information resource? As McNie (2007) notes, useful scientific information will inherently improve environmental decision-making by expanding the range of possible solutions to an issue and enabling policy makers to make informed choices; however, this requires an understanding of how scientific information becomes useful. The system by which traditional academic scientific information is established as useful is well-known: research that passes a rigorous peer review process and is published in a reputable academic journal has been determined to be a useful contribution to scientific enquiry (Cronin 2005). Obviously, to date, no such single system exists to determine the value of SOE reports, which are produced outside of these traditional academic channels, to policy makers and other stakeholders in environmental management, whose information needs are diverse and whose interests in the issues may be divergent or even diametrically opposed (Ernst 2004; Jacobson, Lisel et al. 2013; Shanley and Citlalli 2009). Hence, researchers studying the impact of SOE reporting have sought to establish metrics by which the usefulness of scientific information to decision makers can be evaluated. One common metric considers the balance of a report's salience, legitimacy, and credibility (McNie 2007; Mitchell et al. 2006). McNie defines salient information as information that is relevant to the information needs of users (McNie 2007). In order to be considered legitimate, an information product must be created by a process that users view as free of political bias and that reflects the interests and concerns of all affected stakeholders (Mitchell et al. 2006). Credible information products must be perceived by users as presenting scientific knowledge with accuracy (Mitchell et al. 2006).

A recurring theme in the discussions of salience, legitimacy, and credibility highlighted above is that the process by which a SOE report is developed can have as much impact on the usefulness of its scientific information as the content of the product itself, a point made strongly by Wells (2003). This impact is particularly pronounced if the process involves stakeholders who are in the report's prospective audience. Indeed, Mitchell et al. (2006, p. 308) go so far as to declare that SOE reports are "better conceptualized as social processes rather than published products."

Mitchell et al.'s (2006) conclusion that the process of generating SOE reports is as important, if not more important, than the reports themselves extends from their observations of the interplay between the SOE attributes of salience, legitimacy, and credibility. The most common cause of this interplay resulting in the detriment of a SOE report's usefulness is an overemphasis on credibility, manifested in allowing the process to be exclusively controlled by scientists and attempting to remove all political influence from the process. This attempt to remove political influence from a process that is explicitly intended to influence the public in general, and politicians and policy makers in particular, is antithetical to the success of the report's goals: such a report may very well fail to address the relevant concerns of policy makers and the public (Mitchell et al. 2006). Notably, the reverse can also occur; for instance, in attempts to generate highly salient information without adequately consulting the scientific community, scientists may be forced to make recommendations based on incomplete or premature results, thus raising questions about the credibility of the information contained in a report (Mitchell et al. 2006). Furthermore, attempts to foster legitimacy can come at the expense of credibility. Mitchell et al. (2006) note that involving stakeholders who can represent the views of the report's audience can result in the report being viewed as less than credible by other scientists and even some policy makers.

Mitchell et al. (2006) note that attempts to generate influential reports do not necessarily have to involve trade-offs between attributes. In fact, effective involvement of local stakeholders in the development of SOE reports can create a positive feedback loop among the attributes. For instance, in the case of the movement to combat acid rain, the effort to improve salience and legitimacy by increasing stakeholder participation in SOE creation had a salutary effect of improving the credibility of said SOE reports, as they were able to complement scientific research with quality local knowledge (Andonova 2006). Effective engagement of stakeholders alongside scientists in the process of generating SOE reports can thus improve all three attributes simultaneously, resulting in reports that maximize their value and impact (Mitchell et al. 2006).

As noted above, it is this potential mutual reinforcement of attributes that leads Mitchell et al. (2006, p. 324) to two complementary conclusions: that "influence flows from the process by which it creates knowledge rather than from the reports it may produce" and that, thus, "the content and form of [SOE] reports are poor predictors of their influence." They propose abandoning the existing model of the SOE reporting process that relies on scientists' attempts to communicate the best available scientific information to audiences, and replacing it with a model that views the SOE reporting process as relying on extended interactions focusing on mutual education between scientists and stakeholders.

Stakeholder participation in the SOE reporting process, if conducted effectively, ultimately fosters all of the attributes of successful reports: salience, since the involvement of policy makers and other stakeholders allows scientists to focus their efforts on presenting the scientific information that provides the best decision-making support for the actual decisions under consideration; legitimacy, because extended dialogue between scientists and end-users serves to reassure those users that their concerns are being taken into account in the review of the relevant scientific information; and credibility, because those stakeholders—industrial or otherwise—who are perceived as responsible for the environmental problem(s) being addressed can provide valuable data that is otherwise unavailable to scientists, while the perception that their concerns about the process are being heard reduces their distrust of the knowledge produced by the SOE reporting process (Mitchell et al. 2006). A similar cooperative approach to the production of knowledge in SOE reporting was a key recommendation by Battaglia et al. (2013), which endorsed a growing trend toward hybrid approaches to knowledge production that aim to synthesize expert and local knowledge.

12.4 Methodology

This case study of the State of the Scotian Shelf Report had the benefit of drawing on an established body of work by the EIUI research program for its methodological design (see www.eiui.ca). Previous EIUI studies of the awareness and use of information utilized a range of methodologies: both qualitative—surveys and interviews with relevant individuals in the communities of practice (Cossarini et al. 2014; Soomai, MacDonald and Wells 2011); and quantitative—citation searching and analysis conducted via both academic databases, such as Web of Science, and mainstream search engines, such as Google (Cordes 2004; Hutton 2009).

To best assess the awareness and use of the SoSS Report, a mixed methods approach was adopted, with quantitative data from citation searches and web traffic statistics providing context for the findings of surveys and interviews with key audiences for the report (Brannen 2005). Citation searches in the major academic resources Web of Science and Google Scholar, and web traffic analysis of the report's online home at COINAtlantic (www.coinatlantic.ca) provided the quantitative context for the study by supplying direct evidence of access to and use of the SoSS Report. These quantitative data sources are explained at length by Ross (2015). Online surveys were distributed to two audiences for the report: the subscribers of ACZISC's *Coastal Update* e-newsletter—the primary venue for promotional notices for the report—and the former members of the ESSIM SAC. For additional insight into both the development of the report and its reception among its primary stakeholders, all members of the ESSIM SAC were also invited to participate in semi-structured interviews. Following completion of the interviews, participants were assigned a participant code to conceal their identity. The interviews were transcribed and then coded to identify common themes among respondents.

12.5 Results

In total, 66 *Coastal Update* subscribers and 14 ESSIM SAC members responded to the online surveys discussing their awareness and use of the SoSS Report, while 8 members of the ESSIM SAC participated in interviews focusing on their role in the report's development and their assessments of the report following its publication.

12.5.1 Use of the State of the Scotian Shelf Report

The results of the surveys and interviews provided an excellent demonstration of the conception of information use as a broad continuum as proposed by Nutley et al. (2007). The responses of participants—both interviewees and survey respondents—highlighted uses that fall across the spectrum of conceptual to concrete usage, while also emphasizing that the SOE reports are potentially useful to a variety of audiences in government, education, public advocacy, and industry.

12.5.2 Use of the Report for Government Purposes

Nearly all interviewees identified government officials—both elected politicians and civil servants—as primary audiences for SOE reports. While it was generally assumed that government users would be those individuals with jurisdiction over the ecosystem covered in the report, two survey respondents from outside the Scotian Shelf region indicated that using the SoSS Report would broaden their understanding of ICOM and SOE reporting efforts in other regions. Table 12.2 depicts the range of potential ways in which those individuals perceived reports being used.

The SOE reports are often conceived as primarily intended for use by government officials and managers. As the responses show, the reports are seen as useful throughout planning processes, as planning aids during the conception stage, as argumentative support when presenting plans to decision makers, and, following a plan's execution, as tools to evaluate their success. This last role highlights an observation many participants made regarding the value of SOE reports: reports increase in value over time if they are updated periodically because the succession of *snapshots* can be analyzed to evaluate ecosystem trends.

TABLE 12.2

Uses of the SoSS Report for Government Purposes

Category of Use	Subcategory of Use		
	Strategic	Elaborative	Substantive
Conceptual	Outreach tool from government to stakeholders in the region, demonstrating that the government understands stakeholders' concerns	Reference for governments interested in past ICOM efforts	Compendium of salient policy issues to clarify potential actions to policy makers
Concrete	Argumentative support for managers advocating for or defending management decisions	Information resource to help designate potential species at risk, marine protected areas, and ecologically and biologically significant areas Information resource for consultants providing advice to industry groups. Evaluative tool for measuring the success, or lack thereof, of established environmental management plans	Reference document for policy makers involved in planning in an ICOM context Information resource for conducting risk assessments and environmental impact assessments

12.5.3 Use of the Report for Public Advocacy

Potential users identified in the nongovernmental public sphere included community groups, environmental activists, First Nations groups, nongovernmental organizations with an environmental focus, and members of the interested public. The three potential uses of the report for public advocacy identified by participants were public education efforts by activists raising awareness of environmental issues, similar efforts to raise awareness of actively debated management and policy issues, and efforts by interested individuals to educate themselves regarding salient local issues.

Such uses are difficult to classify using the schematic presented by Nutley et al. (2007) (see Chapter 2 in this volume) because, while they are predominantly substantive-conceptual uses, their application to advocacy efforts lends them a substantive-concrete dimension. Indeed, one interview participant, a member of an advocacy group interested in coastal issues, argued that scientific information presented in language that is comprehensible to nonscientists is not just an aid to effective participatory democracy, but an essential precondition for it. The participant noted that environmental management initiatives typically involve public consultations in affected communities, but that

> there's a whole crowd of [members of the public] that might have an interest, but … they know they don't know anything much about it. They're concerned, but they're not confident about it … [they] don't have the confidence to speak up, to write a letter to the government, or to participate in these things, because they just feel ignorant.

Both interviewees representing community groups specifically identified the Report as a support to democracy; it serves the stated interest of the government to encourage citizen participation and input to democratic decision-making, as well as the interests of members of the public who wish to give input but feel ill-informed to express their opinions to managers and policy makers.

12.5.4 Use of the Report by Industry Groups

One interviewee, who was closely involved in the development of the SoSS Report, expressed skepticism about its potential use by industry stakeholders. Furthermore, questions about the commitment of particular industries to both the use of scientific information and participation in the ICOM process were also raised by other industry stakeholders. Despite this view, participants representing the fisheries, shipping, and oil and gas industries all identified themselves as potential users of the report. Furthermore, one participant representing industry stakeholders directly confirmed use of the SoSS Report in the development of recommendations to industry decision makers.

Echoing themes expressed by the community representatives interviewed for this study, two of the interviewed industry stakeholders noted that SOE reports address an information need that may otherwise go unfilled. One participant stressed the value to industry stakeholders of having baseline information from a credible government source. The participant noted that in the absence of government-provided scientific information, industry organizations may have to rely on less credible information provided by consultants, rendering it more difficult to make effective, evidence-informed decisions. The need for credible scientific information may be even more urgent in the case of industry groups. Whereas the absence of such information for the public results in nonparticipation, the lack of such information for industry may not result in a reduction of their activity, but simply a reduction in the degree to which that activity is being planned with reference to relevant and sound scientific information.

12.5.5 Use of the Report for Educational Purposes

Participants identified a range of potential audiences for SOE reports in the educational sector, most prominently students and educators at the high school and university level, but also academic researchers. Such audiences were seen as including both scientific and environmental management fields. Two survey respondents specifically noted that the SoSS Report was used as assigned reading in environmental management classes where they were students, suggesting that use of the report for educational purposes may also promote awareness and use of the report by future environmental managers. Table 12.3 presents the potential educational uses identified by participants.

Two interviewees expressed the view that the information contained in SOE reports was unlikely to be useful to researchers in scientific fields, as these reports tend to provide broad overviews of established knowledge. However, two other participants argued that such reports could be useful

TABLE 12.3

Uses of the SoSS Report for Educational and Research Purposes

Category of Use	Subcategory of Use	
	Elaborative	Substantive
Conceptual	Starting point for future research, due to comprehensive list of references	Educational tool for science teachers at the high school level Educational tool for university professors in the subjects of marine and environmental management
Concrete	Authoritative reference for scientists and researchers seeking to establish accepted environmental conditions in primary literature	Testing of baseline information in the report in subsequent research

references for scientists looking for authoritative sources describing baseline conditions of an ecosystem as background for a study or experiment. This suggestion is consistent with recent research by Avdić (2013), who found that the UN Food and Agriculture Organization's The State of World Fisheries and Aquaculture biennial report was most likely to be cited in the introduction of a scientific publication to establish the authoritative baseline data.

12.5.6 General User Satisfaction with Form and Content of the Report

The wide range of potential audiences and uses for SOE reports represents a challenge to producers. Teams creating such reports must design a single document (or single collection of documents, in the case of modular reports like the SoSS and the State of the Gulf of Maine reports) that translate scientific information into language that is comprehensible and useful for users that have divergent needs, interests, and familiarity with environmental science. In light of this challenge, it is notable that responses to the surveys and interviews reported a generally high level of satisfaction with the SoSS Report's format and content.

Interview participants generally expressed satisfaction with the writing level and scientific credibility of the report. Five interviewees, representing government, industry, and community stakeholders, explicitly identified the report as being written at an appropriate technical level for its intended audiences: in the words of one participant, the Report was written at a level that the general reader "who doesn't have a scientific background would still understand," while still having enough depth—that is, detail and interpretation—to serve the needs of professionals in coastal and ocean management. To the extent that participants had recommendations for improving the content of the Report, the focus was on expanding the geographic and thematic scope of the theme papers in future editions, rather than making major adjustments to the current information and writing style.

The modular and digital-only format of the Report was praised by its users and producers. Users recognized that a modular format facilitates information retrieval and saves valuable time by allowing readers to quickly identify the theme paper that is relevant to their topic of interest, rather than being forced to skim through a large omnibus document looking for particular sections. For report producers, the modular format facilitates report production by enabling a sectional approach to producing and updating the report, reducing the size of individual outlays, and easing the process by which funding is sought.

Participants generally approved of the digital-only version of the Report for simple reasons: low costs of distribution, ease of accessibility for users, and the enhanced ability to edit or update documents post-publication. While some participants expressed concerns about limited access for users with low technological ability or lack of quality broadband connections, they largely dismissed these as a fading concern in an age where Internet use is widespread and broadband penetration is ever increasing.

12.5.7 Enablers of SOE Report Use

Stakeholder engagement was identified as a major enabler of use of SOE reports. The results of this study confirm findings by Mitchell et al. (2006) that effective engagement of stakeholders in the development of information products enhances the salience, legitimacy, and credibility of those products in the eyes of users. Participants repeatedly emphasized the benefit of the availability of scientific information with a government imprimatur, while many expressed the belief that the product was more relevant to their information needs as a result of SAC's involvement in its development.

Participants further identified the potential of stakeholder engagement to mediate tensions between different stakeholder groups. Several interview participants implied a suspicion of the motives, interests, and commitment of other stakeholder groups involved in the development of the ESSIM Plan and the SoSS Report. However, despite these misgivings, nearly all interviewees (7 of 8, or 87.5%) expressed trust in the value of the ESSIM Plan and the SoSS Report, suggesting that they believed that the sum of multiple dissenting viewpoints constituted an acceptable consensus that reflected the interests and information needs of all stakeholders involved in SAC.

In addition to enabling use by improving perceptions of the report's salience, credibility, and legitimacy, participants suggested that stakeholder engagement in the context of an ICOM initiative directly incentivized use of the report by stakeholders. By confronting stakeholders with the scientific information that was deemed relevant to the integrated management of the Scotian Shelf region, DFO was able to indirectly encourage stakeholders to familiarize themselves with this information and alert their respective organizations to alter their own strategic planning in preparation for the implementation of the ESSIM Plan.

12.5.8 Barriers to SOE Report Use

One of the primary barriers to use of the SoSS Report is simple lack of awareness of it. Half of survey respondents reported not being aware of the SoSS Report prior to participating in the survey. As individuals choosing to complete an online survey on the subject of the SoSS Report are presumably interested in the topic, this is a notable result. Furthermore, one interview participant, who was involved in the ESSIM Stakeholder Advisory Committee and expressed a strong interest in the Report, was unaware that the theme papers were ever published until halfway through the interview!

In addition to those who expressed their own lack of awareness, many participants in the interviews and surveys expressed the view that promotion for the report had been inadequate. One interview participant identified the concentration of promotional efforts solely at the time of publication as the most significant barrier to awareness, arguing that sustained promotion over the Report's life cycle is essential. In this participant's view, individuals with a potential interest in the Report may not be interested at the exact moment of publication, and thus may either ignore the initial promotional efforts or not remember them at a later date when the Report's information might be useful to them.

Two interviewees (2 of 8, or 25%) identified time constraints, particularly those affecting industry stakeholders, as a significant barrier to their use of the report. A participant representing the fishing industry stated that although the report would be helpful for planning and decision-making purposes, the participant's organization lacked the resources to incorporate the Report into its planning process. This perspective emphasizes the identification of ICOM initiatives as enablers to information use; in the absence of an incentive to participate in ICOM planning and implementation, organizations in private industry may not be able to, or may choose not to, allocate the resources to use available scientific knowledge as a planning aid.

A further barrier to use identified by participants was a lack of confidence in DFO's commitment to updating the Report. Again, this view reflects a previously identified strength of SOE reports: the potential, with regular updating, for such reports to demonstrate ecosystem trends and aid in the evaluation of the effectiveness of integrated management actions. However, if users are not convinced that regular updating will occur, they are less likely to use the early information to form baselines in their planning. At present, DFO intends to update the SoSS Report theme papers every five years. Participants agreed that this was an appropriate time frame for updating, with most identifying five years as an appropriate cycle prior to being informed that this was the DFO plan. As such, an approach to address this barrier to use must focus on establishing how genuine DFO's existing plans are to update the Report.

12.6 How Can SOE Report Producers Maximize Use and Awareness?

One of the practical aims of this study was to develop recommendations for how the SoSS Report, and other SOE reports, could be better developed and promoted to maximize their awareness and use. Obviously, producers of SOE reports have an interest in seeing them used, even more so in an age of constrained government budgets and strong desire to account for the value of expenditures. Ensuring that they are being used by a wide range of individuals and organizations in their target audiences is crucial to maintaining funding support to continue these reporting endeavors.

One of the primary barriers identified to the use of the SoSS Report was that potential users may simply be unaware of the report. This lack of awareness is compounded by the findings regarding the diverse nature of potential audiences and uses of SOE reports. The SoSS Report was primarily promoted to an audience of practitioners in the field of coastal and ocean management. As such, promotional efforts may well have overlooked other potential audiences, such as educators, community activists, and private individuals with an interest in the management of local ecosystems. Furthermore, as these promotional efforts were tied to the initial publication of the theme papers, they are not well suited to maintaining awareness of the report. A gap of about three years will exist between the publication of the first edition of the final theme papers to be completed and the second edition of the first theme papers to be completed.

The primary solution to this lack of awareness, put simply, is to improve promotional efforts for SOE reports, rather than relying on potential readers to actively seek out information on the Scotian Shelf via Google searches. Of course, this step may be far from simple in practice. The diversity of potential audiences and uses for SOE reports poses a considerable challenge in identifying appropriate, much less ideal, venues for promotion. Furthermore, with the rise of digital communication technology, the advent of social media platforms, and a decline in the readership of traditional media outlets, promotional efforts must contend with a fractured media environment in which no avenue of promotion is guaranteed to reach all potential audience members.

This challenge was reflected in the responses of interview and survey participants, who were asked to identify appropriate potential venues for promotion of the SoSS Report and other SOE reports. In addition to the current *Coastal Update* promotions, responses suggested press releases to traditional media outlets, such as local newspapers and television programs; enhanced use of social media platforms; and direct promotion of the report at relevant conferences and in university classrooms. Queried about their preferred social media platforms for promotion, respondents were divided between Facebook, Twitter, and LinkedIn, with a smaller number of participants also identifying Yammer, a social networking site for organizations that is owned by Microsoft. Further suggestions included the distribution of limited-run, promotional print copies of digital reports and the repackaging of the content of reports into smaller formats that can be more easily mass-distributed in a digital environment.

The diversity of suggestions for the best potential venues of promotion emphasizes that an optimal promotional effort for a SOE report will not be attained by identifying an "ideal" venue for promotion. Rather, an optimal effort will depend on broad promotional initiatives through multiple channels. Report producers must understand the various potential audiences and uses for their products and target them with the appropriate media in the appropriate venue: for instance, notices on professional social networks like LinkedIn to reach practitioners in coastal and ocean management, releases to local newspapers and other news media to reach interested members of the public in areas affected by environmental management decisions, and direct promotion of the report as an educational resource to university professors to reach students in the sciences and in environmental management.

Maximizing the awareness and use of SOE reports requires not only broadening promotional efforts, but sustaining them over time. Awareness is not permanently achieved at the discrete moment of a report's initial release: in the words of one participant, report producers must keep their product "under the noses of … the people who would have an interest in it." Many interview participants suggested a straightforward solution to increasing the frequency of promotional notices for SOE reports produced in a modular format: updating theme papers on a more staggered schedule. If, instead of updating all theme papers over a two-to-three year period, report producers updated a select number of theme papers each year, with each theme paper updated within five years, there could be a steady stream of new releases offering new occasions for promotion. This result will be particularly beneficial because promotion of one theme paper could serve to maintain awareness of the entire report.

A staggered updating schedule for modular reports will also potentially serve to increase use of the reports. The most obvious mechanism for this potential increase is the correlation between awareness and use: while awareness of an information product does not guarantee use of it, awareness is certainly a precondition of use. Additionally, a staggered updating schedule can potentially address another barrier to use: audience members' lack of confidence in the commitment of producers to regularly revise the report and maintain the currency and salience of its information. Although there are some short-term drawbacks to this approach—some theme papers will initially be updated very shortly after the initial publication and may not reflect much in the way of new knowledge—there is no better demonstration of a commitment to keep a report current than to begin releasing updated theme papers.

In addition to addressing current barriers to use of SOE reports, report producers should seek to capitalize on existing enablers. The most prominent enabler identified in this study was the extent of the benefits of stakeholder engagement in the production of the report. This finding echoes similar discoveries by Mitchell et al. (2006) regarding the benefits of stakeholder engagement in the coproduction of knowledge. However, participants' responses went further, suggesting that the benefits of stakeholder engagement to the value of scientific information products extend beyond the production process. By providing stakeholders from industry, government, and the public with a forum in which they can describe and advocate for their interests and have an opportunity to provide input into the environmental management of the region, SOE report producers offer strong incentives for stakeholders to maintain familiarity with the state of scientific knowledge of the region. One interview participant noted that SOE reports enable discussion between stakeholders because they provide a common baseline of accepted data regarding environmental conditions, ensuring that "everybody's working from the same song sheet." To extend this metaphor, an active ICOM program provides the choir with space to rehearse, thus enhancing the value of the common song sheet. Awareness and use of SOE reports, and thus the ultimate impact of scientific information upon the development of policy and management plans, will be far greater if active efforts are made to engage users and encourage them to recognize the value of SOE reports to policy issues and management functions. The significance of the development process to the usefulness of a SOE report does not terminate when the process is finished, but continues via the processes used in the application of that information to policy issues by stakeholders.

In the particular case of the SoSS Report, the ICOM ship has sailed, as the ESSIM Plan was completed in 2008 but ultimately was not fully implemented by the participants in the process, including DFO, other government departments, and industry sectors (McCuaig and Herbert 2013). Though some participants expressed the view that an ICOM plan for the Scotian Shelf region will eventually come to pass in a different form, any present attempts to further engage stakeholders regarding the SoSS Report will need to take place outside the context of ESSIM. One approach would be to invite former members of SAC to participate in the steering committee's updating process for the SoSS Report. If the earlier recommendation to update the report on a more staggered basis was adopted, this re-engagement could take place on an annual basis and focus on those theme papers that were due to be updated. In addition to complementing promotional efforts by maintaining awareness of the report among key stakeholders, this process would also benefit future editions, as participants offered suggestions for improving its content, particularly by expanding its range of themes and its geographical scope.

12.7 Conclusion

What is the value of a SOE report? They are often developed with an eye toward their application in policy-making and environmental management. Accordingly, studies such as this one have generally focused on their use as decision-making aids by policy makers and managers (e.g., Mitchell et al. 2006; Soomai, Wells and MacDonald 2011). However, while the application of scientific information to policy formation demands further investigation (see Chapters 2, 3, and 9 in this volume), the results of this study demonstrate that the value of SOE reports extends well beyond their direct use in policy- and decision-making.

A major finding was that SOE reports serve a wide variety of uses for a wide diversity of audiences. One interviewee, who represented a community group on SAC, noted that SOE reports contain scientific "information that people—the average person—can read, and after reading several similar things, get a general idea about the status of things" and went on to observe that that kind of information is "not as common as you'd think." The SOE reports address a widely held demand for summary information not met by producers of primary literature or private sector organizations. The wide-ranging demand for salient, credible, and legitimate scientific information that is comprehensible to nonscientists is best addressed by the sorts of governmental, nongovernmental, and intergovernmental organizations that produce SOE reports. As the results of this study show, users consider scientific information provided by the government to be inherently legitimate and credible, particularly if it is developed with the participation of regional stakeholders.

By providing this sort of scientific information to the public, SOE report producers within government fulfill many roles that are secondary to the primary aim of supporting evidence-based policy and decision-making. However, even these secondary purposes may ultimately contribute to the primary objective, as the availability of scientific information that is comprehensible to nonscientists is a prerequisite for constructive engagement in the policy-making process by industry stakeholders, environmental advocates, and the interested public. A SOE report cannot compel a citizen to attend public consultations regarding an environmental management plan, but a citizen who is able to educate him- or herself regarding the status of a local ecosystem is, as the interviewee quoted above observed, more likely to contribute to such a consultation. Assigning a SOE report as course material for a student in an environmental management program will not directly affect policy decisions, but that student in a future career as an environmental manager will be more familiar with such information resources available to support decision-making. Furthermore, as SOE reports focus on particular ecosystems and address policy and management issues alongside scientific information, they are uniquely well-suited vehicles for building interest and awareness in affected communities. The process of preparing a SOE report and making people aware of it can be as important as the final product itself, as indicated earlier in this chapter. Ultimately, environmental management efforts, including ICOM programs, are supported by the availability of scientific information to any individual or institution with a stake in the management of a given ecosystem. Thus, the effective production, publication, and promotion of SOE reports is essential to the goal of sound, evidence-informed policy-making and, ultimately, to effective ICOM.

References

Andonova, L. 2006. Structure and influence of international assessments: Lessons from Eastern Europe. In *Global Environmental Assessments: Information and Influence*, edited by R. Mitchell, W. Clark, D. Cash, and N. Dickson, 307–338. Cambridge, MA: MIT Press.

Avdić, V. 2013. Measuring use and influence: An assessment of the FAO's flagship report The State of World Fisheries and Aquaculture. Master's project report, Dalhousie University.

Battaglia, M., E. Meloni, and A. Cautillo. 2013. Technical assessment and public perception of environmental issues: The case of the municipality of Pisa. *Local Environment* 19 (7): 786–802.

Brannen, J. 2005. Mixing methods: The entry of qualitative and quantitative approaches into the research process. *International Journal of Social Research Methodology* 8 (3): 173–184.

Cordes, R. 2004. Is grey literature ever used? Using citation analysis to measure the impact of GESAMP, an international marine scientific advisory body. *Canadian Journal of Information and Library Science* 28 (1): 49–70.

Cossarini, D. M., B. H. MacDonald, and P. G. Wells. 2014. Communicating marine environmental information to decision makers: Enablers and barriers to use of publications (grey literature) of the Gulf of Maine Council on the Marine Environment. *Ocean & Coastal Management* 94: 163–172.

Cronin, B. 2005. *The Hand of Science: Academic Writing and Its Rewards*. Lanham, MD: Scarecrow Press.

Department of Fisheries and Oceans. 2015. Developing a marine protected area network for the Scotian Shelf bioregion. Draft document. Dartmouth, NS: Fisheries and Oceans Canada, Maritimes Region, Ecosystem Management Branch.

Ernst, M. 2004. A survey of coastal managers' science and technology needs prompts a retrospective look at science-based management in the Gulf of Maine. http://www.gulfofmaine.org/council/publications/coastalmanagerssurveyreport.pdf.

Greenberg, D. H. and M. B. Mandell. 1991. Research utilization in policymaking: A tale of two series (of social experiments). *Journal of Policy Analysis and Management* 10 (4): 633–656.

Gulf of Maine Council on the Marine Environment. 2010. State of the Gulf of Maine report. http://www.gulfofmaine.org/2/sogom-homepage.

Hutton, G. R. G. 2009. Developing an inclusive measure of influence for marine environmental grey literature. Master's thesis, Dalhousie University.

Jacobson, C., A. Lisel, R. W. Carter, et al. 2013. Improving technical information use: What can be learnt from a manager's perspective? *Environmental Management* 52: 221–233.

McCuaig, J. and G. Herbert, eds. 2013. *Review and Evaluation of the Eastern Scotian Shelf Integrated Management (ESSIM) Initiative*. Halifax, Nova Scotia: Department of Fisheries and Oceans, Canada.

McLean, M., H. Breeze, J. Walmsley, et al. 2013. State of the Scotian Shelf report. Canadian Technical Report of Fisheries and Aquatic Sciences 3074. http://www.dfo-mpo.gc.ca/Library/352339.pdf.

McNie, E. C. 2007. Reconciling the supply of scientific information with user demands: An analysis of the problem and review of the literature. *Environmental Science & Policy* 10 (1): 17–38.

Mitchell, R., W. Clark, and D. Cash. 2006. Information and influence. In *Global Environmental Assessments: Information and Influence*, edited by R. Mitchell, W. Clark, D. Cash, and N. Dickson, 307–338. Cambridge, MA: MIT Press.

Nutley, S. M., I. Walter, and H. T. O. Davies. 2007. *Using Evidence: How Research Can Inform Public Services*. Bristol: The Policy House.

Oceans Act, 1996 S.C., c. 31.

Ross, J. D. 2015. What do users want in a state of the environment report? A case study of the "State of the Scotian Shelf Report." Master's thesis, Dalhousie University.

Shanley, P. and C. López. 2009. Out of the loop: Why research rarely reaches policy makers and the public and what can be done. *Biotropica* 41 (5): 535–544.

Soomai, S. S., B. H. MacDonald, and P. G. Wells. 2011. The state of the Gulf of Maine report: An initial study on awareness, use, and influence of the theme papers. Halifax: Dalhousie University.

Soomai, S. S., P. G. Wells, and B. H. MacDonald. 2011. Multi-stakeholder perspectives on the use and influence of "grey" scientific information in fisheries management. *Marine Policy* 35 (1): 50–62.

United Nations. 1972. Declaration of the United Nations Conference on the Human Environment. http://www.unep.org/Documents.Multilingual/Default.asp?documentid=97&articleid=1503.

Ward, B. and R. J. Dubos. 1972. Only one earth: An unofficial report commissioned by the secretary-general of the United Nations Conference on the Human Environment. New York: W.W. Norton.

Wells, P. G. 2003. State of the marine environment (SOME) reports—A need to evaluate their role in marine environmental protection and conservation. *Marine Pollution Bulletin* 46 (10): 1219–1223.

13

The Environmental Effects of Ocean Shipping and the Science–Policy Interface

Elizabeth R. DeSombre

CONTENTS

13.1 Introduction

Ocean shipping is central to the global economy. More than 90% of all goods transported internationally are carried on the ocean by ships (IMO Maritime Knowledge Centre 2012). Shipping and other ship-based activities have major effects on the health of the ocean. Between resource degradation from the overharvesting of fish, to invasive species introduced to waters distant from their native habitat, to various types of pollution discharged—intentional or otherwise—into the water or air as ships transport goods or people, considerable ocean degradation can be traced to the effects of ships.

The interaction between science and policy in addressing environmental effects of ships is complicated by a number of things. As is true of all issues considered in this book, the vastness of the oceans makes information gathering—and policy implementation—difficult. Specific to the issue of shipping is the large number of ocean-going vessels with the potential to cause environmental effects, and the long distances many of them travel; simply finding out which ships are engaging in which activities with what environmental effects can be daunting. It is also the case that, with the exception of major oil spills, which do engage public attention, most environmental problems caused by ships are out of sight and do not attract the attention of environmentalists or regulatory agencies, and thus rarely garner the public participation that can be key to gaining movement on environmental problems.

Communication of scientific information about environmental harm from ships is thus key to the ability to address or prevent these problems. Uncertainty about ship behavior and the resulting environmental effects can influence the type of policy tools used to address these problems. The role of disasters, such as oil spills or problems from invasive aquatic species, has been key in increasing public awareness of environmental problems that often occur far from the public eye. This public concern can help amplify scientific arguments for policy action, but it can also push for action before scientific consensus on the appropriate type of action is clear.

Uncertainty is, in some ways, the genesis of most environmental problems caused by ships. Some of that uncertainty is fundamental: environmental harm from ships began before we had any real understanding that they could be problematic. Addressing environmental problems from ships, therefore, involves addressing ways of doing business that had previously been seen as nonproblematic but that we have come to understand contribute to environmental degradation. When ships are operating on increasingly tight margins, it can be extremely difficult to persuade shipowners to change behavior if they are not legally required to do so.

For that reason, global cooperation—most frequently accomplished through the International Maritime Organization—is also necessary. No shipping company can afford to be the one that undertakes costly changes in operations (for the sake of environmental protection) while its competitors do not. Ratification thresholds in most International Maritime Organization (IMO) treaties—in which treaties do not take effect until a certain number of states, representing a certain percentage of registered shipping, have been ratified—are an important way to garner sufficient participation before behavior change is required.

Issues of international cooperation in general also make political solutions to environmental problems more difficult. The inherently international nature of ocean shipping means that rules that apply to ships are determined by nation-states in a context in which they have to choose to cooperate. Even when states do decide to cooperate, emissions from ships are sometimes omitted from the universe of collaborative agreements on some pollutants (such as the Kyoto Protocol to the United Nations Framework Convention on Climate Change). Moreover, for any state that remains outside the international regulatory system on a given issue, its ships are not bound internationally by the relevant rules. Even when there is agreement on the underlying science, concern about suffering a competitive disadvantage by implementing solutions when other states may choose not to can prevent resolution of these issues. Claiming scientific uncertainty is a time-honored way to avoid regulation by states that would rather not have to change behavior. Major disasters, therefore, in which it suddenly becomes publicly obvious (for instance) that oil spills from ships are possible and environmentally problematic, can be an important trigger to tightening rules.

What makes both science and policy more difficult to conduct specifically with respect to environmental effects of shipping is the broader context of flag of convenience registration. Because of the way ship registration has evolved, shipowners can choose to register their vessels in whichever states offer the least intrusive regulatory environment and the lowest cost. This registration opportunity puts many ships outside of the regulatory reach of states or international organizations attempting to protect ocean resources, and makes collecting even the most basic information about environmentally relevant behavior of ships difficult. It is, therefore, the case that even when sufficient information is available to create a nuanced understanding of the contribution of ships to environmental problems and regulatory processes exist for preventing or mitigating these problems, the number of ships registered in locations that do not require them to take on environmental regulations causes a problem for the actual implementation of environmental policy.

In short, research or information can influence our understanding of what a ship-related environmental problem can be, and contribute to discussion of desired solutions. But the short-term economic interests that drive the shipping industry have led to a wide variety of creative ways to avoid participation in policy efforts to respond to or prevent these environmental problems, and the political reality of efforts to regulate shipping must take account of this incentive structure. Science, even good science, can be powerless in the face of legitimate economic incentives.

There is some hope in this issue area, however. First, the primary international regulatory body, the IMO, has a reasonably good track record of making scientifically sound policy decisions pertaining to shipping, even though it can be slowed down by economic interests. Science has influenced the creation of rules on both accidental and intentional oil pollution through equipment standards and air pollution through efficiency standards, and on process requirements to decrease transport of potentially invasive marine species. Second, and more important, is the fact that approaches taken by states and other actors both within and outside this international regulatory process have made important strides in increasing the percentage of ships subject in one way or another to international rules. Although it remains difficult to monitor, implement, and enforce these regulations, creative efforts have led to improvements on many environmental conditions influenced by shipping.

13.2 Ship Registration

The key aspect of global shipping regulation is how ships are registered. All ships have to have a nationality, which determines the domestic and international rules by which they must abide, and the political entity that has oversight and enforcement authority over them. Historically, this nationality was generally the nationality of the owner or captain of the ship, which was usually the same. Although there were exceptions in times of war or for other political purposes, this system of congruent ownership and nationality operated for centuries.

But beginning in the 1920s, and increasingly after World War II, some states began offering what is known as *open registration*, allowing ships owned or operated by non-nationals to register in these locations. The process began almost as a bureaucratic accident in Panama, but developing states quickly learned that they could earn income from making their registration attractive to shipowners by keeping costs (such as registration fees or taxes) low (DeSombre 2006).

The other way to keep costs down for shipowners was to refrain from requiring onerous regulations, since pollution control mechanisms or processes, or requirements to refrain from discharging pollution, make shipping more costly than operating without concern for the environment. This issue is even more important for protection of fisheries, since states that are not bound by fishing regulations are permitted to catch as many fish as they can by whatever method they choose (DeSombre 2005). Open registries seeking ship registrations thus initially refrained from joining many of the major international agreements to protect the environment (or safety) or labor rights, and their lack of participation in these agreements made operating a ship flagged in these locations much less expensive than flagging elsewhere would be. These registries are known, somewhat derisively, as *flags of convenience* because the registration is chosen for the convenience of the shipowner rather than the international community.

After World War II, the percentage of ships registered in these locations climbed dramatically (Boczek 1962). Liberia, one of the early open registries, became the world's largest in the early 1960s, and it remained so until it was surpassed by Panama in the early 1990s (DeSombre 2006). Although the designation of open registry is not clear-cut and it can thus be difficult to designate a registry as open or closed, it is likely that more than 60% of large commercial ships are currently registered in open registries (George 2013). The IMO, for instance, differentiates between the top "open" registries—Panama, Liberia, the Marshall Islands, Hong Kong, the Bahamas, and Singapore—and the top "controlled fleets," meaning where the parent companies that control the ships are located—Japan, Greece, Germany, China, the United States, and the United Kingdom (IMO Maritime Knowledge Centre 2012).

The broader implication is that ships are registered in locations to which they have little connection otherwise. The most popular registries have both lower regulatory standards to begin with, and laxer enforcement of those standards (Winchester and Alderton 2003). Some of these flag states intentionally refrain from participating in the international regulatory process and others lack the desire or capacity to implement any international rules.

This context sets the stage for any efforts to translate knowledge about environmental effects of ships into effective protection measures.

13.3 International Regulatory Process

Because the oceans beyond the exclusive economic zone are an international space, regulatory action for the high seas undertaken by any one state (or its ships) is doomed to failure. Taking on rules to protect the environment when other ships do not is insufficient for ocean protection and also a competitive disadvantage for those who bear the costs of actions when others refrain from doing so. States have long realized that protection of the oceans is only achievable through international cooperation.

The IMO is the primary international decision-making body addressing environmental problems created by ships. The organization (originally called the International Maritime Consultative Organization) was created in 1948 as a specialized agency of the United Nations. It became fully operational when its founding convention entered into force in 1958. It was initially seen as an organization to address maritime safety and related technical issues. A major coastal spill from an oil tanker (the *Torrey Canyon*) shortly after the organization began its operations drew it into greater involvement with the issue of oil pollution (IMO 2015b).

The main decision-making body of the IMO is the Assembly, which includes all member states. It generally meets every other year. When the Assembly is not in session, governing decisions are undertaken by a Council, whose size has increased over time. Currently, it is composed of 40 member states elected for 2-year terms by the Assembly. It includes the 10 states "with the largest interest in providing international shipping services," as well as another ten most involved in international seaborne trade, and an additional 20 that ensure geographic diversity (IMO 2015f).

The IMO focuses on a wide variety of issues relevant to shipping, including things that are only tangentially related to the environment, such as safety and security. Other treaties, negotiated outside of the auspices of the IMO, have come to be administered by it. The most important IMO entity for environmental protection is the Marine Environmental Protection Committee (MEPC), which focuses on preventing pollution from ships. The Committee was not an original part of the organization, but was added formally via a 1975 amendment to the organization's founding document that entered into force in 1982 and affirmed the environment as co-equal with safety in the IMO's mandate (M'Gonigle and Zacher 1979, p. 48). The Committee, which operates with representation from all member states, has taken the lead in negotiations and is working to encourage the implementation of a variety of treaties and other procedures addressing pollution from ships.

Also key is the Joint Group of Experts on the Scientific Aspects of Marine Environmental Protection (GESAMP), formally established in 1969 within the UN system to advise and coordinate among multiple UN organizations on issues of the marine environment, the IMO central among them. GESAMP was designed specifically "to provide authoritative, independent, interdisciplinary scientific advice" on the marine environment to these UN organizations. GESAMP was instrumental in moving the IMO toward environmental action to prevent oil pollution from ships, and it has played other important roles in assessing and suggesting approaches to addressing other problems of the marine environment (Administrative Secretary of GESAMP 2005).

Structurally, the IMO has struggled because of the flag of convenience phenomenon. The organization's structure, set up to give the greatest decision-making power to important maritime states, involves allocating representation by shipping tonnage registered. For instance, the initial composition of the Maritime Safety Committee—the most important technical committee of the Organization—was set to be 14 states, with eight of them being the eight largest shipowning states, as measured by ship registration. The rise of flag of convenience registration would have disenfranchised the traditional maritime states, as their share of ship registration decreased. Between the creation of the IMO founding documents and the entry into force of the treaty, both Liberia and Panama had risen to be among the eight largest registries, and tried to claim membership on the Committee. The traditional maritime states opposed this attempt, and the case was eventually decided in favor of Liberia and Panama by the International Court of Justice (Knudsen 1973). Perhaps as a result, the IMO expanded the membership on this Committee via amendments to 16 and then to the entire IMO membership (IMO 2015f).

Despite these changes making the IMO more broadly inclusive, the influence of states with large registries and incentives to keep the regulatory environment minimal can lead to a slowness to take action to prevent potential maritime environmental damage. It should be noted, however, that the traditional (closed registry) maritime powers that have a central role in the global economy also seek unencumbered—and inexpensive—shipping, which works against adoption of stringent international standards as well.

The organization also allows involvement of nongovernmental organizations (NGOs), provided they can "reasonably be expected to make a substantial contribution to the work of IMO" (IMO 2013, p. 1). The most numerous and most influential NGOs that have been granted consultative status are organizations of shipowners, operators, or others involved with the shipping industry rather than those concerned with scientific or environmental issues. Environmental NGOs do participate actively in the MEPC, however (Karim 2015, p. 20), including both general purpose environmental organizations and shipping industry organizations, such as the International Tanker

Owners Pollution Federation, which is focused on environmental protection (IMO 2015e).

The IMO undertakes two sorts of regulatory activities. Many of the IMO recommendations are issued as codes, such as the Code of Practice for the Safe Unloading and Loading of Bulk Carriers, or the International Safety Management Code. These are drafted by IMO committees and the Assembly. While not legally binding, these codes represent IMO's wisdom on how best to conduct operations. Most states implement these codes and other recommendations nationally. The IMO also issues recommendations pertaining to issues that have not yet been negotiated internationally, and in that way sometimes serve as a forerunner to negotiation of a legally-binding instrument, or inclusion in modifications of existing agreements.

The primary regulatory activity undertaken by the IMO comes through the negotiation of international conventions, though technically these are simply intergovernmental negotiations that are convened by the IMO, rather than IMO regulations per se. The IMO serves as the secretariat for such instruments and currently oversees more than 50 international agreements and protocols pertaining to ships, plus amendments. The regulation of marine pollution has become one of the most important aspects of IMO governance and this happens primarily through legally-binding agreements.

Most IMO agreements have a double ratification threshold, in which a particular number of states accounting for a certain percentage of registered shipping tonnage must ratify an agreement for it to enter into force. For example, the International Convention for the Prevention of Pollution from Ships (IMO 2015d MARPOL 73/78) required ratification by a minimum of 15 states that collectively accounted for at least 50% of the gross tonnage of the world's registered merchant shipping fleet. Even though members need not adopt IMO-negotiated agreements, there is a high degree of participation in IMO conventions, including by the major shipping states. Nonetheless, there is unevenness in acceptance of international obligations or, more frequently, implementation of adopted agreements.

A major criticism of the IMO, particularly relevant to the science–policy interface, is its lack of transparency. Unlike most international organizations with a regulatory mandate, much of the information it gathers about its conventions and the behavior of its member states is not shared with the public, although committee documents are accessible. Despite the broader international trend for increasing access to data and information from international organizations, the IMO is one of only a very small number of organizations that has decreased public access to information rather than increasing it. The privacy demanded by commercial interests in shipping probably contributes to the closed nature of the organization, but it does not fully explain this trend, which severely hinders academic analysis of the science–policy interface or of the negotiations or effects of IMO rules. This approach also seriously hinders public engagement on the issues, which is often a pathway to bring scientific concerns to broader attention.

The IMO has, nevertheless, had some important successes. An external review of the organization conducted in 2001 pointed to successes in passing increasingly strict regulations in the wake of major accidents, the ability of the organization to adapt to an increasing international focus on environmental issues, and the decreasing amount of oil pollution from tankers due both to accidents and other causes (Campbell et al. 2001; GESAMP 2007).

13.4 Port State Control

The rise of flags of convenience made efforts by the IMO or states acting regionally to protect the oceans from ship-based environmental degradation much more difficult. Collective action among states to protect a common-pool resource is difficult enough (Barkin and Shambaugh 1999); it is much harder for states to agree to (often initially costly) environmental restrictions to achieve collective benefit if others are likely to remain apart from those agreements. The resulting measures bear less benefit for those who have sacrificed just as much to bring them about, knowing that the possible outcome makes them less willing to undertake the measures in the first place. The increase in flag of convenience registration initially dramatically lowered collective levels of environmental (and safety and labor) protections on ships globally, as states offering these flags remained outside of the regulatory system and an increasing number of ships registered in these locations (DeSombre 2006).

States that had adopted higher levels of regulation, however, were reluctant to let others free ride on their action. They were concerned not only about the decreased benefit from collective action, but also about the additional relative cost they would have to bear if some ships did not apply these standards. Those states implemented what has come to be known as port state control (PSC). This is not a new concept: it simply involves officials from a port state boarding, inspecting, and, if necessary, detaining a foreign-flagged vessel in its port. The broad justification for these actions can be found in the United Nations Convention on the Law of the Sea, which lays out the ability of states to "establish particular requirements for the prevention, reduction, and control of pollution of the marine environment as a condition for the entry of vessels into their ports" (UNCLOS 1982, article 211) and requires that flag states hold their ships responsible for providing information to port states that require it for these purposes.

This individual right allows a state to mandate and enforce its own standards. No state acting alone is likely to have a major influence on collective standards. States acting together, however, can do just that. At the regional level, states created a set of memoranda of understanding (MOUs) that coordinate both the standards expected of ships in ports and an inspections process to check that they are being upheld. The MOUs make no new laws pertaining to ships; they refer to existing international agreements on safety and environmental protection that ships must uphold, and they do so largely by incorporating these obligations into domestic law. In the existing MOUs, port state authorities agree to inspect some percentage of ships that enter their ports during the course of a year. As a result of the inspection process, a ship can be found to be "clean" (to pass with no problems), can have some number of recorded deficiencies, and, if there are enough deficiencies or they are serious enough, can be detained in port until the most egregious ones are corrected (Hare 1997).

One of the most important aspects of the PSC system is that it explicitly endorses discrimination based on the state in which a vessel is flagged. Because not all ships can be inspected, inspectors hope to examine those most likely to be problematic. The PSC systems thus keep records of overall detention rates for all inspected ships. Flag states whose ships exceed these average rates are singled out for more frequent inspections. Ships registered in flag states that have not adopted the major international agreements relevant to the standards covered in inspections are also inspected more frequently (DeSombre 2006, pp. 93–94).

Flag of convenience registries use their low environmental, safety, and labor requirements for ships as a way to keep shipping costs low and thus make their registries more attractive to shipowners with many choices about where to register (DeSombre 2006). But that advantage disappears if ships are singled out for inspection based on being flagged in locations with poor reputations or a history of failing inspections. As Julio Sosa, the Panamanian Maritime Consul in Houston explained, "No one wants to be in a flag where the coast guard is going to be fingering you all the time" (Morris 1996, p. 15). Not only do inspected ships face the possibility of being detained for being in poor condition, but the mere process of inspection is time-consuming and thus costly in an industry where turnaround time is key to profitability.

Since the 1980s, this nascent inspection process has had a major effect on ship registration and oil tanker safety. By the 1960s, Liberia had become the largest ship registry, with oil tankers making up a high percentage of registered ships. It had also gained a reputation as a problematic registry for tankers: its ships were older, not well maintained, and more prone than average to accidents. Several ocean disasters involving Liberian-registered oil tankers in the late 1960s and early 1970s led to the demand for a PSC-type inspection regime (and also contributed to the creation of MARPOL).

Owners of Liberian-registered oil tankers, especially Americans, were concerned that these new inspection processes could cause difficulties and perhaps even decrease the advantage of flagging outside of the United States if it meant increased scrutiny for Liberian-flagged vessels. These owners, therefore, persuaded the Liberian government to ratify a number of existing safety and environmental treaties pertaining to shipping (which it did in 1980 and 1981), and persuaded the Liberian registry to implement its own inspections system, to increase the likelihood that any Liberian vessels that were inspected would pass. Liberia agreed to take on these additional regulations because it feared the loss of revenue from ship registrations if many vessels decided to flag elsewhere for fear of being singled out. Shipowners were also supportive because although it would cost more to meet the newly adopted standards, those costs were still outweighed by the tax benefits of continuing to flag in Liberia (compared to standard registries). Those shipowners who already met the standards in question benefited competitively even more (Carlisle 1981, pp. 185–186).

The IMO and its regulatory process translate science into policy action. But the IMO will fail to fully address the problems it takes on unless all ships actually implement these standards, whether or not all flag states have legally adopted and enforced them. It is through the PSC process that many of the environmental standards for ships are propagated through a large percentage of the ocean-going vessels that engage in international trade.

13.5 Environmental Issues

Many environmental problems are created by ships, ranging from overfishing and related ecosystem destruction (addressed in other chapters in this volume and so not covered further here) to intentional or accidental discharge of a variety of pollutants into the air or water during transit or shipping, to ecosystem disruption caused by movement of invasive species. This section covers the main types of discharges from ships that have been identified scientifically and addressed within the policy process; other subcategories, for example, sewage waste and chemical operational waste, are omitted for space reasons. The science–policy interface has varied depending on the issue, the players, and the time at which the problem first emerged.

13.5.1 Accidental Oil Pollution

Oil is transported globally on the oceans, in increasingly large tankers. The dangers of traversing the ocean, with possible storms or unexpected conditions, leaves open the possibility of ship damage or loss and resulting oil spills on the open ocean or nearby coasts.

In addition to the series of IMO operational measures that are intended to decrease the likelihood of accidents through the adoption of operational safety measures, the primary regulatory measure adopted has been the requirement for tankers to be constructed with double hulls. The idea is that if a tanker hull is breached, it has a much smaller chance of leading to a catastrophic oil spill, as the tanker has essentially a second layer of ship between the oil and the ocean.

There was major disagreement about the efforts to require double hulls for tankers. The United States lobbied hard for this requirement to be included initially in MARPOL (both in the original negotiation of the convention in 1973 and in the negotiations for the protocol added in 1978, which became the version of the convention that ultimately entered into force), but other states were unwilling to accept this provision as mandatory (Karim 2015, p. 47). The *Exxon Valdez* oil spill, in which more than 11 million gallons (4.1×10^7 L) of oil were spilled into the ocean near Alaska in 1989, however, redoubled American resolve and prompted unilateral action. The United States adopted the Oil Pollution Act of 1990, which required all US-flagged tankers, as well as all tankers that passed through American waters or entered US ports, to be double-hulled by 2015, whether new or retrofitted (Tan 2006, p. 140).

This unilateral action, applied in a way that affected ships from many registries, increased the willingness of the IMO to adopt such standards itself in the 1992 amendments to MARPOL. These measures required that double hulls be installed on newly built tankers and required the eventual retrofit of pre-MARPOL tankers within 25 years of their delivery date. Tankers previously built to MARPOL standards needed to be retrofitted with double hulls by the time they were 30 years old (Karim 2015, p. 48).

Later oil disasters also influenced movement on tanker standards intended to decrease the likelihood of spills. The *Erika* disaster, in which a single-hulled tanker spilled oil off the coast of France, motivated the 1999 amendments to MARPOL that moved the phaseout date of single-hulled tankers to 2015, with some older tankers required to be retrofitted earlier. Yet another oil disaster—that of the *Prestige,* which sank off the coast of Spain in 2002—accelerated the timeline further to 2010 (MEPC 2003). Demonstration of the fate and effects of oil spills clearly influenced the policy agenda.

There was actually some question about whether double hulls are, in fact, environmentally better than single hulls on tankers. Initially, the argument relied less on clear evidence and more on the logic that a second hull would necessarily reduce the likelihood of spills. Recent studies suggest that having a double hull does reduce the sizes of spills on average (Yip et al. 2011). At minimum, however, this standard would not have prevented some of the most dramatic disasters that motivated the changed legal rules. The *Prestige,* having literally split in half and sunk, could not have been saved by a double hull. Nonetheless, major changes in ship design and construction can be pointed to as clear signs of action, whereas operational changes may not be as visible. The double-hull requirement was thus perhaps useful, but at the time it was enacted, it was not scientifically supported as necessary to reduce the threat of oil spills.

Additional IMO agreements address aspects of oil pollution. Among these are provisions for coastal states to take emergency action "to prevent, mitigate, or eliminate danger to its coastline" when faced with the dangers of an oil spill nearby, and the requirement that states create emergency response plans for use in case of oil spills under the International Convention on Oil

Pollution Preparedness, Response and Co-operation (1990). Later agreements included the same measures for noxious and hazardous substances other than oil—for instance, the Protocol on Preparedness, Response and Co-operation to pollution Incidents by Hazardous and Noxious Substances (2000).

The oil spill provisions of MARPOL and related agreements have likely had a beneficial effect. The incidence of large oil spills from tankers decreased from an average of 25 per year in the 1970s to fewer than four per year currently, with the most dramatic decrease in the 1970s and 1980s when MARPOL provisions took effect (Psarros et al. 2011). Although it may have taken major oil spills to create and strengthen the rules, those rules—along with improvement in navigation aids—are now more likely to prevent that sort of disaster.

13.5.2 Intentional Oil Pollution

Although accidental pollution is what we most frequently think about when we think of oil pollution in the oceans, for quite some time, more of the oil in the water was discharged intentionally than accidentally by ships (Mitchell 1994, p. 70). This process of intentional oil pollution was part of how tankers operated, especially as they were working on increasingly narrow profit margins and had to maximize speed in loading and unloading their shipments of oil.

Oil is a heavy cargo, and vessels constructed to ship it have large tanks aboard in which to carry it. But once tankers have offloaded the oil in the location to which it is being shipped, they then have to travel back to the oil-producing location to pick up the next shipment. A ship that was safe to navigate with tanks full of oil will ride far too high in the water without the oil. These vessels would, therefore, take on water in their tanks for ballast. Tanks that had previously been filled with oil will instead be filled with sea water, providing the additional weight to balance the ship.

Unloading and reloading at port must be done as quickly as possible, thus, the ship would traditionally discharge the water in its tanks before arriving at its destination. Because the water had been transported in oil tanks, it would have oil mixed in when discharged. In other words, it is not that tanker operators intended to discharge oil, but that oil was necessarily a part of the water they discharged during standard operations.

From any one ship the amount of oil may seem insignificant, but in the context of the global ocean transport of oil, the amounts added up. Oil was estimated to constitute between 0.3% and 0.5% of a tank of ballast water discharged. While that might seem inconsequentially small, it could actually constitute 300–500 tonnes of oil per tanker voyage (Mitchell 1994). This intentional discharge constituted up to five million tonnes annually (Wardley-Smith 1983; Mitchell 1994). The public experienced this type of pollution through oil patches and tar balls washing up on beaches (United Kingdom Ministry of Transport 1953).

Initially, in addition to rules prohibiting discharges close to land (Mitchell 1994), this problem had been addressed internationally via regulations that required ships to limit the percentage of oil that could be discharged with ballast. Eventually the set of agreements landed on the need to avoid creating a "visible trace" of oil pollution (Mitchell 1994, p. 95). In other words, they were allowed to discharge ballast water that had been stored in oil tanks so long as the water discharged did not include oil, determined by a visual inspection. Most of the water discharged from a tank would not have noticeable oil; it was only at the end of the discharge process that greater amounts of oil would be mixed in.

But these approaches ran into several aspects of uncertainty. At the level of individual behavior, it was difficult for ship operators to determine the point before which their discharges would create a visible trace of oil. Their interest was in discharging as much of their ballast water before entering a port (the less water that remained, the more quickly they could re-fill the tank with oil), so they did not want to stop the discharge any sooner than they needed to. But the only way to know where the end point of safely discharging without oil came was to pass it. In other words, ship operators were deciding when to stop the discharge based on when they first saw oil emerge, which led to continuous discharges of oil, even if they were smaller in amount than had previously been the case.

The second uncertainty came from knowing, when an oil trace was found in the ocean, which ship it came from. One of the difficulties of the ship-based behavior rules was that there were hundreds of tankers taking trips across the ocean at any given time. In the first place, that means that there are many opportunities to not comply with the regulations. Even a ship that generally follows the rules may (intentionally or accidentally) not do so on one or more of the trips that it takes. It is extremely difficult to know at any one time if a rule that requires keeping track of many ships on many trips is being followed, especially in an area as vast at the global oceans. Similarly, when an oil trace was found (indicating that a ship had indeed violated the rules, assuming it was bound by them), the number of ships and the vastness of the oceans meant that it was nearly impossible to trace that evidence back to an individual ship that might have caused the problem.

The new rules, negotiated under MARPOL 1973/78, focused instead on equipment-based approaches to address the problem of intentional oil pollution. The primary effort was to require that ships be built with segregated ballast tanks (SBT) so that the water taken on for ballast could not be carried in the same tank that had carried the oil. In addition to providing an equipment-based way to ensure that ships would know that they were not discharging water that contained oily residue, it also made information gathering about ship behavior much easier. Instead of determining which ships may or may not have discharged oil and on which trips, ships could be inspected and certified at the point of construction; if they had the requisite equipment, they could be assumed to not discharge oil. This agreement has fundamentally changed the way ships are built and has dramatically decreased the extent of oil pollution (Mitchell 1994).

MARPOL was originally negotiated in 1973 but failed to achieve sufficient ratification to enter into force quickly; negotiation of a protocol in 1978 was what allowed the combined agreement to become legally binding. The same set of oil disasters that motivated US unilateral action on double hulls also persuaded the United States to indicate that it would set unilateral SBT requirements (to be imposed on all ships in U.S. waters and ports) if the IMO did not take decisive action to implement strong equipment-based rules to prevent intentional oil discharges, which were written into the revised agreement (Mitchell 1994).

One final point worth noting in the discussion of oil pollution of the oceans more generally is that although science certainly played a role in the policy approaches adopted, scientific viewpoints on the effects of oil pollution on ocean ecosystems are mixed. Expert scientific panels reviewing the problem have concluded that oil pollution does cause serious problems, primarily in the short term (i.e., less than 5–10 years) and on coastlines (Camphuysen 1989; National Research Council 1985, 2003; GESAMP 1993, 2007). The early conclusion that oil spilled in the open ocean, that is, offshore, does not have a long-term permanent effect on marine ecosystems is largely upheld (GESAMP 1990), although population effects on pelagic sea birds and other animals of the open sea remain unclear. Where oil is spilled in areas of abundant natural seeps, such as in the 2010 Deep Water Horizon blowout in the Gulf of Mexico, conclusions about long-term effects have yet to be reached (Cornwall 2015). Ultimately, action to prevent oil pollution from ships has been driven by public perception, even more than scientific facts, especially in the wake of visible oil disasters or other signs of oil pollution in coastal areas.

13.5.3 Air Pollution

For the most part, the focus on pollution from ships has been on emissions that harmed the water, but ship-based operations can also affect the air. Because ships burn fossil fuels (various weights of oil), they have similar air pollution effects to any fossil fuel–burning operation, resulting in particulate pollution, sulfur oxides (SO_x), and nitrogen oxides (NO_x) emissions that contribute to acid rain, and carbon dioxide (CO_2) emissions that contribute to climate change. The fuel oil most frequently used by ships historically had a high sulfur content, leading to higher SO_x emissions than would have been the case with different fuel oil.

Concern about emissions of these substances on land has a long history. Long before anyone considered addressing emissions from ships, the basic scientific understanding had been worked out. That is not inconsequential; for an issue like acid rain, environmental damage was present before we understood the long distances that acidifying substances could travel and the (often nonlinear) route by which effects transpired. Because the issue had been initially addressed in a land-based context, considerable uncertainty had been resolved by the time focus turned to emissions from ships. Concern about air pollution from ships, however, was initially limited to emissions close enough to affect land.

Although discussions about air pollution began at the IMO in the 1980s, on the heels of efforts in Europe to address acid rain, it did not adopt rules under MARPOL to address the issue until 1997. It did so via a new annex to MARPOL, Annex VI, which entered into force in May 2005 and limits air pollution (sulfur oxide, nitrogen oxide, and ozone depleting substances) from ships (IMO 2015e). The rules are weak, however. They initially required fuel oil used on ships to have no more than 4.5% sulfur content (1997 Protocol to Amend the Convention for the Prevention of Pollution by Ships 1973/1978, Regulation 14), which was no lower than the average sulfur content of fuel oil used on ships at the time (IMO 1997; Karim 2015). Oil exporting countries and oil industry groups lobbied against measures that would have reduced the sulfur content of marine fuels by any greater amount (Tan 2006).

Despite this pressure, subsequent amendments to Annex VI, Regulation 14, mandated increasingly stringent rules: beginning in 2012, sulfur content could be no more than 3.5%, and by 2020 it may be no more than 0.5% (IMO 2015g). In addition, in a compromise for those who wanted more stringent rules, states were allowed to create SO_x emission control areas. Ships in these areas are required to have a lower sulfur content of fuel, an exhaust gas cleaning system, or some other technological approach that limits the SO_x emissions to an equivalent amount (IMO 2015a). Europe and the United States, which had hoped for stronger measures under MARPOL, have instead worked to impose domestic requirements limiting oil to a much lower sulfur content and to impose those rules on ships visiting their ports, so there were ways around the unwillingness to impose stricter fuel-content regulations. Similar approaches were taken to prevent NO_x emissions from fuel oil.

More recently, focus has shifted to the climate change implications of ship operations. Managing climate change globally has been far from a simple task, and whole books have been written on the science–policy interface on that issue generally. For the most part, the broader climate change discussions initially did not focus on emissions from ships, and fuel for ships was even excluded from calculations of national greenhouse gas emissions (United Nations Framework Convention on Climate Change Secretariat 2014). It was expected that the IMO would take up this issue (Kyoto Protocol to the United Nations Framework Convention on Climate Change, Article 2(2), 1997).

The IMO Assembly adopted a resolution in 2003 directing MEPC to determine what would be necessary for the adoption of greenhouse gas emission reduction measures (IMO 2003). An internal IMO study in 2009 concluded that the maritime sector was responsible for 3.3% of global greenhouse gas emissions in 2007, with most emissions resulting from international shipping.

Emissions were expected to grow dramatically (IMO 2009), although the global economic downturn and resulting decrease in shipping trade that followed reduced maritime greenhouse gas emissions in the years immediately after that study (IMO 2015c).

Oil-producing states (led by Kuwait and Saudi Arabia) and large developing countries (including India, China, Brazil, and Chile) strongly resisted taking meaningful action on these issues (Karim 2015). They were concerned more about the extent to which the obligations would have a negative effect on their profits than about the issue of uncertainty. Their opposition delayed and weakened the resulting policy decisions, and they voted against the measures. India spoke against the measures but was ineligible to vote because it was not a party to Annex VI of MARPOL. Other developing countries, including St. Vincent and the Grenadines—a major flag of convenience—abstained (Karim 2015).

The 2009 IMO energy study concluded that major greenhouse gas emission reductions were possible from operational and design measures taken to increase energy efficiency. Following extensive negotiations within the MEPC, the IMO adopted further amendments to Annex VI of MARPOL in 2011, which require energy efficiency measures on all new ships and an energy management plan for all ships of a certain size (IMO 2015a). To ensure that ships are indeed following the new policies, this amendment created a certification process. It is notable that these regulations are framed in terms of efficiency, which has a plausible cost advantage even if there are upfront costs to ships for the measures taken to achieve it. The IMO predicts significant savings in fuel costs because of the efficiency measures (IMO 2011). Straight emission reduction obligations were not seriously considered. Developing states were nevertheless concerned that the principle of common but differentiated responsibilities, put forward in other international negotiations about greenhouse gas emissions, was not applied in this context (Karim 2015, p. 109).

Although some states are likely to refuse to adopt these amendments to the regulations, the port state control measures discussed above are likely to be a powerful tool to increase ship adherence to them. If port states mandate the use of these efficiency measures for ships allowed to enter their ports—especially if, as has generally been the pattern, major trading states participate in the agreement—then ships that intend to travel to those locations will be forced to adhere to the measures regardless of whether their flag states require it.

13.5.4 Invasive Species Transport

Ocean-going vessels have unintentionally transported species from one habitat to another since they began to move across the seas. Initially, the damage done by this process was to land ecosystems. For example, goats transported to the Galapagos Islands by early whalers or pirates multiplied on the islands in the absence of natural predators and affected the local eco-systems, causing the extinction of native species (Carrion et al. 2011).

The same use of ballast water that was rendered less problematic in terms of oil content by earlier MARPOL regulations still involved moving water from one part of the ocean to another, along with species small enough to be taken up. Ballast water can be taken up when any heavy cargo has been offloaded, so it is not unique to oil transport. When these species are deposited along with the water discharged before the ship enters a new port, many of them do not survive in the new environment, but some that do survive can thrive where their natural predators may be absent. In some cases, organisms transported with ballast or bilge water may also be threats to human health (McCarthy and Khabat 1994).

The United States and Canada first brought the issue of ballast water transferring invasive aquatic species before the IMO in the 1980s because of concern about foreign species being introduced into the Great Lakes. More recently, the issue has been driven by Australia's concern for native species in its coastal waters, which are threatened by species from Asian waters, where shipping to Australia frequently originates (Hayes and Sliwa 2003). Scientific studies sampled ballast water and determined that large numbers of species could survive in ballast water. One meta-study determined that more than 1000 species are transported in ballast water (Gollasch et al. 2002).

One way to reduce the likelihood of invasive species from ballast water is to exchange ballast water closer to the place it was picked up; ships have been encouraged to take up new ballast water repeatedly as they transit the ocean. In particular, the general scientific understanding has been that nearshore organisms where ballast water is generally taken on initially, are unlikely to survive in the deep ocean, so replacing ballast water further into the voyage can decrease the risk of species contamination (Tan 2006). However, doing so is operationally risky: when ships offload ballast they are less stable and more prone to accident. Similarly, exchanging ballast water takes time, up to several days, depending on the size of the ship (Gollasch et al. 2007), and thus entails cost that ships are unwilling to take on voluntarily.

In the wake of this concern, IMO first adopted operational guidelines in 1993. These have been updated and expanded multiple times since then, including guidelines for how to study ballast water species (Gollasch et al. 2007), culminating in the adoption of the 2004 International Convention for the Control and Management of Ship's Ballast Water and Sediments, known informally as the BWM Convention. The Convention requires all ships to have a ballast water management plan approved by its flag state government, as well as a record book to record its ballast water actions.

In particular, the BWM Convention adopts two approaches to ballast water management. The exchange standard requires ships to exchange more than 95% of their ballast water at least 200 miles away from land and in water at least 200 m deep, with some additional provisions for trips in which these distances or depths are not possible. The performance standard requires specified concentration limits for organisms in water that must not be exceeded in ballast water discharged close to land. The size of the ship and its construction date determine the date by which either of these standards must be implemented, with full implementation required by 2016.

These measures, particularly the ballast water exchange, are understood to be imperfect compromises between what is environmentally desirable and what is practically achievable. Exchange of even 95% of ballast water has been shown not to reduce organisms by that amount. In some cases, an increase in organisms may even be detected after exchange (McCollin et al. 2001). In addition, given that on some shipping routes ships may never be sufficiently far from land to fully meet the highest exchange standards, ballast water exchange will still result in the movement of organisms from one place to another.

The performance standards also involved a compromise based on political acceptability rather than scientific assessment of the most useful standards. In fact, few delegations to the negotiations of the convention contained sufficient scientific expertise to be able to have a scientific discussion of the appropriate level. Nevertheless, the standards agreed to are likely to achieve a greater reduction than would be achieved by ballast water exchange requirements alone (Gollasch et al. 2007). Even a dramatic reduction in the number organisms per cubic meter of ballast water discharged near land, however, will still result in some movement of organisms and thus some risk of species invasion.

Actual treatment of ballast water on board before it is discharged would decrease further the risk of invasive species transport. Although this option is foreseen by the convention, it is not required under it, other than to allow ships to participate in programs by their flag states to test ballast water treatment in lieu of other ballast water measures. The convention also includes a process of review within its operational requirements to examine the effectiveness of existing measures and the possibility of new requirements.

As of this writing, the convention has not yet entered into force. Shipping companies strongly oppose mandatory measures that are costly, risky, and give them no operational advantage, and so have attempted to persuade states not to ratify the agreement. In the interim, many states are imposing convention-level standards on their own ships or in their waters, but ships are able to avoid implementing them depending on registry state and shipping route, creating a global double standard on ballast water management (Karim 2015).

13.6 Conclusion

Ships operate far from shore in great numbers on a vast ocean. It can be difficult to track exactly how they are operating on any voyage or the cumulative effects of their operations. Even when a collective understanding can be reached of the environmental effects from the operations of ships, changing this behavior is costly and requires widespread international agreement.

The trickiness of this political compromise, especially when the realm of actors involved is expanded by the role of flags of convenience states whose primary goal is to keep shipping costs low, means that science frequently takes a backseat to political calculations. It is, nevertheless, the case that major progress has been made: operational and accidental discharge of oil from ships has been reduced, air pollution limits are starting to have a positive effect, and strong efforts to decrease the risk of transport of organisms in ballast water are underway.

Several cross-issue conclusions can be drawn about the science–policy interface on these issues. First, uncertainty about behavior made an important case for ship-related standards that are equipment based rather than behavior based. In some cases, operational rules, if followed, could possibly have a stronger effect than equipment changes, but the uncertainty about whether all ships would follow procedures on all trips argued strongly in favor of changing ship construction to remove that uncertainty. Similarly, public demand for action, especially to prevent oil spills, can be more clearly satisfied by pointing to changes in ship design than by discussing the value of operational procedures.

Second, it is worth noting the role of disasters or other high-profile and visible problems from ship pollution in garnering political action. Oil spills in the 1970s provided the impetus for the creation of the initial rules under MARPOL for ship-related improvements to prevent them, and in the late 1980s and early 1990s for shifting the timeline for requiring these changes. Other visible evidence of environmental damage, such as tar balls on Mediterranean beaches or invasions of jellyfish or zebra mussels, galvanized public demand for action, whether (in the case of invasive species) or not (in the case of operational oil pollution) science was unified in its concern about the issue.

Finally, although the IMO faces constraints from its member states and the shipping industry in making publicly available the information it generates or works with, it has done a reasonably good job of bringing scientific expertise to bear in the negotiation, operation, and revision of agreements designed to reduce the environmental effects of shipping. That is all the more remarkable given the percentage of ships registered in locations that allow them to skirt some international regulations. Action taken by high-standard states to apply their domestic rules to any ships that transit their waters or visit their ports helps expand the application of newly-negotiated measures to protect the ocean environment from the operations of ships. This chapter shows that while scientific information is important, it is only one factor in decision-making on shipping issues in international waters as well as in coastal areas.

References

Administrative Secretary of GESAMP. 2005. The new GESAMP: Science for sustainable oceans: A strategic vision for the IMO/FAO/UNESCO-IOC/WMO/WHO/IAEA/UN/UNEP Joint Group of Experts on the Scientific Aspects of Marine Environmental Protection. London: IMO.

Barkin, J. S. and G. E. Shambaugh, eds. 1999. *Anarchy and the Environment: The International Relations of Common Pool Resources*. Albany: SUNY Press.

Boczek, B. A. 1962. *Flags of Convenience: An International Legal Study*. Cambridge, MA: Harvard University Press.

Campbell, P., J. Hushagen, and D. Sinha. 2001. Challenges, opportunities and evolution: Review of the Secretariat of the International Maritime Organization, MAMNET, Switzerland, 26 March 2001. London: IMO.

Camphuysen, C. J. 1989. *Beached Bird Surveys in the Netherlands, 1915–1988: Seabird Mortality in the Southern North Sea Since the Early Days of Oil Pollution*. Amsterdam: Werkgroep Noordzee.

Carlisle, R. 1981. *Sovereignty for Sale*. Annapolis, MD: United States Naval Academy Press.

Carrion, V., C. J. Donlan, K. J. Campbell, et al. 2011. Archipelago-wide island restoration in the Galápagos islands: Reducing costs of invasive mammal eradication programs and reinvasion risk. *PLoS One* 6 (5): e18835 (1–7).

Cornwall, W. (2015). Deepwater horizon. After the oil. Five years on, the world's largest accidental marine spill has left subtle scars on the Gulf of Mexico. *Science* 348 (6230): 22–29.

DeSombre, E. R. 2005. Fishing under flags of convenience: Using market power to increase participation in international regulation. *Global Environmental Politics* 5 (4): 73–94.

DeSombre, E. R. 2006. *Flagging Standards: Globalization and Environmental, Safety, and Labor Regulations at Sea*. Cambridge, MA: MIT Press.

George, R. 2013. *Ninety Percent of Everything*. New York: Picador.

GESAMP (Joint Group of Experts on the Scientific Aspects of Marine Environmental Protection). 1990. The state of the marine environment. GESAMP Reports and Studies No. 39. New York: United Nations.

GESAMP (Joint Group of Experts on the Scientific Aspects of Marine Environmental Protection). 1993. Impact of oil and related chemicals on the marine environment. Reports and Studies No. 50. London: IMO.

GESAMP (Joint Group of Experts on the Scientific Aspects of Marine Environmental Protection). 2007. Estimates of oil entering the marine environment from sea-based activities. Reports and Studies No. 75. London: IMO.

Gollasch, S., M. David, M. Voigt, et al. 2007. Critical review of the IMO international convention on the management of ships' ballast water and sediments. *Harmful Algae* 6 (4) (2007): 585–600.

Gollasch, S., E. Macdonald, S. Belson, et al. 2002. Life in ballast tanks. In *Invasive Aquatic Species of Europe: Distribution, Impacts and Management*, edited by E. Leppäkoski, S. Gollasch, and S. Olenin, 217–231. Dordrecht: Kluwer Academic.

Hare, J. 1997. Port state control: Strong medicine to cure a sick industry. *Georgia Journal of International and Comparative Law* 26: 571–594.

Hayes, K. R. and C. Sliwa. 2003. Identifying potential marine pests—A deductive approach applied to Australia. *Marine Pollution Bulletin* 46 (1): 91–98.

IMO (International Maritime Organization). 2003. Resolution on IMO policies and practices related to the reduction of greenhouse gas emissions from ships. IMO Doc. A. 963 (23).

IMO (International Maritime Organization). 2009. Second IMO GHG study. IMO Doc. MEPC 59/4/7 (9 April).

IMO (International Maritime Organization). 2011. Main events in IMO's work on limitation and reduction of greenhouse gas emissions from international shipping. Accessed 7 March 2015. http://www.imo.org/MediaCentre/resources/Documents/Main%20events%20IMO%20GHG%20work%20-%20October%20 2011%20final_1.pdf.

IMO (International Maritime Organization). 2013. Rules and guidelines for consultative status of non-governmental international organizations with the International Maritime Organization. Accessed 6 March 2015. http://www.imo.org/About/Membership/Documents/RULES%20AND%20GUIDELINES%20 FOR%20CONSULTATIVE%20STATUS.pdf.

IMO (International Maritime Organization). 2015a. Air pollution, energy efficiency and greenhouse gas emissions. Accessed 24 September 2015. http://www.imo.org/en/OurWork/Environment/PollutionPrevention/AirPollution/Pages/Default.aspx.

IMO (International Maritime Organization). 2015b. Brief history of IMO. Accessed 6 March 2015. http://www.imo.org/en/About/HistoryOfIMO/Pages/Default.aspx.

IMO (International Maritime Organization). 2015c. Greenhouse gas emissions. Accessed 6 March 2015. http://www.imo.org/ourwork/environment/pollutionprevention/airpollution/pages/ghg-emissions.aspx.

IMO (International Maritime Organization). 2015d. International convention for the prevention of pollution from ships (MARPOL). Accessed 19 September 2015. http://www.imo.org/en/About/Conventions/ListOfConventions/Pages/International-Convention-for-the-Prevention-of-Pollution-from-Ships-%28MARPOL%29.aspx.

IMO (International Maritime Organization). 2015e. NGOs in consultative status. Accessed 6 July 2015. http://www.imo.org/en/About/Membership/Pages/NGOsInConsultativeStatus.aspx.

IMO (International Maritime Organization). 2015f. Structure of IMO. Accessed 6 March 2015. http://www.imo.org/About/Pages/Structure.aspx#2.

IMO (International Maritime Organization). 2015g. Sulphur Oxides (SOx)—Regulation 14. Accessed 24 September 2015. http://www.imo.org/en/OurWork/Environment/PollutionPrevention/AirPollution/Pages/Sulphur-oxides-(SOx)-%E2%80%93-Regulation-14.aspx.

IMO Maritime Knowledge Centre. 2012. *International Shipping Facts and Figures—Information Resources on Trade, Safety, Security, Environment*. London: IMO.

International Convention for the Control and Management of Ship's Ballast Water and Sediments. 13 February 2004. IMO Doc BWM/CONF/36, 16 February 2004.

International Convention on Oil Pollution Preparedness, Response and Co-operation. 30 November 1990. 1891 UNTS 51. As amended by the Protocol on Preparedness, Response and Co-operation to Pollution Incidents by Hazardous and Noxious Substances. 15 March 2000. Australian Treaties Library [2003] ATNIF 9.

Karim, M. S. 2015 *Prevention of Pollution of the Marine Environment from Vessels: The Potential and Limits of the International Maritime Organisation.* Dordrecht: Springer.

Knudsen, O. 1973. *The Politics of International Shipping.* Lexington, MA: Lexington Books.

Kyoto Protocol to the United Nations Framework Convention on Climate Change. 11 December 1997. UN Doc. No. FCCC/CP/1997/L.7/Add.1 (10 December 1997).

MARPOL (International Convention for the Prevention of Pollution from Ships). 2 November 1973. 1340 UNTS 184. As amended by the protocol of 1978, 17 February 1978, 1340 UNTS 61 and the Protocol of 1997, 26 September 1997, Can TS 2010 no 14.

McCarthy, S. A. and F. M. Khambaty. 1994. International dissemination of epidemic *Vibrio cholerae* by cargo ship ballast and other nonpotable waters. *Applied and Environmental Microbiology* 60 (7): 2597–2601.

McCollin, T., E. M. Macdonald, J. Dunn, et al. 2001. Investigations into ballast water exchange in European regional seas. In *Proceedings of the Second International Conference on Marine Bioinvasions,* 94–95. New Orleans, 9–11 April 2001.

MEPC (Marine Environment Protection Committee), International Maritime Organization. 2003. Annex 1: Resolution MEPC.111 (50). Report of the Marine Environment Protection Committee on its Fiftieth Session. IMO Doc. MEPC 50/3, (8 December).

M'Gonigle, M. R. and M. W. Zacher. 1979. *Pollution, Politics, and International Law.* Berkeley: University of California Press.

Mitchell, R. B. 1994. *Intentional Oil Pollution at Sea.* Cambridge, MA: MIT Press.

Morris, J. 1996. Lost at sea. *Houston Chronicle,* 22 August, 15.

National Research Council. 1985. *Oil in the Sea. Inputs, Fates and Effects.* Washington, DC: National Academy Press.

National Research Council. 2003. *Oil in the Sea III: Inputs, Fates, and Effects.* Washington, DC: National Academies Press.

Oil Pollution Act. 1990. Pub L No 101–380, 104 Stat 484 (codified as amended at 33 USC ch 40 §§ 2701–2762).

Psarros, G., R. Skjong, and E. Vanem. 2011. Risk acceptance criterion or tanker oil spill risk reduction measures. *Marine Pollution Bulletin* 62 (1): 116–127.

Tan, A. K-J. 2006. *Vessel-Source Marine Pollution: The Law and Politics of International Regulation.* Cambridge: Cambridge University Press.

UNCLOS (United Nations Convention on the Law of the Sea). 10 December 1982, 1833 UNTS 3.

United Kingdom Ministry of Transport. 1953. Report of the committee on the prevention of pollution of the sea by oil. London: Her Majesty's Stationery Office.

United Nations Framework Convention on Climate Change Secretariat. 2014. Emissions from fuel used for international aviation and maritime transport (international bunker fuels). Accessed 6 March 2015. http://unfccc.int/methods/emissions_from_intl_transport/items/1057.php.

Wardley-Smith, J., ed. 1983. *The Control of Oil Pollution.* London: Graham and Trotman, Publishers.

Winchester, N. and T. Alderton. 2003. *Flag State Audit 2003.* Cardiff: Seafarers International Research Centre.

Yip, T. L., W. K. Talley, and D. Jin. 2011. The effectiveness of double hulls in reducing vessel-accident oil spillage. *Marine Pollution Bulletin* 62 (11): 2427–2432.

14

Just Evidence: Opening Health Knowledge to a Parliament of Evidence

Janice E. Graham and Mavis Jones

CONTENTS

14.1 Introduction

The chapters in this volume are prefaced by a common understanding that the health of our oceans matters. The collection provides rich accounts dealing with how scientific information is used to build a research base and collaboratory networks to exchange, manage, signal risk, influence, and govern policy- and decision-making. For these authors, water matters in an ecological sense, in the same way that other components of the environment, that is, land and air, matter. Individually and collectively, they constitute the "one health" we all share and should not take for granted (One Health Global Network 2015; One Health Initiative n.d.; Centers for Disease Control and Prevention 2013; Public Health Agency of Canada 2015). A one health approach disrupts arguments that reduce the environment to health, health to environment (Burger 1990), or human behavior to some purported set of rational acts of self-interest, for example, *Homo economicus*. Taking into account ecosystems and social networks, for example, one health reorients policy to accommodate both human and nonhuman indicators of health (Rabinowitz and Conti 2012). It opens a space to consider shared reciprocal relations in the exchange of goods that are more complex than that explained by an anthropocentric rational market model (Sahlins 1972; Maurice 1999; Graham and Bassett 2006). Situated between these points of reference, between private corporatist strategies for profit and natural (presumably, though contestably public) resources, government regulatory policies and practices across all sectors are intended to safeguard citizens against undue harm.

In this chapter, we present a complementary perspective to the oceanic theme of this volume by ethnographically engaging the circulation of scientific knowledge and evidence in a different but comparable policy decision-making environment. We will use two case studies in *health regulation* that explore (1) national regulatory practices and policies for emerging health products, and (2) global vaccine development and implementation platforms. These case studies and the conclusions that we draw illustrate the role(s) that information plays in decision-making processes at the science–policy interface in regulatory contexts; this parallels the observations and conclusions of the authors of other chapters in this book.

The determination of evidence deemed valuable along the health regulatory pipeline for emerging pharmaceuticals and biologics (including vaccines) is based on a range of explicit and tacit knowledge. Regulatory science, as in all science platforms, relies on the construction of standards, instruments, and guidelines that order certain types of evidence, exclude other types, and shape our lives (Collins and Evans 2002; Wynne 1996; Lampland and Star 2008; Bijker et al. 2009). Institutions define and determine expertise, evidence, and its interpretation; expert elites authorize what can and cannot be considered in order to balance technical, cultural, and political considerations. The policies and practices of the individuals and organizations that decide whose information counts, and what information is used, matter. However, the processes by which results are interpreted, and conclusions are made, remain obscure. Clinical trial study protocols, for example, are designed by drug developers who are interested in producing data that will result in approval of their products. As a result, industry studies, compared to trials with any other source of funding, are more likely to favor the sponsor's product. These biases, however, cannot always be explained by standard assessment tools, for example, randomization or blinding (Lundh et al. 2013). We know, for example, that financial conflicts of interest can sway opinion unconsciously (Kassirer 2007; Sismondo 2008). The importance of making primary data available for independent review cannot be lost on government scientific regulators. This was made evident in 2015 with the reanalysis of GlaxoSmithKline (GSK)'s paroxetine trial, which showed that the antidepressant was neither safe nor effective in adolescents (Le Noury et al. 2015).

Regulatory advice sought early by product sponsors improves marketing authorization success (Hofer et al. 2015), and has been encouraged in recent regulatory modernization policies. But how close should relationships be between the sponsors of products and the regulator? Both healthcare providers and the public trust that there is no conflict of interest and that the mechanisms and instruments of the regulatory process ensure that the drugs and vaccines provide more benefit than harm. The trustworthiness of this evidence requires building a framework for accountability (O'Neill 2014). What questions should we be asking to ensure credibility, legitimacy, and public trust in health regulation?

The determination of regulatory policy, the reach of regulatory activity, and the scientific and ethical competencies of regulators are central to the debate about the nature of a just society and the relative importance of public health issues. For most people, however, regulatory processes are obscure, unclear, and even unfathomable. Citizens do not often think about the safety and effectiveness of the products they consume. They assume government regulators do that for them, until there is a crisis. While emergency preparedness occupies more and more of national state and international multilateral agency activities, the role of good regulation is to cut crises off at the pass.

As an anthropologist of science, technology, and medicine, the first author's research (Graham) on health regulatory activities has taken her to the shores of Canada's three oceans as well as to the land-locked sub-Saharan African Sahel, where desertification and the recent drying trend from warming African waters are contributing to societal and health consequences. These range from respiratory infections irritated by the pervasive dust-carrying winds, to malnutrition from the agricultural crises brought on by scarce water resources (Giannini et al. 2003; van Eeckhout 2015). Humans have fairly predictable ways to address crises. When faced with environmental, social, economic, political, or health challenges, humans react. They respond. They move. They innovate. Incentivizing the tangible products of innovation has become a key objective of most governments and across several sectors.

Unfortunately, not all innovations improve health. Indeed, government's dual role, as both incentivizer of new health products and protector of the public health, puts it in a potential conflict of interest. Regulators fall prey to claims of regulatory capture when governments are seen to advance commercial or lobbyists, interests, while their agencies are mandated to act in the public interest (Carpenter 2013; Lexchin 2012). Showing that their products are novel, safe, and effective is the goal of the product sponsors. Ensuring these claims are true is the responsibility of our regulators. What happens in between is the correspondence at the interface of science and politics.

14.2 Weaving the Technical, Relational, and Political into a Parliament of Evidence-Based Knowledge

Innovation inspires; it drives humans beyond static *being* into dynamic *becoming*. It brings new solutions to old problems, new values to tired tenets, and opens up new markets, needs, and desires. Human evolution maps to the creation, replacement, and communication of new ideas and artifacts. We evolve with novel technologies that simultaneously change us culturally and biologically. The anthropologists Augustín Fuentes (2013) and Tim Ingold (2013a) suggest an intertwined, woven "correspondence" between biology and social relationships that places human *beings* perpetually in the process of *becoming* human. In this view, where we are not born but become, genetics and social identities "mix and mingle with one another in that zone of interpenetration we are used to calling the 'environment'" (Ingold 2013b, p. 16). Relationships with one another and with other things are formational to humanity and to most material achievements. Our relational accomplishments, for these must be acknowledged as innovations too, develop, perform, inform, and transform along intersecting social and technical pathways. We coalesce around new things and we make friends and enemies, colleagues, and competitors around ideas that change us. Biology, in this view, might be seen to be more complicated than genetics, acting on our genes and composed of complex synergistic epigenetics and behavioral and symbolic inheritance systems that can radically transform us. We are continuously *becoming* human in our interactions throughout our lifespan. We become, as Fuentes has said, "what we eat, who we meet, how we use our feet, and how we perceive the world" (Fuentes 2010).

People are won over by the enthusiasm surrounding new things. While we innovate to make living better, our best intentions can go awry; novel products build and sustain, but they can harm and destroy too. Whether innovations are used to feed our families, kill our enemies, clear an oil slick, or prevent, detect, treat, and manage disease and sickness, they unfold into unknown future ecologies as expectations at first and then material accomplishments or detriments that make up our individual and collective becoming. Inevitably, the products of biosocial relations are fraught with risk and uncertainty and with benefits and harms that can surprise even their developers.

It is the task of regulators to be on guard before a product is approved, and remain so afterwards for the identification, assessment, communication, and response to real risks in the world. The post-market world holds uncertainties that cannot be contained in the controlled clinical trials of the pre-license process. The adverse events identified in small clinical trials needed to attain a product's approval cannot foresee population effects brought on by adverse events, viral type replacement, declining immunogenicity, herd immunity, epidemics, climate change, tsunamis, droughts, crop failures, forced migrations, and relocations. Cascading unknowns can upset the fine balance upon which the original regulatory decisions were based. While synthetic pharmaceuticals, protein targeted radiopharmaceuticals, vaccines, and other biologics can make us more comfortable, prevent disease, and even cure us, they also disrupt.

How might it be possible to widen input into decision-making to include more diverse communities, a broad range of expertise beyond the regulatory scientists required to meet both government policies, for example, faster access, and rigorous critical scientific appraisal?

Isabelle Stengers (2005) proposes a cosmopolitical future that allows for the deliberative engagement of all "constituents" who share a common goal that implicitly involves social justice and generational equity. This vision aims to benefit more than harm. If that common goal is secured through improved health of our bodies, populations, and environments, then the avenue to that end must include full and open disclosure of all potential conflicts of interest and all research data, including untampered clinical study reports (Doshi et al. 2012; Doshi 2015; Jefferson et al. 2014). Different constituencies build different evidence bases and explanations for their interests, for what matters to them, and the kinds of facts they need to gather and manipulate in order to be convinced. If regulators have only partial access to data, or to only one or two sectors within a potential range of constituencies, the impartially of their decisions in applying the best of scientific rigor is left open to doubt. If knowledge and beliefs are constructed and communicated in the everyday practices of science and medicine, regulation, and markets, then tools need to be developed that open and make transparent all sources of data and study design, analytical interpretations, and regulatory decisions, to avoid the perception of conflict of interest, lack of transparency, and regulatory capture. What Gluckman and Allen have referred to in Chapter 10 (this volume) as "the balancing act of science in public policy," and what Sarkki et al. (2014) describe as "balancing credibility, relevance and legitimacy" might be developed into a decision-making framework that, drawing from the works of Bruno Latour (1993) and Isabelle Stengers (1997), would be a cosmopolitical parliament of drug evidence. Such a platform would involve open access to all data for independent analysis, and a transparent platform for engaged and reflexive deliberation and decision-making with mechanisms to prevent more powerful actors from influencing the process.

Multiple constituents would be included in openly determining the safety, effectiveness, and quality of a health product. Routes to follow-up studies that are relevant to constituents post-market would be made available through a lifecycle approach that can introduce and address new information from all communities.

14.3 Indication and Intellectual Property Creep

Clinicians often prescribe drugs developed originally for one medical indication—for instance, a biologic for non-Hodgkin's lymphoma—for a different condition, such as treating sufferers of rheumatoid arthritis. This introduces uncertain considerations surrounding safety and effectiveness. Does the product qualify as new? Can it then be privileged for extended patent protection? A synthetic drug said to be moderately effective for the treatment of people with Alzheimer's disease is prescribed for the "worried well" for mild memory loss (Graham 2008). Is there a problem with that? Health technologies are commonly prescribed for conditions other than their original intention, often on the fly, with no record of experimentation or clinical trials. Weapons were transformed into surrogate limbs when Afghani amputees adapted used missile casings for prosthetics. Lifesaving therapies can turn into killers when off-label indication creep unknowingly captures those at risk. Before its withdrawal in 2004, the COX-2 non-steroidal, anti-inflammatory drug Rofecoxib was approved and aggressively marketed. Notoriously, Merck withheld evidence of increased risk for heart attacks and strokes for over 5 years, resulting in an estimated 88,000–140,000 deaths (Graham et al. 2005; Bhattacharya 2005). Both the withholding of data for safety and efficacy and its exceptionally aggressive marketing contributed to the large number of deaths through misinformation and therapeutic creep (Wright et al. 2001; Therapeutics Initiatives 2001, 2001–2002, 2004). Similarly, the recombinant glycoprotein hormone, erythropoietin, useful in cancer care treatment, can also cause lethal thrombotic complications (Hébert and Stanbrook 2007). Therapeutic or indication creep commonly comes from information seeded by industry to clinical scientists conducting late Phase III and Phase IV post-marketing studies. It is enabled by prescribing clinicians (Fugh-Berman and Melnick 2008; Djulbegovic and Ash 2011; Kesselheim, Meloo, and Studdert 2011; Riggs and Ubel 2015).

Misinformation seeded by other groups, motivated politically, religiously, or maliciously, takes on a different sort of threat. Often grounded in local logics, anxieties and rumors continue to derail vaccine campaigns (Leach and Fairhead 2007). Clinical researchers and public health vaccine campaigns have begun to pay attention to the fact that ignoring local understandings and explanations is at an immunization campaign's peril (Ghinai et al. 2013; Larson 2014).

How does the knowledge of scientists, health providers, policy makers, and citizens—whether that information sits as data in scientific repositories or in citizens' collective thoughts and actions—get equitably configured into evidence databases? Logical systems, no matter whose logic, are not immutable (Longino 2002). Sometimes new studies bring to light old (folk) remedies. Local knowledge that may have been dismissed by experts as anecdotal, or folklore, or gossip, reappears later with scientific recognition and potential market value (CBC 2015). The intellectual property rights for new medicines can be fought over in highly contested legal fields and are unlikely to provide the same gains to the original creators as they do for larger, more powerful industry interests (Hayden 2003).

Considerable public and private efforts are put into incentivizing and supporting the development of new technologies to address the matrix of multidimensional factors that contribute to and threaten the one health we all share. Indeed, incentivizing the development of global health technologies has become the goal of a growing cadre of billionaire philanthropists. Their foundations advance the principles and ideologies that brought them their wealth in the first place and provide them with the resources to set the research agendas of their hearts' desires, ranging from agricultural and health technologies to ocean sciences (Broad 2014). Philanthrocapitalists, rather than public agencies and independent experts, have increasingly directed strategic planning for global health and environment. Much of the money put forth in the strategies advanced by philanthrocapitalist groups is directed from public funds, commonly in the guise of public–private initiatives (Mazzucato 2011; Light 2009; Lezaun and Montgomery 2015). By the time the private sector becomes "technically" involved (political involvement is integral to the philanthropic strategic plan), there are few risks for an already advanced product. The return on (private) investment at the end stage of development has been assured by the public coffers. The Ebola vaccines developed by the Public Health Agency of Canada and the National Institutes of Health in the United States, now referred respectively as the Merck and GSK Ebola vaccines, are exemplary cases in point.

14.4 Moving to a Solution: Some Questions First

So far, the dual role of governments and the relational reality of cozy regulatory–industry activities has been discussed, which put forth a rhetoric of fireguards between industry and regulator, but nonetheless include opportunities for bilateral meetings to introduce new evidence to persuade a hesitating regulator. What would prevent trial design and research evidence from being gamed by industry? What if evidence of therapeutic improvement had to be agreed upon by an independent body of evaluators representing diverse backgrounds, rather than fast-tracked through a regulatory pipeline increasingly compromised by a government advocating and creating policies for industry partnerships, commodity fetishism, and corporate drivers (Graham 2001)? How might the independence of evaluators be as integral as industry imposed regulatory time limits? What would evidence of value-added health improvements look like in a setting where the greater push for newer drugs, that are not always better, could be reset (Graham and Nuttall 2013)?

Central to this inquiry would be the development of techniques to demand that new health technologies contribute significantly to value-added health improvement (not all new therapies work better). In liberal democracies, it is important to find out how the actors and practices that command techno-scientific authority sometimes hold sway, and sometimes do not, in matters of decision-making, governance, and the determination of what matters. It is worthwhile to unpack the disproportionate roles and interests different actors have in determining what information matters, where it comes from, who it is passed to, who *gives* and who *receives* knowledge, training, and treatment, and whose metrics are used to measure and declare the success of interventions. Information flows in many directions, and decision-making is often more political than scientific (Bishop and Lexchin 2013; Burchett et al. 2012). Brian Wynne has argued that we should be critically engaged in "the enrollment of science in global economic and political forms" (Weiner 2011). To that end, we might consider systematically unpacking the circulation of expertise (and interests) that contributes to the approval of health products. At issue, in the governing of the public's health, is whether it is possible to gather together a panel of truly independent evidence-based evaluators whose expertise in research design, methodological rigor, and clinical experience is not compromised by some form of conflict of interest. Central to the work of several science and technology studies scholars has been an examination of the information that scientific and political actors use to build evidence. In addition, the degree to which authorities listen to and involve diverse communities in building the knowledge base, then reach decisions using those data, and the role of scientific advice in democracies generally, have been ongoing questions (Bijker et al. 2009).

We shall briefly present two case studies on how the information from the best-made science can be diverted by practices that prevent knowledge from being fully realized in the world. We will conclude the chapter with a prolegomenon of what we might do to resolve this problem. Our recommendations will emphasize a close parallel to issues that confound decision-making at the science–policy interface described in other chapters in this volume.

14.4.1 Case 1: International Regulatory Practices and Policies for Emerging Health Products: Efficacy and Safety

Beginning in 2001, Graham became engaged in several years of participant-observation in a regulatory platform (the Canadian federal department Health Canada). This research was pursued in order to describe the regulatory actors and their tasks, map the regulatory territory of scientific evidence and policy decisions, and illustrate how a regulatory system adapts in response to contingency and rapidly emerging scientific and policy changes. This study followed the step-by-step process of product submission and regulatory review as teams of research scientists, biologists, medical officers, and technicians, equipped with state-of-the-art technologies and instrumentation, evaluated clinical science trial data and inspected manufacturing sites. Scientists, clinical evaluators, and policy advisors reviewed regulatory submissions, sampled consistency, conducted extensive chemistry and manufacture confirmatory tests, reanalyzed data, and checked back with the sponsors for missing data or for any queries they might have had about the submitted evidence. Decision-making frameworks were established by the various parties, but decisions to submit, resubmit, or finally withdraw an application were in the hands of the sponsor. Inevitably, the actors on both sides of the product decision must balance legislated deadlines with partial data, and weigh individual and public health safety against public and industry desires.

Beginning in the early 2000s, the Health Products and Food Branch (HPFB) of Health Canada established a series of initiatives to "ensure that Canadians have faster access to the safe drugs they need" (Government of Canada 2002). Focusing originally on smart regulation (Graham 2005), HPFB moved to a more acceptable language of a "lifecycle approach" as part of the regulatory modernization at Health Canada (Health Canada 2015). In keeping with government policy, Health Canada developed policies and instruments to open up access to new drugs. While regulatory work up to 2004 had mostly concentrated on the assessment of pre-market pharmaceuticals and biological therapies—that is, isolated from natural sources such as living cells or tissues—for market approval, the lifecycle approach was intended to manage the approval of drugs for market placement more quickly, through a progressive licensing strategy. Although a post-market approval authority to follow the products in their application was part of the scheme, health advocates were concerned that funding and enforcement would lag behind approval, compromising the safety of Canadians prescribed by these early licensed products (Graham and Nuttall 2013). Internationally, there has been wide adoption of regulatory modernization across all government sectors such that the parallels in how this development has been carried out among these sectors, for example, health and environment, will become apparent to the reader. As in all such processes, intricate convergences of human and nonhuman environments and technological and cultural ecologies have occurred.

Clinical trial evidence that is not open or transparent harms everyone (Muir Gray 2012). Graham was fortunate to have been a student at McMaster University in Hamilton, Canada, in the early 1980s, when the innovators of what became known as evidence-based medicine (EBM) were tutors in the graduate clinical epidemiology and biostatistics course and were testing their systematic review methodologies. She learned to analyze the clinical trial evidence by critical appraisal of the methodological designs, data, and interpretations of medical studies. Graham believed, as acolytes do, in the potential of evidence-based approaches to open up and make transparent clinical study data so that critical appraisal could be carried out by anyone curious enough to care about the results. How disillusioning, then, to watch the sleight of hand as these evidence-based standards for the scientific stewardship of clinical trials research were undermined by interests other than science and by consensus panels and expert advisory committees that sometimes exercised authority without attending necessarily to the evidence. Things are not always as they appear (Gilbert 2006). The keepers of best practice in health care miss the integrative thinking needed for health systems (or, for that matter, integrated coastal and ocean management).

David Sackett et al. (1996) described EBM as the integration of "individual clinical expertise and the best external evidence." A problem occurs, however, when the best evidence is limited. The gold standard of the blinded randomized controlled trial (RCT) is ideal in theory, but has corroded in practice. The costs of conducting sound EBM trials have restricted it largely to private firms, which control the data in and the analysis out. Field biologists as well as economists know how difficult it is to control for external conditions in the laboratory, let alone the natural world (see Chapters 8, 13, and 16 in this volume). If externalities can be controlled, taking account of every known contingency, the unknowns will still rule the day. This is why slow cautious longitudinal research in natural conditions is invaluable, if for nothing more than to remind us of the damage wrecked by Frankenstein's hubris.

Items missed in the data collection in an RCT cannot always be accounted for afterwards. The best studies for the best external evidence do not necessarily see the light of day. Expensive to conduct, most randomized controlled trials are industry sponsored, whose objective is to produce evidence that will see their products approved for market. The sponsors are most often the pharmaceutical industry hoping to make profits. So, many types of drugs that are already past patent protection, products like aspirin, for example, have been largely neglected in clinical trials while the hope and money have been placed on much more profitable *innovative* new drug products, because they are patentable. Only about half of all RCT studies are ever published, and negative studies, that is, research that shows no improvement of treatment in comparison to the control group, are seldom published at all (Maund et al. 2014; Scherer et al. 2007; Chan et al. 2004). Why? Because interested sponsors fund trials, and often industry-backed researchers carry out the research. You are not likely to sell a car if you tell someone it is a lemon. Therefore, we mainly see partial and *interested* information directed at selling a drug as a commercial commodity rather than therapies and services for the public good.

The randomized controlled trial is a standard that misses an important component, *clinical meaningfulness*, advanced by Alvan Feinstein (1987) in response to statistical dominance in medicine, though not without critique

(Hobart 2007). In the 1990s, as a naïve postdoctoral fellow, Graham thought that "meaningfulness" would provide an avenue to tie patient and care-giver experiences into a truly integrated approach to evidence for treatment outcomes for clinical trials. She developed a qualitative methodology that would take into account everyday symptoms of decline and improvement from a patient's, caregiver's, and doctor's points of view. To her mind, this approach would provide a valuable humanistic and personal component for ascertaining the effectiveness of potential treatments. She thought that this symbolic local ecological knowledge (as an anthropologist, and local knowl-edge matters) (Geertz 1983) could augment the materialist statistical signifi-cance of clinical trial studies (Graham 2008). Unfortunately, and predictably (as time taught healthy skepticism), the manufacturer who sponsored the study selected only the positive results from the database and ignored the not so positive cases in order to make its argument for inclusion of the drug into provincial formularies.

Quick to catch on that personal testimonials matter more than statistics, industry carefully selected the particular data from the study to sell their product. The company cherry-picked the best evidence to tell a different story. A profit-incentivized pharmaceutical company captured Graham's method, but used only the positive accounts to promote the drug. If the stories of decline had been included, the minimal effectiveness of the drug would have been shown. Furthermore, by placing that drug in the provincial formularies, its costs were charged to the public health care system. Several years later, Graham witnessed the last province to resist allowing that drug into its formulary, based on the paucity of evidence, fold under political pres-sure from an aggressive campaign of "expert" clinical researchers, namely, the same folks who had conducted the industry's studies, as well as assem-bled the industry-funded patient groups.

Industry pays for the research, the researchers, and the evidence that most advances their interests. Personal testimonials from select actors trumped the minimal evidence for therapeutic improvement. Profits (in a country where natural resource extraction often overrides the best evidence of declining supplies and catastrophic environmental consequences) do not always have public health among their interests. The use of scientific evidence and the regulators who protect good science need governance. The independence of science and education of policy analysts to recognize its importance warrant continuing attention.

In case this account seems a testimonial in itself, it is not uncommon. The *British Medical Journal's* "open data campaign" defended key Cochrane reviewers who demanded to see company-protected data in order to assess the efficacy of the influenza antiviral Tamiflu (oseltamivir) medica-tion sufficiently. Reviewers from the internationally recognized Cochrane Collaboration, who conduct systemic reviews of primary research in health care and health policy using evidence-based approaches, were denied access to clinical study reports held by the manufacturer, Roche (Doshi et al. 2012).

These reviews address important questions such as "Does treatment X work better than Y and will it do more good than harm?" Through "sophisticated marketing rather than verifiable evidence," countries around the world stockpiled Tamiflu, costing billions of dollars; purchasers believed that Tamiflu would suppress the threat of pandemic H1N1 influenza. There was no reliable evidence to confirm this position. Regulators failed to appraise the full data; they failed the public trust.

14.4.2 Case 2: Global Vaccine Development and Implementation Platforms. Equity. Developing Vaccines for the Global South

In 2001, the Gates Foundation provided seed funding to develop a new meningococcal serogroup A conjugate vaccine, MenAfriVac™ for endemic and repeated epidemics of meningitis A in sub-Saharan Africa. The vaccine had to be affordable, to cost around 50¢ a dose. While a meningitis C vaccine was developed within months during an outbreak in the United Kingdom in the 1990s that killed one thousand people, in Africa during the same period, some 700,000 people were affected by Group A *Neisseria meningococcus*, the most prevalent meningitis strain in sub-Saharan Africa. Meningococcal serogroup A infection claimed 100,000 lives and left 600,000 others with life-long morbidity. The vaccine promised to save hundreds of thousands of people devastated by periodic meningitis outbreaks. Other meningitis vaccines existed, but patents make them unaffordable and no manufacturer was interested in developing a vaccine with limited potential for large profit. Fueled by a feasibility study of existing intellectual property, a multilateral partnership under the umbrella of the World Health Organization/Program in Appropriate Technologies (PATH), the Meningitis Vaccine Project arranged for the technology transfer, clinical trials, regulatory approval, and implementation of MenAfriVac (LaForce et al. 2007; LaForce and Okwo-Bele 2011; LaForce et al. 2009). The vaccine worked, and meningitis A has been controlled in vaccinated populations.

But, while hundreds of millions of dollars spent to build capacity for disease and safety monitoring, surveillance, and training was directed to the epidemiological center in Burkina Faso's capital, little knowledge filtered into (or out of) the other communities (Graham et al. 2012; Mounier-Jack et al. 2014). When meningitis W-135 and X and *Streptococcus pneumonia* meningitis popped up in epidemic clusters, the year after the campaign, just as Graham's Burkinabé colleagues told her would happen 4 years earlier, people who thought they had been immunized against meningitis were infected. The capacity and knowledge for a single disease targeted vertical vaccination program was not integrated, it did not filter down to real people or local health care workers. Worse still, local knowledge, scientific, medical, and lay, was not engaged. The monovalent vaccine, rather than a quadrivalent to protect against the other meningitis subtypes, was slated to be adopted into the routine immunization program. In a country that spends only \$9/day/capita on health services, fees are still charged for hospital and clinic visits, illiteracy is around 26%, and child mortality remains one of the highest in world. Despite strengthened surveillance, mass campaigns, such as the Meningitis A vaccine introduction, remain missed opportunities to strengthen health systems because they lack "integration with other health systems" (Mounier-Jack et al. 2014; Sanou et al. 2009). Vertical global health programs, even successful ones, miss or strategically ignore everyday reality and significant local knowledge (McGoey 2012a, 2012b, 2014).

14.5 Integrating Knowledge from All Levels in a Parliament of Evidence

Within the social studies of science, risk regulation regimes are characterized as dominated by a technocratic approach and as neglecting publicly located, socially situated epistemological standpoints, that is, the real world. Several nations have taken this critique on board through regulatory modernization, where strategic efforts are being directed to open up, enable scrutiny, and solicit input into decision-making from a broad range of citizens. We have suggested that the evidence base for risk regulation could benefit from accommodating more ways of knowing (Graham and Jones 2010; Jones and Graham 2009). We have argued that it is not only *lay* public knowledge, but also *expert* scientific understanding that is neglected in modern risk regulation regimes. A symmetrical approach to evidence-based risk regulation is needed which draws from ethnographic studies, the literature on risk, regulatory science, and science and technology studies. Drawing from the work of Bruno Latour and Isabelle Stengers, this framework can be developed as a parliament of evidence for decision-making.

Policies might be created that solicit, even promote, wide public dialogue that could generate broad information exchanges across public platforms—that is, those outside official science-based regulatory offices—as evidence that could be included in evidence-based decision-making. These policies could encourage citizens with specialized knowledge to contribute to regulatory decision-making. Among Canada's regulatory comparators—the United Kingdom, the United States, France, and Australia—public input is generally sought in cases where the government seeks policy direction or approval for decisions. However, few mechanisms exist to consider public input in a similar manner to scientific evidence.

International efforts to consider other types of citizen evidence are part of regulatory modernization. It keeps pace with political neighbors in terms of policy, economy, and science and technology, using such harmonizing tools as memoranda of understanding, trade agreements, accords, and other means of operating at a supra-state level. To the extent that it resembles the late twentieth-century project of modernity, regulatory modernization authorizes scientific knowledge to be the principal informant for evidence-based decision-making, characterizing the mutually-dependent features of innovation and economic growth as essential goods. In this configuration, modernization prioritizes narrowly construed definitions of expert rationality over open, democratized forms of decision-making.

Yet, openness and democratization feature centrally in governments' expressed vision of regulatory modernization. Where unresolved uncertainties about risk proliferate, regulators operate in conditions where international trends lean toward public participation in technology governance. This is particularly true of regulatory systems designed to protect citizens from high-profile risks when lives are at stake, such as those connected with therapeutic health. When regulatory failure results in compromised health or death among members of the public, trust in the regulatory system is compromised. This is an important implication of modernization: it works to reduce not only technological risks, but also political ones. States not actively engaging with their citizens risk being characterized as out-of-touch at best, and illegitimate at worst. Structuring in public participation is a symbol of good governance, of the state's capacity for the social distribution of expertise. The challenge for the modern regulator is to create a system capable of pre-empting the critique that this democratic version of regulatory modernization is merely rhetorical, a way to enhance the legitimacy of the regulatory regime while devolving responsibility for detecting and assuming risks onto members of the public, in the name of citizenship.

Although modernization lends itself as a topic for science studies researchers interested in the transition from knowledge to practice, much scholarship in this area suffers from an incomplete understanding of the requirements of on-the-ground regulatory practice. Regulators, industry, expert advisors, and citizens are all regulatory actors engaged in risk governance. On a daily basis, these actors encounter elements of their environment that both constrain and enable transformation. They engage in practices that help them make sense of their environment; considering their differing epistemological positions, these practices often lead to contests over meaning and significance.

The determination of evidence is a prominent site of contest in risk regulation. Different actors may entertain different perceptions of what is and is not appropriate evidence for regulatory decision-making. Growing public awareness of the role of industry in shaping scientific evidence, from the tobacco lobby to clinical trials and global warming, means that few disagree that politics can affect the production and dissemination of scientific research (Oreskes and Conway 2010).

In a changing paradigm of risk, regulators attempt to address the tension between *perceived* (culturally constructed) and *objective* (identified through expert measurement) risk (Doern and Reed 2000, p. 10). However, the ideal of regulatory objectivity is performed differently in political cultures (Jasanoff

2011). The distinction between *objective* and *perceived* risk, while analytically useful, is loaded with inequity when perceptions and experiences that count for some are not taken into account by others. This distinction reproduces objectivity as an achievable criterion for evidence assessment. It neglects the significance of values, power, and culture within scientific decision-making, as well as the widespread acknowledgment of conflicts of interest and bias buried in evidence-based, and in particular, industry-sponsored studies. This critique of objectivity familiar to science studies scholars is gaining ground in scientific communities, which are forced to acknowledge how evidence has been compromised through conflicts of interest, and indeed, how even the term "sound science" could be appropriated and used by the tobacco lobby. With the recognition that technologies developed to create science are equally fallible to human foibles (biases and conflicts of interest), the need for alternative paths and mechanisms to assess evidence for risk regulation has emerged.

The approach we proposed would enable qualitatively different kinds of evidence to be assessed, evaluated, and judged to be valid via distinct, identifiable, and transparent techniques (Graham and Jones 2010). We considered how to arrive at a modernized regulatory framework that accounts for both the need to assess risk objectively through measures of safety and efficacy, *and* the need to include local understandings and experiences as relevant, valid contributions to the evidence base. The policy features of *accountability* (accepting responsibility for the consequences of decisions), *openness* (willingness to consider input from public sources), *transparency* (making available study data and information about decisions), and *flexibility* (recognizing that a one size fits all approach to regulatory decisions is not always appropriate) figure centrally. While the inclusion of "timeliness" is a clear address to the ubiquitous industry and patient-group complaints of the slowness of regulatory decision-making, the "open" form of modernization proposed presents a gentler, more democratic, pluralist version in contrast to the innovation-friendly, technocratic form of modernization. This *symmetrical* framework proposed for transforming evidence-based risk regulation would expand on international trends for transparency and accountability, rather than endorse drivers for economic innovation alone.

14.6 Modernization, Risks, and Regulatory Science

Objective and perceived risk remain in hierarchical tension, the consequence of state reliance on (scientific) evidence-based decision-making to the exclusion of pragmatic citizen knowledge. First, regulators have excluded important information based on local ways of knowing and social context; second, they have risked fostering public cynicism by maintaining a hold on access to proprietary data, thereby denying independent review despite high-profile exposures of gross misrepresentations and misjudgments in scientific advice; and third, they have neglected the role of values in shaping scientific knowledge. As a result, science and technology policy tends to suffer high levels of critical attention (and, therefore, politicization) as the public and scientists alike query what exactly is going on in these closed regulatory circles. The response to political pressure often taken by official decision makers is to "give the people what they want": open up the system to accountable practices, set up mechanisms for participation, and enhance goodwill (and legitimacy) by demonstrating a commitment to meeting public demands. However, the adoption of such measures without critical reflection and a carefully thought out methodology and vision is unlikely to accomplish what it sets out to do. Moreover, such an approach risks eroding the legitimacy already held by state expert systems, as well as compromising the credibility of the regime by investing time and taxes in consultations and similar activities that may result in very little visible change in the trajectory of decisions. The National Institute for Health and Clinical (changed to "Care" in 2012) Excellence (NICE) is an example of an agency that works hard to incorporate best science with a deliberative process of citizen engagement. Despite its attempts, it is constantly under assault by industry and patient lobbyists whenever it arrives at recommendations (Graham 2008).

Regulators are placed in a dilemma. If they retain their reliance on extant expert systems to produce the evidence for decision-making, they risk further destabilization from public demands associated with growing distrust in science. If they bow to demands for greater public participation, they risk eroding the existing strengths of their system, that is, efficiency of systematic evaluation and risk assessment in the vast majority of reviews. The problem facing risk regulation regimes engaged in modernization is how to find an acceptable medium that does not compromise safety and efficacy along that spectrum of choices. To that end, symmetrical regulation would require accountability through constructivist realism (*not* accountability through objectivity alone); openness (*not* just transparency); and reflexivity (*not* flexibility).

14.7 A Symmetrical Approach: Constructivist Accountability, Openness, and Reflexivity

14.7.1 Accountability through Both Independent Scientific Assessment and Constructivist Realism

Objectivity (along with value neutrality) is a defendable aspiration of expert systems of scientific advice supporting regulatory frameworks; its intent of application of rigorous science and methods is a necessary aim. The independence (value neutrality) in which objectivity claims are associated are, however, widely critiqued. The consequence is that the legitimacy of scientific knowledge as the primary authority for policy advice is questioned along with the legitimacy of the policy decision. In the cause of symmetrical regulation, constructivist realist knowledge might be adopted to address this legitimacy gap, thereby acknowledging partial perspectives, social contexts, and shaped standpoints.

How would a constructivist–realist accountability look in practice? Consider, for example, the accumulation of physical and social facts, call them symptoms, which mark a neuro-social degenerative condition such as Alzheimer's disease. The decision as to which particular constellation of symptoms and signs one calls upon to understand this illness depends on whether one is a clinician or a family member. Political, social, and physical-pathology are flexible factors affecting diagnosis (Graham and Ritchie 2006). Building both clinical and social outcomes and regulatory mechanisms to accommodate these varying data sources involves necessarily both constructivist *and* positivist analyses. In a similar way, regulators and indeed, the clinical research community could apply their awareness of interpretive relativism to their regulatory outcomes. They could push for more rigorous clinical trial design and methodologies, including the independent analysis of research data and results in order to detect methodological and interpretive bias. Consultation with independent (nonsponsor) clinical researchers could be used to balance the data and interpretation provided by sponsor-supplied researchers. Legislation to control more comprehensively the premature marketing and hyping of new products could be enforced.

14.7.2 Openness, Not Just Transparency

Transparency is about provision of detailed information through one-way communication. It can be seen, however, as a photo-op for deliberative democracy in a political climate of gag orders, as a way of overloading pressure groups seeking information on opaque policy processes. Transparency devolves responsibility onto citizens without giving them real opportunities to contribute to decisions. Openness, on the other hand, is two-way, where information flows in multiple directions through engaged exchange and discussion.

How might openness look in practice? The inclusion of a broader range of constituents on official bodies would gain ground as a trust-building measure. Bringing different perspectives to the same table is one way to support the coproduction of a symmetrical evidence base.

14.7.3 Reflexivity for a Symmetrical Evidence Base

Finally, the third feature of a symmetrical evidence approach is reflexivity, *not* flexibility. Flexibility is a feature of "smart" risk regulation regimes, and is critiqued elsewhere (Graham 2005). Reflexivity, instead, recognizes and builds a dialogue between conflicting systems of knowledge, for example, experts, and experienced and concerned citizens. Conflicting expert advice leaves decision makers with the task of determining which expert advice to follow.

Reflexivity recognizes that not all types of evidence are assessed in the same way. Standardization is an essential part of the process. The dilemma of regulation is that standardization—seen as a way to ensure both fairness and rigor—often accomplishes just the opposite. Regulators need to be prepared to enact flexibility, not through a predetermined kit of approaches from which they can draw "the perfect tool," but rather, by assessing each case according to the best way to deal with its particularities.

Reflexivity requires vigilance by regulators regarding methods and outcome. If regulatory scientists discuss a "risky" (uncertain, potentially unsafe) product together, they should be able to identify common questions and approaches to answer them. This process should not be a systematic wearing down of evaluators' queries by industry-sponsored scientific teams. Regulators need to define what outcomes are appropriate, and sponsors must provide those outcomes. The decision on acceptable outcomes should not be a negotiated benchmark between sponsor and regulator, but a carefully determined outcome from several expert (independent and non-conflicted) sources.

14.8 Conclusion

The three features described here—accountability through constructivist realism, openness, and reflexivity—are not necessarily new in the recommendations sections of scholarly critiques of risk regulation regimes. They can be seen to overlap with features that Sarkki et al. (2014) refer to as "trade-offs" between credibility, relevance, and legitimacy and in what Gluckman and Allen (Chapter 10 in this volume) call the "the balancing act of science in public policy." In Canada, accountability, openness, and reflexivity have been actively employed in regulatory policy-making and practice. In low income and emerging countries, as we have seen, local knowledge (even scientific and clinical knowledge) may provide only minimal input in targeted disease initiatives. It is the way these features are operationalized in regimes that will have an impact on not only the power of the evidence base, but also the effectiveness of the regulatory regime as a whole. Worldwide, policy makers have turned their attention to post-approval regulatory activities, emphasized as a more holistic, real world lifecycle approach in regulatory renewal frameworks. How regulators will reinforce the integrity of pre-market assessment in a post-market environment remains to be seen in practice. Whether subsuming commercial technical drivers of innovation and economics under principles of timeliness can satisfy all the actors, scientific and other citizens alike, calling for accountability, openness, and reflexivity, also remains a question. This is, perhaps, especially the case in the growing global health economies.

A symmetrical approach to regulatory decision-making that provides mechanisms to hear and assess different types of evidence and multiple epistemologies would begin to address the decline in trust of regulatory decisions brought about by highly publicized product withdrawals and exacerbated by the close relationship between regulator and industry, and by the preponderance of industry-sponsored evidence. It would preempt the need for reanalyses by independent reviewers that result in findings of unsafe and ineffective therapies. A symmetrical approach would bridge rigorous scientific evaluation and public input, providing the best evidence from all available sources to be discussed and contested by a diverse range of actors toward the goal of arriving at a common agreement. The question, paraphrasing Latour (2003, p. 4), should not be whether the conclusion has been constructed, for of course it has been, but whether it is based on an accountable, open, and reflexive process that can "differentiate good and bad construction" in order to arrive at the optimal decision to approve or reject a new health product.

In a symmetrical approach to regulation, scientists, policy makers, and all citizens have the opportunity to modestly witness, as Donna Haraway (1997) calls it, the interpenetration of capitalism and technoscience. We have seen where clinical research, health technology assessment, and global health initiatives do not follow citizen driven models of horizontal alliances in the type of deep democratic manner involving longitudinal community engagement and input considered, for example, by Arjun Appadurai (2001). Nor do they adhere to the plea for slower science proposed by Isabelle Stengers, where "we slow down, that we don't consider ourselves authorized to believe we possess the meaning of what we know" (Stengers 2005, p. 2). Instead, we witness the power of financial and corporate elites to control interests and to "favour the 'project' model, in which short-term logics of investment, accounting, reporting and assessment are regarded as vital" (Appadurai 2001, p. 30). At stake are precautionary consideration, democratic engagement, and sustainable health delivery systems. In the years leading up to the implementation of the meningitis A vaccine, African scientists and clinicians recognized and expressed an array of concerns (to J. Graham) surrounding the meningitis vaccine project, including a fear of new outbreaks of *S. pneumonia*, meningitis W 135 and meningitis X. Their knowledge, though overlapping with the Meningitis Vaccine Project's scientists, clinicians, and policy workers, was deeper in contextualized understanding of the landscape of diseases and the availability of resources to address them.

A symmetrical approach to decision-making would provide mechanisms and a platform to hear, assess, and incorporate diverse methodologies and understandings. A Latourian parliament of things would bring science and politics together to address the decline in trust of regulatory decisions exacerbated by anti-vaccine groups, highly publicized product withdrawals, and by the preponderance of industry-sponsored evidence and regulatory capture.

The principles of accountability through constructivism, openness (two-way information exchange between engaged actors), and reflexivity (where all types of evidence are not assessed in the same way) could provide a space whereby citizens, scientists, regulators, and the private sector could each express a common value that would filter into their engagement. In line with post-normal science approaches to wicked issues, extending the expertise on which decisions are based offers a path to respect the political commitments that are at stake (for stakeholders). Such innovations would be a force for social and environmental change (Turnpenny et al. 2011) rather than for individual interests and desires. Through accountable, open, and reflexive science with public deliberation, we could see our innovations working toward a common future where human becomings are socially, ethically, and ecologically transcendent. In a parliament of evidence, power elites alone would not drive decision-making (Wynne 1996); vulnerable groups could be heard and conflicts of interest addressed. In a parliament of evidence, the cost of not attending to our water, our environment, and our health would matter more than financial profit (Stern 2006).

Acknowledgments

This work was supported by the Canadian institutes of Health Research grant OGH-111401, Articulating standards: Translating the practices of standardizing health technologies.

References

Appadurai, A. 2001. Deep democracy: Urban governmentality and the horizon of politics. *Environment & Urbanization* 13 (2): 23–44.

Bhattacharya, S. 2005. Up to 140,000 heart attacks linked to Vioxx. *New Scientist*, 25 January. https://www.newscientist.com/article/dn6918-up-to-140000-heart-attacks-linked-to-vioxx.

Bijker, W. E., R. Bal, and R. Hendriks. 2009. *The Paradox of Scientific Authority: The Role of Scientific Advice in Democracies*. Cambridge, MA: MIT Press.

Bishop, D. and J. Lexchin. 2013. Politics and its intersection with coverage with evidence development: A qualitative analysis of expert interviews. *BMC Health Services Research* 13 (88): 1–10. http://www.biomedcentral.com/1472-6963/13/88.

Broad, W. J. 2014. Billionaires with big ideas are privatizing American science. *New York Times*, 15 March. http://www.nytimes.com/2014/03/16/science/billion-aires-with-big-ideas-are-privatizing-american-science.html.

Burchett, H. E., S. Mounier-Jack, U. K. Griffiths, et al. 2012. New vaccine adoption: Qualitative study of national decision-making processes in seven low- and middle-income countries. *Health Policy and Planning* 27 (Supplement 2): ii5–ii16.

Burger, E. J. 1990. Health as a surrogate for the environment. *Daedalus* 119 (4): 133–153.

Carpenter, D. 2013. Corrosive capture? The dueling forces of autonomy and industry influence in FDA pharmaceutical regulation. In *Preventing Regulatory Capture: Special Interest Influence and How to Limit It*, edited by D. Carpenter and D. Moss, 152–172. Cambridge: Cambridge University Press.

CBC (Canadian Broadcasting Corporation). 2015. Medieval potion kills antibiotic-resistant MRSA superbugs. *CBC News*, 3 April. http://www.cbc.ca/news/tech-nology/medieval-potion-kills-antibiotic-resistant-mrsa-superbugs-1.3020735.

Centers for Disease Control and Prevention. 2013. One health. Accessed 15 September 2015. http://www.cdc.gov/onehealth/.

Chan, A. W., A. Hróbjartsson, M. T. Haahr, et al. 2004. Empirical evidence for the selective reporting of outcomes in randomized trials: Comparison to published articles. *Journal of the American Medical Association* 291: 2457–2465.

Collins, H. M. and R. Evans. 2002. The third wave of science studies. *Social Studies of Science* 32 (2): 235–296.

Djulbegovic, B. and P. Ash. 2011. From efficacy to effectiveness in the face of uncertainty: Indication creep and prevention creep. *Journal of the American Medical Association* 305 (19): 2005–2006.

Doern, G. B. and T. Reed. 2000. Canada's changing science-based policy and regulatory regime: Issues and framework. In *Risky Business: Canada's Changing Science-Based Policy and Regulatory Regime*, edited by G. B. Doern and T. Reed, 3–30. Toronto: University of Toronto Press.

Doshi, P. 2015. No correction, no retraction, no apology, no comment: Paroxetine trial reanalysis raises questions about institutional responsibility. *British Medical Journal* 351: h4629.

Doshi, P., T. Jefferson, and C. Del Mar. 2012. The imperative to share clinical study reports: Recommendations from the Tamiflu experience. *PLoS Medicine* 9 (4): e1001201.

Feinstein, A. 1987. *Clinimetrics*. New Haven: Yale University Press.

Fuentes, A. 2010. Blurring the biological and social in human becomings. Paper presented at the 11th European Association of Social Anthropologists Meetings, Maynooth, Ireland, 24–27 May.

Fuentes, A. 2013. Blurring the biological and social in human becomings. In *Biosocial Becomings: Integrating Social and Biological Anthropology*, edited by T. Ingold and G. Palsson, 42–58. Cambridge: Cambridge University Press.

Fugh-Berman, A. and D. Melnick. 2008. Off-label promotion, on-target sales. *PLoS Medicine* 5 (10): e210.

Geertz, C. 1983. *Local Knowledge: Further Essays in Interpretive Anthropology*. New York: Basic Books.

Ghinai, I., C. Willott, I. Dadari, et al. 2013. Listening to the rumours: What the Northern Nigeria polio vaccine boycott can tell us ten years on. *Global Public Health* 8 (10): 1138–1150.

Giannini, A., R. Saravanan, and P. Chang. 2003. Oceanic forcing of Sahel rainfall on interannual to interdecadal time scales. *Science* 302 (5647): 1027–1030.

Gilbert, D. 2006. *Stumbling on Happiness*. New York: Vintage Books.

Government of Canada. 2002. The Canada we want: Speech from the throne to open the second session of the Thirty-Seventh Parliament of Canada. 30 September. http://publications.gc.ca/site/eng/236442/publication.html.

Graham, D. J., D. Campen, R. Hui, et al. 2005. Risk of acute myocardial infarction and sudden cardiac death in patients treated with cyclo-oxygenase 2 selective and non-selective non-steroidal anti-inflammatory drugs: Nested case-control study. *The Lancet* 365 (9458): 475–481.

Graham, J. E. 2001. Harbinger of hope or commodity fetishism: Re-cognizing dementia in an age of therapeutic agents. *International Psychogeriatrics* 13 (2): 131–134.

Graham, J. E. 2005. Smart regulation: Will the government's strategy work? *Canadian Medical Association Journal* 173 (12): 1469–1470.

Graham, J. E. 2008. Facilitating regulation: The dance of statistical significance and clinical meaningfulness in standardizing technologies for dementia. *BioSocieties* 3 (3): 241–263.

Graham, J. E. and R. Bassett. 2006. Reciprocal relations: The recognition and co-construction of caring with Alzheimer's disease. *Journal of Aging Studies* 20 (4): 335–349.

Graham, J. E., A. Borda-Rodrigeuz, F. Huzair, et al. 2012. Capacity for a global vaccine safety system: The perspective of national regulatory authorities. *Vaccine* 30 (33): 4953–4959.

Graham, J. E. and M. Jones. 2010. Rendre evident: une approche symetrique de la réglementation des produits thérapeutiques (Determining evidence: A symmetrical approach to the regulation of therapeutic products). *Sociologie et societies* 42 (2): 153–180.

Graham, J. E. and R. Nuttall. 2013. Faster access to new drugs: Fault lines between Health Canada's regulatory intent and industry innovation practices. *Ethics in Biology, Engineering & Medicine – An International Journal* 4 (3): 231–239.

Graham, J. E. and K. Ritchie. 2006. Mild cognitive impairment: Ethical considerations for nosological flexibility in human kinds. *Philosophy, Psychology and Psychiatry* 13 (2): 31–43.

Haraway, D. 1997. *Modest_Witness@Second_Millennium.FemaleMan_Meets_OncoMouse: Feminism and Technoscience*. New York: Routledge.

Hayden, C. 2003. *When Nature Goes Public: The Making and Unmaking of Bioprospecting in Mexico*. Princeton: Princeton University Press.

Health Canada. 2015. Initial development. Accessed 8 September 2015. http://www.hc-sc.gc.ca/ahc-asc/activit/strateg/mod/ini/index-eng.php.

Hébert, P. C. and M. Stanbrook. 2007. Indication creep: Physician beware. *Canadian Medical Association Journal* 177 (7): 697.

Hobart, J. 2007. A brief critique of clinimetrics. *The Lancet Neurology*. http://www.thelancet.com/cms/attachment/2000991592/2003658649/mmc1.pdf.

Hofer, M. P., C. Jakobsson, N. Zafiropoulos, et al. 2015. Regulatory watch: Impact of scientific advice from the European medicines agency. *Nature Reviews Drug Discovery* 14: 302–303.

Ingold, T. 2013a. *Making Anthropology, Archaeology, Art and Architecture*. New York: Routledge.

Ingold, T. 2013b. Prospect. In *Biosocial Becomings: Integrating Social and Biological Anthropology*, edited by T. Ingold and G. Palsson, 1–21. Cambridge: Cambridge University Press.

Jasanoff, S. 2011. The practices of objectivity in regulatory science. In *Social Knowledge in the Making*, edited by C. Camic, N. Gross, and M. Lamont, 307–337. Chicago: University of Chicago Press.

Jefferson, T., M. Jones, P. Doshi, et al. 2014. Oseltamivir for influenza in adults and children: Systematic review of clinical study reports and summary of regulatory comments. *British Medical Journal* 348: g2545.

Jones, M. and J. E. Graham. 2009. Multiple institutional rationalities in the regulation of health technologies: An ethnographic examination. *Science and Public Policy* 36 (6): 445–455.

Kassirer, J. P. 2007. Financial conflicts in the medical profession: An ongoing, unresolved problem. *Open Medicine* 1 (3). http://www.openmedicine.ca/article/view/133/64.

Kesselheim, A. S., M. M. Meloo, and D. M. Studdert. 2011. Strategies and practices in off-label marketing of pharmaceutical: A retrospective analysis of whistleblower complaints. *PLoS Medicine* 8 (4): e1000431.

LaForce, F. M., K. Konde, S. Viviani, et al. 2007. The meningitis vaccine project. *Vaccine* 3 (25) Suppl. 1: A97–100.

LaForce, F. M. and J. M. Okwo-Bele. 2011. Eliminating epidemic Group A meningococcal meningitis in Africa through a new vaccine. *Health Affairs* 30 (6): 1049–1057.

LaForce, F. M., N. Ravenscroft, M. Djingarey, et al. 2009. Epidemic meningitis due to Group A *Neisseria meningitidis* in the African meningitis belt: A persistent problem with an imminent solution. *Vaccine* 24 (27) Suppl. 2: B13–19.

Lampland, M. and S. L. Star. 2008. *Standards and Their Stories: How Quantifying, Classifying, and Formalizing Practices Shape Everyday Life.* Ithaca: Cornell University Press.

Larson, H. 2014. Underlying issues are key to dispelling vaccine doubts. *Bulletin World Health Organization* 92: 84–85.

Latour, B. 1993. *We Have Never Been Modern.* Cambridge, MA: Harvard University Press.

Latour, B. 2003. The promises of constructivism. In *Chasing Technoscience: Matrix for Materiality*, edited by D. Ihde and E. Selinger, 27–46. Bloomington: Indiana University Press.

Leach, M. and J. Fairhead. 2007. *Vaccine Anxieties: Global Science, Child Health and Society.* London: Earthscan.

Le Noury, J., J. M. Nardo, D. Healy, et al. 2015. Restoring study 329: Efficacy and harms of Paroxetine and Imipramine in treatment of major depression in adolescence. *British Medical Journal* 351: h4320.

Lexchin, J. 2012. Those who have the gold make the evidence: How the pharmaceutical industry biases the outcomes of clinical trials of medication. *Science Engineering Ethics* 18 (2): 247–261.

Lezaun, J. and C. M. Montgomery. 2015. The pharmaceutical commons: Sharing and exclusion in global health drug development. *Science, Technology & Human Values* 40 (1): 3–29.

Light, D. W. 2009. Advanced market commitments. Current realities and alternate approaches. Health Action International (HAI) Europe. Paper series reference 03-2009/01. http://haieurope.org/wp-content/uploads/2010/12/27-Mar-2009-Report-AMC-Current-Realities-Alternate-Approaches.pdf.

Longino, H. E. 2002. *The Fate of Knowledge.* Princeton: Princeton University Press.

Lundh, A., S. Sismondo, J. Lexchin, et al. 2013. Industry sponsorship and research outcome. *Cochrane Database of Systematic Reviews 2012* (12): Art. No. MR000033.

Maund, E., B. Tendal, A. Hróbjartsson, et al. 2014. Benefits and harms in clinical trials of Duloxetine for treatment of major depressive disorder: Comparison of clinical study reports, trial registries, and publications. *British Medical Journal* 348: g3510.

Maurice G. 1999. *The Enigma of the Gift.* Chicago: University of Chicago Press.

Mazzucato, M. 2011. *The Entrepreneurial State.* London: Demos.

McGoey, L. 2012a. Strategic unknowns: Towards a sociology of ignorance. *Economy and Society* 41 (1): 1–16.

McGoey, L. 2012b. Logic of strategic ignorance. *British Journal of Sociology* 63 (3): 553–576.

McGoey, L. 2014. The philanthropic state: Market hybrids in the philanthrocapitalist turn. *Third World Quarterly* 35 (1): 109–125.

Mounier-Jack, S., H. E. D. Burchett, U. K. Griffiths, et al. 2014. Meningococcal vaccine introduction in Mali through mass campaigns and its impact on the health system. *Global Health: Science and Practice* 2 (1): 117–129.

Muir Gray, J. A. 2012. Address to the APAC forum on quality improvement in health care. Accessed 19–21 September 2015. http://www.vimeo.com/49892096.

One Health Global Network. 2015. What is One Health? Accessed 15 September 2015. http://www.onehealthglobal.net/what-is-one-health/

One Health Initiative. n.d. About the One Health initiative. Accessed 4 February 2015. http://www.onehealthinitiative.com/about.php.

O'Neill, O. 2014. Trust, trustworthiness, and accountability. In *Capital Failure: Rebuilding Trust in Financial Services,* edited by N. Morris and D. Vines, 172–192. Oxford: Oxford University Press.

Oreskes, N. and E. M. Conway. 2010. *Merchants of Doubt. How a Handful of Scientists Obscured the Truth on Issues from Tobacco Smoke to Global Warming.* New York: Bloomsbury Press.

Public Health Agency of Canada. 2015. One Health. Accessed 8 September 2015. http://www.phac-aspc.gc.ca/owoh-umus/index-eng.php.

Rabinowitz, P. and L. Conti. 2012. Links among human health, animal health, and ecosystem health. *Annual Review of Public Health* 34: 1–16.

Riggs, K. R. and P. A. Ubel. 2015. The role of professional societies in limiting indication creep. *Journal of General Internal Medicine* 30 (2): 249–252.

Sackett, D. L., W. M. C. Rosenberg, J. A. Muir Gray, et al. 1996. Evidence based medicine: What it is and what it isn't. *British Medical Journal* 312 (7023): 71–72.

Sahlins, M. 1972. The original affluent society. In *Stone Age Economics.* London: Routledge.

Sanou, A., S. Simboro, B. Kouyate, et al. 2009. Assessment of factors associated with complete immunization coverage in children ages 12–23 months: A cross sectional study in Nouna District, Burkina Faso. *BMC International Health and Human Rights* 9: S10. http://www.biomedcentral.com/1472-698X/9/S1/S10.

Sarkki, S., J. Niemelä, R. Tinch, et al. 2014. Balancing credibility, relevance and legitimacy: A critical assessment of trade-offs in science–policy interfaces. *Science and Public Policy* 41: 194–206.

Scherer, R. W., P. Langenberg, and E. von Elm. 2007. Full publication of results presented in abstracts. *Cochrane Database Systematic Review* 2: MR000005.

Sismondo, S. 2008. How pharmaceutical industry funding affects trial outcomes: Causal structures and responses. *Social Science & Medicine* 66: 1909–1914.

Stengers, I. 1997. *Power and Invention: Situating Science.* Minneapolis: University of Minneapolis Press.

Stengers, I. 2005. The cosmopolitical proposal. In *Making Things Public: Atmospheres of Democracy,* edited by B. Latour, 994–1003. Cambridge, MA: MIT Press.

Stern, N. 2006. *Stern Review on the Economics of Climate Change.* London: HM Treasury. http://webarchive.nationalarchives.gov.uk/+/http:/www.hm-treasury.gov.uk/sternreview_index.htm.

Therapeutics Initiative. 2001. Selective COX-2 inhibitors: Are they safer? January–February 2001, *Newsletter* 39. Accessed 8 September 2015. http://www.ti.ubc.ca/newsletter/selective-cox-2-inhibitors-are-they-safer.

Therapeutics Initiative. 2001–2002. COX-2 inhibitors update: Do journal publications tell the full story? November 2001–January 2002, *Newsletter* 43. Accessed 8 September 2015. http://www.ti.ubc.ca/newsletter/cox-2-inhibitors-update-do-journal-publications-tell-full-story.

Therapeutics Initiative. 2004. Rofecoxib (Vioxx®) withdrawal generates uncertainty about COX-2s—Do product monographs adequately inform? July–October 2004, *Newsletter* 53. Accessed 8 September 2015. http://www.ti.ubc.ca/newsletter/rofecoxib-vioxx%C2%AE-withdrawal-generates-uncertainty-about-cox-2s-%E2%80%93-do-product-monographs-ad.

Turnpenny, J., M. Jones, and I. Lorenzoni. 2011. Where now for post-normal science? A critical review of its development, definitions, and uses. *Science, Technology and Human Values* 36 (3): 287–306.

van Eeckhout, L. 2015. Winds of climate change blast farmers' hopes of sustaining a livelihood in Burkina Faso. *The Guardian,* 7 July, http://www.theguardian.com/world/2015/jul/07/winds-climate-change-blast-burkina-faso-farmers.

Weiner, T. 2011. A report on the 2010 4S conference in Tokyo. 4 January. http://somatosphere.net/2011/01/report-on-2010-4s-conference-in-tokyo.html.

Wright, J. M., T. L. Perry, K. L. Bassett, et al. 2001. Reporting of 6-month vs 12-month data in a clinical trial of celecoxib. *Journal of the American Medical Association* 286 (19): 2398–2400.

Wynne, B. 1996. May the sheep safely graze? A reflexive view of the expert–lay knowledge divide. In *Risk, Environment and Modernity: Towards a New Ecology,* edited by S. Lash, B. Szerszynski, and B. Wynne, 186–226. London: Sage.

15

Information Matters: The Influence of the Atlantic Coastal Zone Information Steering Committee on Integrated Coastal and Ocean Management in Atlantic Canada

Andrew G. Sherin and Alexi Baccardax Westcott

CONTENTS

15.1 Introduction

The Atlantic Coastal Zone Information Steering Committee (ACZISC) is an informal and neutral forum for information sharing and collaboration between government agencies, university researchers, and civil society for the integrated coastal and ocean management (ICOM) community of practice (CoP) in the four provinces of Atlantic Canada. A CoP is defined as "a group of people who share a concern or a passion for something they do and learn how to do it better as they interact regularly" (Wenger 1998, p. 226). According to Wenger, a CoP requires a domain (i.e., coasts and oceans), a community (i.e., members engaging in joint activities and discussions, e.g., meetings, and workshops, and sharing information, e.g., newsletters and websites), and a practice (i.e., coastal and ocean management). A CoP is a key component of ICOM because people are at the root of ICOM, people with a shared concern or mandate for the effective management of coastal and ocean resources. A principal characteristic of a CoP is information sharing. For the ocean and coasts, no single organization has all of the information necessary for integrated management, thus making information sharing mandatory.

Established in 1992, the ACZISC has helped build the ICOM CoP in Atlantic Canada, promoted data and information sharing between members of the CoP, and influenced the development of ICOM policies and data and information products to support it. This chapter provides a brief history of the ACZISC. It describes the products and services the ACZISC has provided to build and sustain the ICOM CoP and promote the sharing of data and information. It also describes how the ACZISC has influenced coastal and ocean management in Atlantic Canada, using selected examples of its work.

15.2 About ACZISC

The ACZISC has members from federal and provincial government departments and agencies, nongovernmental organizations (NGOs), industry associations, and academe. Its activities have been supported by financial and in-kind contributions from the four Atlantic provincial governments and several federal government departments with a mandate and interest in ICOM. The ACZISC has worked with partners to provide a basis for federal–provincial NGO–community cooperation and exchange of information in the region (Rounce and Beaudry 2002, p. 40).

The Land Registration and Information Service (LRIS), an agency of the former Council of Maritime Premiers, established the ACZISC in 1992 to develop a regional strategy for information management in the Atlantic coastal zone and to identify and coordinate coastal zone information programs that would form the basis of a region wide program (LRIS 1991). Its first meeting was held in Halifax, Nova Scotia, on 15 January 1992, with twelve people representing nine federal and provincial organizations. The Secretariat at that time was provided by LRIS. Three to four ACZISC meetings per year have been held since its inception, rotating between the provincial capitals of Atlantic Canada. The ACZISC has developed other products and services beyond the regular meetings to build and sustain the ICOM CoP; these are described later in this chapter.

Two major strategic planning exercises have been conducted during the tenure of the ACZISC. The most recent exercise in 2011–2012 progressed from developing a consensus on the major issues impacting the coastal zone through to the definition of priorities and strategies (ACZISC 2012). The three priorities resulting from this planning process are (1) encouraging action on the implementation of ICOM as a tool to realize environmental, economic, and social sustainability, (2) collaborative sharing of data and information between members and with the wider ICOM CoP on the management issues of priority to members, and (3) encouraging the engagement of organizations in the ACZISC that is reflective of the diversity of the ICOM CoP. Examples of activities supporting these priorities are described here.

15.3 Influencing Integrated Coastal and Ocean Management

The ACZISC Strategic Plan Priority 1 is "encouraging action on the implementation of ICOM as a tool to realize environmental, economic, and social sustainability" (ACZISC 2012).

15.3.1 Promoting ICOM

The ACZISC has promoted ICOM through its ICOM Working Group, organizing workshops, inviting presentations on management initiatives in Atlantic Canada to the ACZISC meetings, and contributing to advisory committees. The discussion paper on its role in the development and implementation of ICOM in Atlantic Canada suggested that its strength is in providing an apolitical and inclusive forum for the sharing of ideas, information, and data, leading to a consistency of approach to policy development and implementation (ACZISC 2009). Coastal strategies and policies for New Brunswick, Nova Scotia, Newfoundland and Labrador, and Prince Edward Island have been discussed regularly at its meetings. The ACZISC has conducted workshops and webinars on ICOM topics, including a 2003 workshop on large marine ecozone boundaries in Atlantic Canada, and a 2014 workshop and webinar on marine and coastal protected areas. Members have been active at all Coastal Zone Canada Conferences. Recently, the ACZISC contributed to two significant advisory bodies, the State of the Scotian Shelf Report Steering Committee (see Chapter 12 in this volume) and the Advisory Committee for the National Coastal Climate Change Adaptation Assessment.

15.3.2 Building the Community of Practice for ICOM in Atlantic Canada

One aspect of a CoP is building relationships between the stakeholders in the ICOM arena. The face-to-face meetings held regularly by the ACZISC provide an important and regular opportunity to develop relationships between individual practitioners and organizations. The 74 ACZISC meetings were attended by 1858 participants from January 1992 to February 2015. These meetings had 25–30 participants, many of them regular attendees. Due to constraints on travel in both provincial and federal governments, participation has declined since October 2011, reducing interprovincial interactions. To partially compensate for this decline, the ACZISC has been using teleconference and webinar technology and videos of presentations on YouTube. The website and newsletter are also important tools for building and sustaining a CoP by maintaining communication between face-to-face meetings.

A 2011 survey of members asked them to assess the value of the group's products or output. Figure 15.1 summarizes the results of this survey, covering meetings, the Coastal Activities Report, the *Coastal Update* newsletter, and the COINatlantic.ca website (which replaced ACZISC's). From the survey, the meetings were considered by all respondents to be valuable (ACZISC 2011). Three other products, the Coastal Activities Report (part of the minutes from each meeting), the *Coastal Update* e-newsletter, and the COINatlantic.ca website, were considered to be valuable by the majority of respondents. A survey of e-Newsletter recipients indicated that most people use it for their work in

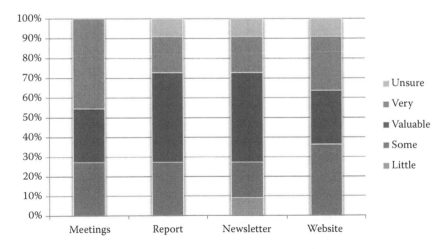

FIGURE 15.1
The value of four ACZISC products to ACZISC members, based on a survey.

support of research, policy development, conservation, community development, and general awareness (ACZISC 2013). It is reasonable to assume that the other products were used in the same way.

15.3.3 Contributions of the Coastal and Oceans Sector to Local Economies and Well-Being

To inform and influence investment by governments in the management of coastal and ocean resources, it is important to document the contribution of the oceans to the economies of Atlantic Canada and the well-being of Atlantic Canadians. The ACZISC initiated and coordinated the first economic study for Nova Scotia, which was eventually repeated in the other Atlantic provinces. Two particular contributions in this area are described in this chapter. An initial study, "Estimating the Economic Value of Coastal and Ocean Resources: The Case of Nova Scotia," using statistics from 1994, was published in February 1998. The report concluded that coastal and ocean resources including both the private sector, for example, fishing and fish processing, and the public sector, for example, the Department of National Defense, have a significant impact on the economy (i.e., total gross domestic product (GDP) impact of $2.8 billion, or 17.5%), well-being (i.e., total household income of $2.6 billion, or 24.8 %), and total employment (i.e., 93,507 jobs, or 24.8%) of Nova Scotians (Mandale et al. 1998).

In 2006, the ACZISC conducted a study jointly with Canmac Economics Limited for the Atlantic Canada Opportunities Agency (ACOA), resulting in the report "The Ocean Technology Sector in Atlantic Canada" (ACOA 2006). The oceans technology sector consisted of 137 firms and contributed 5298 person years of employment, $201.8 million of household (labor) income, and $280.9 million of GDP on an annual basis, in direct, indirect, and induced (spin-off) regional economic activity.

Although no direct evidence exists of the impact of these studies, the four provinces embarked on coastal strategies after the publication of the economic studies. At the federal level, Canada's Ocean Action Plan states, "our oceans have been a very dynamic growth sector for Canada's economy, and currently generate more than $22 billion directly through ocean-related industries" (DFO 2005).

15.4 Influencing Accessibility to Coastal and Ocean Data and Information for ICOM

The ACZISC Strategic Plan Priority 2 is "collaborative sharing of data and information between members and with the wider ICOM CoP on the ICOM issues of priority to members" (ACZISC 2012).

The Commissioner of the Environment and Sustainable Development, in the Office of the Auditor General of Canada, recognized the importance of accessibility to information in the Fall 2010 report to the Parliament of Canada (Vaughan 2010). This report states that "solid, objective, and accessible information is essential to identify and respond to the quickening pace and complexity of environmental change, in Canada and globally" (Vaughan 2010). The ACZISC has always promoted data and information accessibility in its meetings, workshops, and other products and services, and worked to bring together and increase understanding of the complementary roles between two parts of the CoP, the ICOM policy developers and managers, and the geospatial data and information specialists.

15.4.1 Mapping the Coast

The ACZISC has influenced coastal mapping in Atlantic Canada through working groups, projects, and workshops. Six workshops from 1996–2015 contributed to the integration of geospatial data into mapping information to support ICOM. The workshops were designed to share information of mapping activities to members of the ICOM CoP, to disseminate information about mapping best practices including accessibility to mapping data and information, to encourage adoption of common mapping standards, and to promote collaboration. Three of the workshops discussed laser induced detection and ranging (LiDAR) for coastal mapping. The most recent were organized in conjunction with the CoastGIS series of symposia on geographic information systems and computer mapping for coastal zone management in collaboration with the Province of British Columbia (in 2013), the United States Army Corp of Engineers, and South Africa's Council for Scientific and Industrial Research (in 2015). Atlantic Canadian agencies initially expressed reluctance to adopt LiDAR technology for acquiring elevation information over traditional photogrammetric techniques. However, after the 1999 workshop and improvements in LiDAR technology and the demonstration of the efficacy for coastal management purposes, many LiDAR surveys have been conducted in Atlantic Canada, although no organized regional campaign for planning or cost sharing has yet to be established. In 2015, the ACZISC recommended that the federal–provincial Regional Committee for Coastal and Ocean Management (RCCOM) initiate a collaborative project to document existing remote sensing data sets, adopt a policy of data sharing between all RCCOM member agencies, and develop a plan for priority product development and data acquisition to support ICOM based on a risk assessment exercise (ACZISC 2015).

In 1994, the ACZISC was selected as the Canadian coordinator for the East Coast of North America Strategic Assessment Project (ECNASAP), conducted in collaboration with the United States National Oceanic and Atmospheric

Administration's (NOAA) Strategic Environmental Assessment Division. The ECNASAP was a groundbreaking project. It was designed to maximize use of existing data, information, and knowledge by developing comprehensive information and map products on living resources, environmental characteristics, and anthropogenic impacts on the environment and its resources, for the east coast of North America (Brown and Butler 1994, pp. 860–874). Inshore and offshore case studies were developed. Its major contribution to ICOM was an increased awareness of the benefits of ecosystem-based multiple species analysis of fish abundances over single species stock assessment (Brown et al. 1996; DFO 1996).

15.4.2 Managing Data for ICOM: Building the Marine Geospatial Data Infrastructure

Building a marine geospatial data infrastructure (MGDI) for the sharing of data to support ICOM has been an objective of the ACZISC from its inception. The MGDI is a "system of data/information products and enabling technologies that are critical to sustainable development and management of freshwater, coastal and ocean areas" (Gillespie et al. 2000, pp. 15–24).

The ACZISC published its first database directory in October in 1992, a major accomplishment. The database directory became an online resource in 1996, and was followed by the Atlantic Coastal Information Portal (ACIP) in May 2005. The ACZISC established the Coastal Information Technology Architecture Plan and Standards Working Groups and supported the longer-term deliberations of the Geomatics Working Group to advance the concepts originating in the Coastal and Ocean Information Network (COIN) and Inland Waters Coastal and Ocean Information Network (ICOIN) initiatives (Butler et al. 1988; Hamilton 1989). The concepts were also advanced by an initiative on the west coast of Canada called COINPacific (BCMSRM 2003).

The GeoConnections Program of Natural Resources Canada was established in 1999 to play a key role by building the Canadian Geospatial Data Infrastructure (CGDI). In the second phase of the program, focused on developing and expanding the CGDI for users, the ACZISC was active in the deliberations of the marine user node of GeoConnections to build an MGDI. However, the marine user node was disbanded in 2010, with new emphasis being placed by the program on policy, standards, outreach, and CGDI integration. Consequently, the development of the MGDI has been dispersed and slowed.

In 2008, with GeoConnections funding, the ACZISC developed COINAtlantic, web-based tools to facilitate the discovery and visualization of spatial data. The first deployment of the COINAtlantic Search Utility (CSU) tool, that replaced ACIP, had a search functionality that was linked to the GeoConnections Discovery Portal, a metadata database developed and promoted by Natural Resources Canada (Sherin et al. 2010, pp. 73–85). Subsequent versions of the tool have used the ubiquitous Google search application program interface (API) to discover and display spatial data resources that include web mapping services (WMS) and keyhole markup language (KML) files to build maps to support ICOM practitioners (Boudreau and McKenna 2014). In addition to the CSU, the ACZISC developed the COINAtlantic Geocontent Generator (CGG). The CGG is a web-based tool that allows users, particularly smaller organizations contributing to ICOM, to build basic metadata for an organization, publication, project, or data set along with a spatial feature (polygon, line, point, etc.). It saves the information in a web-accessible KML file, and facilitates the discovery of the metadata by the Google search engine. These tools have been expanded, modified, and enhanced, and continue to contribute to building a MGDI for Atlantic Canada and to improve the discoverability and accessibility of data needed for ICOM.

The MGDI can only be effective if data-providing agencies that hold spatial data relevant to ICOM deliver data to the Internet using accepted standards. In 2014, the ACZISC published a data accessibility self-assessment tool (CDAST) to encourage data-providing agencies to improve the effectiveness of data sharing, as well as data-dissemination policies and processes. The CDAST is based upon four sets of principles: the Organisation for Economic Co-operation and Development's (OECD) Principles and Guidelines for Access to Research Data from Public Funding; the Operating Principles for Canada's Open Government site; the principles from the United States' Open Data Policy-Managing Information as an Asset; and the G8 Open Data Charter and Technical Annex. In mid-2015, the CDAST was in a pilot phase for testing, and initial users were being asked for their comments, to assist with future modifications and improvements.

Through its COINAtlantic project, the ACZISC has become active in the International Coastal Atlas Network (ICAN), a project of the International Ocean Data Exchange (IODE) program of UNESCO's Intergovernmental Oceanic Commission (IOC). Participating in this network provides access to worldwide experts in delivering spatial data and information to ICOM practitioners (Wright et al. 2010, pp. 229–238). The ACZISC has also joined the Data Sources Working Group of the Canadian Geomatics Community Roundtable (now GeoAlliance Canada) as an additional vehicle to influence the national availability of marine spatial data resources in support of ICOM.

In its recent report on ocean science in Canada, the Council of Canadian Academies found "that challenges exist in achieving geographical coverage and integration of data management" for Canada's ocean observing systems (CCA 2013, p. xiv). The ACZISC is working with the Marine Environmental Observation Prediction and Response (MEOPAR) network to address these challenges (MEOPAR 2014). A recent MEOPAR data management workshop recommended the establishment of an ocean observation–data management CoP, an essential building block toward a functional MGDI and access to ocean observation data and information for ICOM.

Canessa et al. (2007) concluded that the vision and concept of the MGDI have been consistent for many years and that local and regional projects, such as COINAtlantic, have been bringing the implementation of the MGDI closer to completion. However, they also concluded that impediments to institutional collaboration are still present. More recently, commitments made to "Open Government and Open Data" by the Government of Canada to the Open Data Partnership (Government of Canada 2014), and the Government of Newfoundland and Labrador (GNL n.d.), as well as discussions and initiatives in the other Atlantic Provinces, suggest these impediments may be disappearing.

15.5 Facilitating Relationships within the CoP

The ACZISC Strategic Plan Priority 3 is "encouraging the engagement of organizations in the ACZISC that is reflective of the diversity of the ICOM CoP."

Since its inception, the ACZISC has worked to bring CoP members with a *science and technology* interest together with CoP members with a *policy* interest. In addition, coastal and ocean management often has a strong need for data and information to be presented geospatially, especially with the growth of marine spatial planning as a tool for ICOM. The development of both the ICOM and Geomatics Working Groups was an indication of the range in interests and expertise in the CoP. The challenge was to keep the attention of those in the CoP who may not be as interested in the technical aspects of ICOM if their focus is policy issues, and vice versa.

Through meetings, workshops, training sessions, and working groups the ACZISC has engaged the diversity of the ICOM CoP. Varying interests, people, perspectives, policy priorities, departments, and mandates have been brought to the ACZISC. "Breaking down silos" between practitioners is one of the goals of ICOM and ACZISC. ACZISC's success in this area is difficult to measure. However, in a 2011 survey of members, 84% of respondents considered the ACZISC to be effective at building relationships (25% somewhat effective, 34% effective, and 25% very effective) (ACZISC 2011). Examples of this effectiveness include data, information, and funding sources being shared at meetings, facilitated by direct contact. Partnerships and collaborations have begun between members attending meetings, illustrating the influence of the ACZISC process.

15.6 Discussion

15.6.1 The Changing Organizational Environment

The organizational environment has changed over time for the ACZISC. The ACZISC and its Secretariat are no longer an official organ of provincial governments, that is, the Council of Maritime Premiers, as the interest in joint enterprises among the provinces has waned. The Committee still exists and meets, but the Secretariat is now employed by an incorporated society, the ACZISC Association. The ACZISC addresses issues important to ICOM through building awareness, sharing information, education, and influence. Indeed, the ACZISC has become less of a *committee* with a mandate to *steer*, and more of a network of persons and organizations with responsibilities for ICOM and an interest in information sharing to improve the effectiveness of ICOM practices.

15.6.2 Shifting Coastal and Ocean Policy

With the proclamation in Canada of the Oceans Act in 1996 and establishment of the Oceans Action Plan in 2005, more formal federal–provincial venues for consultation were established, with regional committees on coastal and ocean management for the Maritimes and Newfoundland and Labrador. This meant less emphasis on federal–provincial policy conversations at the ACZISC.

In addition, the implementation of policy within the federal government with respect to ICOM has changed (Ricketts and Hildebrand 2011, pp. 4–19). For example, the Eastern Scotian Shelf Integrated Management (ESSIM) Plan developed over a number of years in collaboration with a Stakeholder Advisory Committee was not endorsed by the Minister of Fisheries and Oceans (McCuaig and Herbert 2013; see also Chapter 12 in this volume). Although some progress has been achieved on ESSIM Plan objectives, "many of the management strategies that focused on the ESSIM area specifically, and which would have required targeted resources and multi-sector collaboration within the collaborative governance framework, were not pursued" (Ricketts and Hildebrand 2011, p. 54). Indeed, the most recent management plans developed for larger bioregions rather than large ocean management areas (LOMAs) are constrained to planning solely within the mandate of the Department of Fisheries and Oceans (Canada) (DFO). This situation contrasts with initiatives in the United States and Europe where integrated and cross–jurisdictional approaches are supported by governmental policies for coastal and ocean management.

The formation of the regional federal–provincial committees in the Atlantic Provinces has been an important contribution to coastal and ocean management in Canada. However, its focus is only on cooperation among federal and provincial government bodies, with less planning among multiple sectors that are part of the CoP and the ACZISC. Nonetheless, the federal–provincial committees have benefited from the experience of interdepartmental and intergovernmental discussion and collaboration in the ACZISC, whose focus remains on ICOM.

15.6.3 Data Accessibility for ICOM

Despite the slow progress in building a MGDI, the data accessibility policy has been strongly influenced by the "open government and open data" movement, nationally and internationally. Canessa et al. (2007) concluded that the vision and concept of the MGDI has been consistent for many years and that local and regional projects, such as COINAtlantic, have been bringing the implementation of the MGDI closer to reality. However, Canessa et al. (2007) also concluded that impediments to institutional collaborations still remain.

The commitments made to open government and open data by the Government of Canada to the Open Data Partnership (Government of Canada 2014), the Government of Newfoundland and Labrador (GNL n.d.), and discussions and initiatives in other Atlantic Provinces and at the municipal level, suggest that these impediments may be disappearing. More recently, the CGDI and the open data initiative became more strongly linked with the development of the Federal Geospatial Platform (Moore 2015). These developments, along with similar initiatives within provincial and municipal governments, should make more data accessible for coastal and ocean management in the near future. The ACZISC will continue to encourage its members to share data and work within the GeoAlliance Canada's Data Sources Working Group and MEOPAR to encourage this trend and the interoperability between the approaches adopted by each province.

15.6.4 Embracing the Diversity of the CoP

The ACZISC has been effective in building and sustaining relationships between federal and provincial agencies, universities, and nongovernmental organizations, as part of the ICOM CoP. However, there are components of the CoP that are not effectively engaged. Aboriginal organizations in Atlantic Canada have been present at many ACZISC meetings but an ongoing relationship with this part of the CoP has been tenuous. Although there has been some engagement of industry and industry associations, this has been primarily with the geospatial and environmental consulting sectors. Engagement with marine industries and their associations such as fisheries, aquaculture, oil and gas, and so on, has been lacking and is also a recognized gap. There is no ongoing relationship yet developed with the municipal level of government, particularly through their associations in Newfoundland

and Labrador and Nova Scotia. Initiatives to engage these sectors of the ICOM CoP are planned.

15.6.5 From Data to Wisdom for Decision Support

As shown in Figure 15.2, modified from Canessa et al. (2007), there are two ways to view how data supports the decision-making process: a *data view* and a *decision view*. The levels progress to higher and higher levels of integration, rising from disconnected data sets, to interpretations of that data (becoming information), to integrated comprehensive pictures or knowledge, and finally to enabling decisions, each level building on the one below. Selected ACZISC activities and products that have contributed to each level are displayed at each level of the diagram (Figure 15.2).

At the data and data-gathering level, activities are shown for providing accessibility and discoverability of data needed to support ICOM. These include the various forms of database directories from paper to web-based metadata tools, the promotion of good data management, and data accessibility through the MGDI User Node for GeoConnections, GeoAlliance Canada, Data Sources Working Group, MEOPAR Ocean Observation Data Management CoP, and CDAST.

At the information and interpretation level, there are the presentations and workshops organized by the ACZISC, the *Coastal Update* newsletter, and the

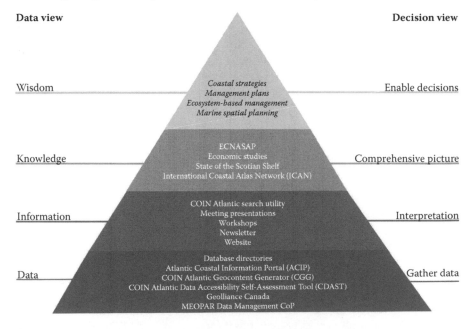

FIGURE 15.2
From data to wisdom for decision support in ICOM. (Adapted from Canessa, R. et al. 2007. *Coastal Management* 35: 105–142.)

COINatlantic.ca website that provide information on a wide range of topics related to ICOM. The CSU is included at this level because it provides the user with the opportunity to pull many layers of data together into a map, enabling basic interpretation of the relationships between the data layers.

Finally, the last layer that the ACZISC directly contributes to is the knowledge and comprehensive picture. Included here are accomplishments of the ECNASAP project that demonstrated a real world implementation of ecosystem-based analysis, the comprehensive coastal pictures obtainable from the coastal web atlases developed by the members of ICAN, the economic and social impact of oceans and coasts demonstrated by the economic studies, and finally state of the environment reporting in the "State of the Scotian Shelf Report." The top layer of wisdom and enabling decisions is outside the control of the ACZISC. However, it has indirectly influenced the development of coastal zone strategies, ecosystem-based management implementation, and management plans for bioregions and LOMAs, and laid a foundation for marine spatial planning.

15.7 Conclusion

An overwhelming amount of data and information for ICOM is potentially available from the Internet, and this information resource is growing rapidly. The ACZISC program strives to aid its members and the CoP in dealing with this challenge within the domain of information for use in ICOM. The ACZISC has coined this issue the *billion URL challenge* (Sherin et al. 2011, pp. 123–131). For over 23 years, the ACZISC has fostered cooperation in Atlantic Canada in the areas of integrated coastal and ocean management, coastal mapping, and geomatics. The nature of the ACZISC has changed over time from an "official" committee of an interprovincial agency to a more loosely organized network, keeping the original goals and objectives. The ACZISC has delivered valuable products to the ICOM CoP in Atlantic Canada that have directly and indirectly influenced strategy and policy development for oceans and coasts in Atlantic Canada. It has been effective in facilitating productive relationships between members of the CoP. The ACZISC's work has adapted to the changing coastal and ocean management and data accessibility policy environments in Canada.

As mentioned, the Commissioner of the Environment and Sustainable Development in 2010 stated that "solid, objective, and accessible information is essential to identify and respond to the quickening pace and complexity of environmental change, in Canada and globally" (Vaughan 2010). The ACZISC will heed the Commissioner's message and continue to facilitate accessibility to data and information essential for effective and inclusive management of coastal and ocean resources and ecosystems.

Acknowledgments

We are very grateful to the reviewers and editors for their extensive reviews of this chapter. We also acknowledge the contributions of ACZIS members and their organizations to the success of ACZISC initiatives since its inception.

References

ACOA (Atlantic Canada Opportunities Agency). 2006. *The Ocean Technology Sector in Atlantic Canada Volume 1: Profile and Impact. Volume 2: Potential Public Sector Demand.* http://publications.gc.ca/collections/collection_2010/apeca-acoa/Iu89-4-41-1-2006-eng.pdf and http://publications.gc.ca/collections/collection_2010/apeca-acoa/Iu89-4-41-2-2006-eng.pdf.

ACZISC (Atlantic Coastal Zone Information Steering Committee). 2009. The role of the ACZISC in integrated coastal and ocean management policy development and implementation in Atlantic Canada. http://coinatlantic.ca/images/stories/documents/role%20of%20aczisc-icom.pdf.

ACZISC (Atlantic Coastal Zone Information Steering Committee). 2011. Preliminary results from ACZISC member consultations. http://coinatlantic.ca/documents/aczisc_meeting_presentations/63ACZISC_PRIORITIES_2011.pdf.

ACZISC (Atlantic Coastal Zone Information Steering Committee). 2012. ACZISC strategic plan 2012–2017. http://coinatlantic.ca/documents/aczisc_miscellaneous%20_documents/ACZISC%20Strategic%20Plan%202012-2017%20FINAL.pdf.

ACZISC (Atlantic Coastal Zone Information Steering Committee). 2013. ACZISC coastal update newsletter survey results. http://coinatlantic.ca/documents/aczisc_meeting_presentations/69Coastal_Update_Survey_Results_2013.pdf.

ACZISC (Atlantic Coastal Zone Information Steering Committee). 2015. Inventory of remote sensing technologies for coastal and ocean management: Final report and recommendations. Report prepared for Environment Canada on behalf of the Regional Committee on Coastal and Ocean Management. Halifax, NS: ACZISC.

BCMSRM (British Columbia Ministry of Sustainable Resource Management). 2003. Benefit analysis: Towards a cooperative ocean information network for the Pacific region, COINPacific. http://coinatlantic.ca/images/stories/documents/COINAtlantic/other%20coins%20-%20coinpacific%20benefit%20analysis. pdf.

Boudreau, P. and J. McKenna. 2014. COINAtlantic—A better way to implement a geospatial portal. *Directions Magazine.* http://www.directionsmag.com/entry/coinatlantic-a-better-way-to-implement-a-geospatial-portal/425845.

Brown, S. K. and M. J. A. Butler. 1994. ECNASAP–Towards international collaboration in strategic environmental assessment. In *Coastal Zone Canada 1994 Conference Proceedings*, Halifax, Nova Scotia, 20–23 September 1994, Volume 1, 860–874. Dartmouth, NS: Coastal Zone Canada Association.

Brown, S. K., R. Mahon, K. C. T. Zwanenburg, et al. 1996. East coast of North America groundfish: Initial explorations of biogeography and species assemblages. Silver Spring, MD: National Oceanic and Atmospheric Administration, and Dartmouth, NS: Department of Fisheries and Oceans Canada. http://www.dfo-mpo.gc.ca/Library/229282.pdf.

Butler, M. J. A., C. LeBlanc, and J. M. Stanley. 1988. Coastal and Ocean Information Network (COIN). Report submitted to the Canadian Hydrographic Service, Department of Fisheries and Oceans by Maritime Resource Management Service Inc.

Canessa, R., M. J. A. Butler, C. LeBlanc, et al. 2007. Spatial information infrastructure for integrated coastal and ocean management in Canada. *Coastal Management* 35: 105–142.

CCA (Council of Canadian Academies). 2013. Ocean science in Canada: Meeting the challenge, seizing the opportunity. The Expert Panel on Canadian Ocean Science. Ottawa, ON: CCA. http://www.scienceadvice.ca/uploads/eng/assessments%20and%20publications%20and%20news%20releases/oceans_2/oceans_fullreporten.pdf.

DFO (Department of Fisheries and Oceans Canada). 1996. East coast of North America strategic assessment project: Groundfish atlas. http://coinatlantic.ca/documents/aczisc_miscellaneous%20_documents/96ecnasap_groundfish.pdf.

DFO (Department of Fisheries and Oceans Canada). 2005. Canada's oceans action plan for present and future generations. Ottawa, ON: Fisheries and Oceans Canada. http://www.dfo-mpo.gc.ca/oceans/publications/oap-pao/pdf/oap-eng.pdf.

Gillespie, R., M. Butler, N. Anderson, et al. 2000. MGDI: An information infrastructure to support integrated coastal management in Canada. *Geocoast* 1: 15–24.

GNL (Government of Newfoundland and Labrador). n.d. Open Government Initiative Framework. St. John's, NL: GNL. http://open.gov.nl.ca/pdf/OpenGovernmentInitiativeFramework.pdf.

Government of Canada. 2014. *Canada's Action Plan on Open Government 2014–2016:* Ottawa, ON: Government of Canada.

Hamilton, A. C., ed. 1989. *ICOIN: Proceedings of a Forum on the Inland Waters, Coastal and Ocean Information Network.* Fredericton, New Brunswick, 13–14 June 1989. Fredericton, NB: Champlain Institute.

LRIS (Land Registration Information System, Council of Maritime Premiers). 1991. Summary report coastal zone information management workshop, Amherst, Nova Scotia. http://coinatlantic.ca/documents/aczisc_workshop_reports/Summary_Report_Coastal_Zone_Info_Mang_Wrkshp_1991.pdf.

Mandale, M., M. Foster, and J. Plumstead. 1998. *Estimating the Economic Value of Coastal and Ocean Resources: The Case of Nova Scotia.* Prepared for the Oceans Institute of Canada and the Atlantic Coastal Zone Information Steering Committee.

MEOPAR (Marine Environmental Observation Prediction and Response Network). 2014. MEOPAR data management workshop executive summary. Montreal, Quebec, 24–25 March 2014.

McCuaig, J. and G. Herbert, eds. 2013. *Review and Evaluation of the Eastern Scotian Shelf Integrated Management (ESSIM) Initiative.* Dartmouth, NS: Fisheries and Oceans Canada.

Moore, A. 2015. The federal geospatial platform–An update. Presentation to ACZISC Meeting #75, Moncton, New Brunswick, 20 May 2015. http://coinatlantic.ca/documents/aczisc_meeting_presentations/75FGP.pdf.

Ricketts, P. J. and L. Hildebrand. 2011. Coastal and ocean management in Canada: Progress or paralysis? *Coastal Management* 39: 4–19.

Rounce, A. and N. Beaudry. 2002. Using horizontal tools to work across boundaries: Lessons learned and signposts for success. CCMD roundtable on horizontal mechanisms, chaired by James Lahey, Canadian Centre for Management Development. http://publications.gc.ca/Collection/SC94-90-2002E.pdf.

Sherin, A., P. Boudreau, and A. Baccardax Westcott. 2011. Meeting the billion URL challenge of the coastal and ocean manager. In *CoastGIS 2011 Conference Proceedings, Oostende*, Belgium, 6–8 September 2011, 123–131.

Sherin, A., M. Butler, C. LeBlanc, et al. 2010. Coastal ocean information network (Atlantic): From concept to reality: A status report. In *Coastal and Marine Geo-Information Systems: Applying the Technology to the Environment*, edited by D. R. Green and S. D. King, 73–85. London, UK: Springer.

Vaughan, S. 2010. 2010 Fall report of the Commissioner of the Environment and Sustainable Development. Ottawa, ON: Office of the Auditor General of Canada. http://www.oag-bvg.gc.ca/internet/English/parl_cesd_201012_e_34435.html.

Wenger, E. 1998. *Communities of Practice: Learning, Meaning, Identity*. Cambridge, UK: Cambridge University Press.

Wright, D. J., V. Cummins, and E. Dwyer. 2010. The international coastal atlas network. In *Coastal Informatics: Web Atlas Design and Implementation*, edited by D. J. Wright, V. Cummins, and E. Dwyer, 229–238. Hershey, PA: IGI Global.

16

A Career-Based Perspective of Science–Policy Linkages in Environment Canada: The Role of Information in Managing Human Activities in Our Ocean Spaces

Peter G. Wells

CONTENTS

16.1 Introduction

Scientific careers in applied science within government, that is, public service, inevitably involve working at the interface of science and its applications in program management, decision-making, and policy formation. My career as an aquatic (marine) biologist, scientist, and science manager for the federal government in Canada spanned the period of 1969–2006. Studies conducted for Environment Canada (herein called EC) covered a wide range of aquatic environmental problems, as the responsibilities of the department expanded with the country embracing its environmental responsibilities. As shown in Section 16.4, most of the applied projects and research were linked, directly and indirectly, to management, policy directives, and legislation. The data and information were used to better manage the coastal zone, through a scientific understanding and enhanced control of water pollution. This occurred despite the fact that the practice of integrated coastal management/integrated coastal and ocean management (ICM/ICOM) in Canada was a vague concept, one newly introduced through the Oceans Act (1996), or one partially in place in Canada through this time period (see Chapter 15 in this book).

This chapter briefly describes examples of aquatic research in support of coastal management conducted under the mandate of EC. Hence, it starts with a description of EC's responsibilities, followed by a description of programs in which I was directly involved, to give examples of the linkage(s) between applied public service science and management needs under current legislation and policies. Much of this work focused on the coastal issues of pollution control, including in watersheds, and marine environmental quality of the Maritime provinces of Canada. Recent activity in the department on the science–policy interface in its science and technology programs is then described. The chapter concludes with a summary of how public service science, in this case with EC, interfaces with policy in the broadest sense and relates to and contributes to coastal management.*

---

* Coastal management is considered to be a component of integrated coastal management (ICM), integrated coastal and ocean management (ICOM), or integrated coastal zone management (ICZM). All deal with coastal waters, and are defined and discussed earlier in this book.

16.2 Environment Canada

This section covers several topics—the early history of EC; its mandate and legislation, early and evolving; its national coastal and marine mandate, including the role of the Environmental Protection Service (EPS) and its water pollution control mandate; the significance of the Government Organization Act of 1979; its international roles and mandate regarding ICM/ICOM; the onset of the national Marine Environmental Quality (MEQ) Program, 1982–1994; and contributions to the National Program of Action (NPA) for the Protection of the Marine Environment from Land-Based Activities (LBA) and the Health of the Oceans (HOTO) initiative. Recent efforts to place more emphasis on the science–policy interface in the department are described in Section 16.6.

EC was established as a new federal department in Canada in 1971, with the merging of existing departments and agencies (e.g., Canadian Wildlife Service (CWS), the Inland Waters Directorate, parts of the Fisheries Research Board of Canada, the Canadian Meteorological Service), and the establishment of the new unit, the EPS. The Fisheries Act, dating from 1867, was strengthened in 1972 with its pollution provisions (section 36) enabling the establishment of regulations and guidelines for industrial pollution control on a sectoral basis. With the stronger Fisheries Act, considerable work commenced in a cooperative way with industry to control air and water emissions of polluting substances on an industrial sector basis, as defined under the Act (see Section 16.4; Pessah and Cornwall 1980; Fisheries Act 1985; Taylor et al. 2013; Wells and Doe 2014). Work also commenced on completely new legislation, the Ocean Dumping Control Act (1975), to prevent pollution from disposal at sea through permits (www.ec.gc.ca), and the Environmental Contaminants Act (1975), to assess and control hazardous chemicals, both Acts being "subsumed into the Canadian Environmental Protection Act in 1988, together with the Clean Air Act, the nutrient provisions of the Canada Water Act, and certain provisions of the Department of the Environment Act" (www.ec.gc.ca/lcpe-cepa/). The Department was EC until 1976, then the Department of Fisheries and Environment to 1979, when the Government Organization Act (1979) was passed, forming the Department of Fisheries and Oceans (DFO) and a new Environment Canada. At that time, all the fisheries and associated research activities, including those dealing with the fate and effects of chemical pollutants in the marine environment, were split off into the new DFO, complicating joint research and monitoring at the scientific level. That event notwithstanding, considerable cooperation and joint projects continued between the two departments in coastal and ocean science, contributing to nascent efforts in ICOM.

EC has also had many international roles and responsibilities to monitor, prevent, and control pollution. For the marine environment, there was the London Dumping Convention (eventually the London Convention 2006) (J. Karau, personal communication) and the MARPOL 73/78 Convention

(MARPOL 1978), coordinated by the International Maritime Organization, and considerable input to the United Nations Convention on the Law of the Sea (UNCLOS) (1982), notably through Part XII on marine pollution. Concern about the effects of land-based pollution on coastal estuaries and waters led to Canada and EC being involved in developing the 1985 United Nations Environment Programme (UNEP) Montreal Guidelines for the Protection of the Marine Environment against Pollution from Land-Based Sources (UNEP 1985; Wells and Gratwick 1988).

It was during this period that the department initiated the national MEQ Program, led by the Ottawa, Ontario, and Dartmouth, Nova Scotia, offices. The MEQ program was important to Canada's contribution to ICM/ICOM. It attempted to tie together the various departmental (EC) activities on marine research, contaminants and wildlife monitoring, state of environment reporting, and interdepartmental cooperation on marine matters. The program made significant progress with its emphasis on reporting on health of the oceans, increasing the profile of the issue of land-based marine pollution, the establishment of the Atlantic Canada Action Program, and contributions to the Interdepartmental Committee on Oceans (Wells et al. 1987; Wells and Rolston 1991; Wells 1996; Vandermeulen 1998; Vandermeulen and Cobb 2004; Wells 2005; L. Hildebrand, personal communication). Despite these successes, in the changing federal landscape of fiscal resources and shared ocean responsibilities, the MEQ program failed to evolve into a fully integrated marine program for the Department. However, it contributed significantly to considerations of what to include in the newly proposed Oceans Act (Wells 1996), led by DFO and passed into law in 1996–1997. MEQ became one of the three pillars of the Oceans Act; with its other objectives of establishing large ocean management areas (LOMAs) and marine protected areas (MPAs) for Canada's coastal areas, the Oceans Act became a key component of Canada's effort to demonstrate ICM/ICOM in waters under its jurisdiction.

During the 1990s, following on from its earlier involvement with UNEP on the land-based marine pollution issue, EC was active as a lead for Canada's NPA for the Protection of the Marine Environment from LBA, formally started in 2000. According to J. Karau (personal communication),

> the NPA was an early attempt at applying ICZM, and Canada was the first country to develop such a plan under UNEP. The early MEQ program and its Health of the Oceans report (Wells et al. 1987; Wells and Rolston 1991) were critical building blocks for the NPA.

The NPA program of UNEP continues, but Canada's participation has waned in recent years, despite the importance of managing land-based marine pollution or LBA as a component of an effective ICOM program.

In 2007, another HOTO initiative of the Government of Canada began, with EC working together with various other federal departments (DFO,

Transport Canada, Parks Canada, and Indian and Northern Affairs Canada). HOTO was one of the new Oceans Act initiatives (the others being MPAs and Integrated Coastal Marine Areas or Areas of Interest (ICMAs/AOIs)). EC was formally recognized by DFO as having a role in developing the national network of MPAs. Such areas were now considered more broadly to mean all coastal and oceanic areas that offer some level of protection to ecosystems, wildlife, and living resources (Stewart 2010), given the CWS and EC's long history with wildlife conservation areas and refuges.

Not to be forgotten is the long-standing contribution to marine environmental protection of EC's Arctic and Marine Oilspill Program (AMOP) conference series. This series is now over 30 years old; through new research, accumulated information, and networked experts from many sectors, it has contributed significantly to Canada's ability to respond to maritime emergencies (spilled oil and chemicals, gas blowouts, and so on) in ocean areas under its jurisdiction, both nearshore and offshore.

As a federal agency, EC has a complex and continuing responsibility regarding Canada's ocean spaces, especially for its watersheds and estuaries, and along its coasts. Its place among the other departments and jurisdictions for matters dealing with ICOM has become clearer with time, although perhaps not with the urgency needed. In areas ranging from pollution control to climate change to wildlife conservation, EC will continue to contribute to the science, information, policy, and management initiatives underlying ICOM for Canada in the twenty-first century (also see Ricketts and Harrison 2007; Ricketts and Hildebrand 2011).

16.3 Career-Based Perspective of Science–Policy Linkages and the Role of Information

As shown previously, the science–policy interface(s) was operationalized early on in specific programs of EC and considered critical to the function of the department. There were many functions and governance levels in EC in which science interacted with policy, especially at its headquarters in Ottawa. As well, there has been a more recent emphasis on the importance of such interactions (see Section 16.6).

By analyzing selected case studies based on one's own work experience in EC and admittedly limited perspective, largely dealing with Canada's coasts and oceans, what can be learned specifically about the use of scientific information in program management, decision-making, and policy formation? Do any new insights emerge about the function of the so-called *science–information–policy* or *science–policy universe*? Much of my career was spent working on applied coastal and marine problems under a defined mandate in EC. In this context, some light may be shed on the importance of understanding the science–policy interface, not only for this department, but, importantly, for advancing ICM and ICOM.

My research (alone and with coworkers) in EC spanned a period overlapping the personal computer and information revolution: typewriters, secretaries, fax machines, mainframe computers, and labor-intensive laboratory and field techniques eventually gave way to desk and laptop computers, smart phones, e-mail and instant communication, and fully automated experimental techniques. This revolution has changed the generation, flow, and accessibility of data and information and the way the users of information (in this case, environmental managers, decision makers, and policy makers) can both acquire information and influence the generation of new information in science programs within their purview. The past few decades (1980s onwards) have been a transitional period for high-level decision makers who are charged with acquiring and applying specific information. The dimensions and dynamics of the science–policy interface or science–management–policy interface have changed as a result of these events.

16.3.1 Case Study Methods

The information sources for the selected case studies were papers and reports from the different programs and projects under analysis, documented in a curriculum vitae (Wells 2015). Each case study is summarized, considering the following questions: (1) Which event, directive, or question led to the study? (2) What did the study consider? and (3) How was the new information used, once published, within the federal government system, with insights on science–policy linkages? A limitation of this approach is that the observations, experiences, and outputs of only one individual were used for the analysis, that is, it is a limited analysis and perspective, being more illustrative than definitive of the use and influence of scientific information in the fields of water pollution, marine environmental quality, and marine environmental management.

16.4 Results and Discussion

Eight case studies (project areas) are presented, spanning the period 1970–2007 and addressing the questions posed in Section 16.3.1. The studies were in the areas of marine oil pollution and water pollution/aquatic toxicology. Table 16.1 illustrates how the various studies link to management and policy and provides an overview and subjective assessment of the operational importance of the science–policy interface. References to the many papers

TABLE 16.1

The Linkages between Science, Management, and Policy Based on Eight Studies Conducted with Environment Canada 1970–2006

Topic of Studies	Scientific Research	Scientific Review	Management-Driven Research	Policy-Driven Research	Specific Outcome of the Research	Importance of the Science-Policy Interface[a]		
						H	M	L
Oil Pollution								
1. Toxicity of crude oil	✓				Fundamental understanding of oil toxicity		✓	
2. Toxicity of refinery effluent	✓		✓	✓	Regulations and guidelines put into place under Fisheries Act	✓		
3. Toxicity of oil dispersants		✓	✓	✓	Dispersant use guidelines for Canada	✓		
4. United Nations reviews		✓	✓	✓	Basic understanding of oil pollution in the oceans	✓		

(Continued)

TABLE 16.1 (CONTINUED)

The Linkages between Science, Management, and Policy Based on Eight Studies Conducted with Environment Canada 1970–2006

Topic of Studies	Scientific Research	Scientific Review	Management-Driven Research	Policy-Driven Research	Specific Outcome of the Research	Importance of the Science–Policy Interface[a]		
						H	M	L
Water Pollution								
5. Toxicity of industrial effluents	✓		✓	✓	Industrial pollution controlled by Fisheries Act	✓		
6. Toxicity of pesticides	✓		✓	✓	Risks associated with pesticides reduced		✓	
7. Sediment toxicity	✓		✓		Basic understanding of sediment quality and effects			✓
8. Gulfwatch Program: chemical monitoring	✓		✓	✓	Data used for shellfish safety and water quality assessment	✓		

[a] H, high; M, medium; L, low.

and reports generated from the studies have been limited for reasons of space but are exhaustively recorded in Wells (2015).

16.4.1 Oil Pollution in the Ocean and Oil Spill Countermeasures (1970–2006)

16.4.1.1 Effects of Crude Oil on Lobsters

This study on the sublethal effects of Venezuelan crude oil on lobsters was initiated due to concerns about the fate and effects of oil pollution in Canadian marine waters, especially the effects on valuable fisheries resources and marine wildlife (Wells 1976). The tanker *Arrow* had spilled Bunker C oil into Chedabucto Bay, Nova Scotia, in early February 1970, threatening local coastlines and fisheries. Due to this spill and earlier ones (*Torrey Canyon* 1967; *Florida* 1969; Santa Barbara blowout 1969), both government and industry expressed a need for a greater understanding about the fate and effects of oil in marine waters, as well as for information to strengthen the Fisheries Act pollution provisions and to protect important fisheries. Hence, the research was driven by practical concerns of the oil and shipping industries, including refineries, the IMO, the federal government, and academic interest. Fundamental scientific questions and policy needs drove the research and interacted in the development of the research.

On the Atlantic coast, one of the most valuable fisheries is for lobsters (*Homarus americanus*). As little was known at the time about the susceptibility of the lobsters' young life stages (larvae and post-larvae) to aquatic pollutants, research on the acute and sublethal effects of crude oil on these organisms was conducted during 1970–1976 as a doctoral study (Wells 1976; Wells and Sprague 1976). It was supported by the Canadian Natural Sciences and Engineering Research Council, Imperial Oil Limited, the Huntsman Marine Laboratory (now the Huntsman Marine Science Centre), and the Department of Fisheries and Forestry (subsequently Fisheries and Environment Canada). Knowledge gained during the research, especially pertaining to the high sensitivity, that is, low toxicity thresholds, of the larvae, contributed to later toxicity studies on oil spill dispersants (see below), as well as to reviews on oil pollution and spill control agents requested by the United States National Research Council (Washington, DC) and the IMO. The science, though largely undirected or basic research, was very tightly fitted to policy and legislative needs in Canada on pollution control and fisheries conservation. This work contributed to both national and international policies on the response to, and management of, marine oil spills.

16.4.1.2 Oil Refinery Toxicity Study

The Fisheries Act was revised in 1972 with the addition of pollution control provisions that included the need for aquatic toxicological information on industrial effluents at source, that is, before the effluents are released into receiving waters. Under the Act, oil refinery regulations and guidelines were being developed; information was needed on the detailed chemical characteristics of the various refinery effluents that operationally were discharged into receiving waters, for example, the Bay of Fundy or Halifax harbor, including a description of their acute toxicity to fish as protected under the Act. The established legislation and the policy for requiring site-specific toxicity data on effluents were directing the science. Hence, a study was conducted for Fisheries and Environment Canada in 1972 on the toxicity of final effluents from the Saint John Irving Oil refinery on lobster larvae, using the techniques developed for the crude oil studies (see Section 16.4.1.1). The results were published in a technical report (Wells 1973) and reported in government correspondence to a government–industry technical committee addressing details of the regulations and guidelines proposed for this industry. The science, in this case applied, was very tightly fitted to legislative needs, the policy of limiting pollution from this industry already having been made.

16.4.1.3 Oil Spill Dispersant Toxicity Program Studies, Environment Canada

Research was conducted from 1973 to 1985 as part of the development of dispersant use guidelines and product approval for Canada (Environment Canada 1984). A policy decision had been made by oil spill emergency personnel within Fisheries and Environment Canada (Ottawa) to consider industrially produced dispersants as a countermeasure option at marine oil spills, but to only use products known to be effective at oil dispersal and of low acute toxicity to fish, as broadly defined in the Fisheries Act. Hence, the research was conducted in response to a policy decision, though no formal written policy for dispersant use was in place until 1984.

Results from a formal program on dispersant effectiveness and toxicity testing were considered crucial to the development of an acceptable policy. This work was conducted in the EC toxicity laboratory at the Bedford Institute of Oceanography (BIO), Dartmouth, Nova Scotia, in the 1970s (Doe and Wells 1978; Wells 1984; Wells and Doe 2014). The initial research results were reported in internal reports to EC managers, who then compiled a preliminary list of dispersant products approved for use at Canadian spills. As well, a summary paper was written for an international conference (Doe and Wells 1978), an invited talk was given at a Gordon Research Conference on oil spills by Wells (1984), and the development of the Canadian dispersant use guidelines was co-led by the BIO research team (Environment Canada 1984). Several other research studies on dispersants were conducted and advice on testing procedures was provided to governments in the United Kingdom and Brazil (Araujo et al. 1987). As a result of the collective studies, a definitive list of acceptable dispersant products was made for Canada, giving guidance to authorities in the Canadian Coast Guard and oil industry charged with stockpiling them for possible use at spills. Hence, there was a direct link between the science on dispersant effectiveness and toxicity, and the proposed management of marine oil spills, both of which influenced the emerging policy on dispersant use.

16.4.1.4 United Nations Reviews on Marine Oil Pollution

Several UN agencies and the headquarters in New York participate in the Joint Group of Experts on Scientific Aspects of Marine Environmental Protection (GESAMP) and have an interest in the fate and effects of marine oil pollution. Oil is a recognized marine pollutant under MARPOL 73/78 and the London Convention. The IMO has a special interest in marine oil pollution, as it is responsible for all aspects of the safety and functioning of ships and international shipping. The GESAMP Secretariat at IMO requested a general study of global oil pollution in the late 1980s, as an update to an earlier one, and stimulated by the massive *Exxon Valdez* spill of March 1989 in Alaska. Canada, that is, EC, was invited to participate in and lead the science-based review, which was published in 1993 (GESAMP 1993). A second study was requested in the late 1990s to describe all sources of oil pollution from sea-based activities, including operational inputs, and to describe the global spatial and temporal trends in oil pollution (GESAMP 2007). How most of the UN member agencies of GESAMP used the data and reports in decision-making is not directly known, but the 1993 report is well cited in the literature (MacDonald et al. 2004). As well, both reports were tabled at the IMO Marine Environmental Protection Committee (MEPC) meetings, the key decision-making arm of IMO regarding prevention and control of pollution from ships. From the 1980s onwards, many efforts were made by the shipping industry to reduce oil discharges in the ballast water of tankers, operational discharges of waste oils from all types of ships, and accidents in coastal waters, thus showing a direct science–policy–management linkage at the international level.

16.4.2 Water Pollution and Aquatic Toxicity Studies

16.4.2.1 Aquatic Toxicity Testing: Methods Development and Application to Industrial Effluents and Related Chemicals (1974–Present)

Several events led to the extensive toxicity testing of industrial effluents and their component chemicals and fractions in the 1970s and 1980s at the Toxicity Evaluation Section, EPS-EC, BIO. As mentioned previously, the Fisheries Act (now 1985) had been revised in 1972 with section 36 now covering new pollution control requirements. Industrial effluent regulations and guidelines were being developed on a sectoral basis, with provisions for effluent quality. Information on the chemical and biological characteristics of the effluents was required, including a description of their acute toxicity to fish. The established policy for controlling pollution and resulting federal legislation were directing the applied toxicology (science). Importantly, the bioassay provisions of each guideline or regulation were established due to a winning argument that every industrial effluent being discharged into waterways should be acutely non-toxic, that is, nonlethal, to fish (Pessah and Cornwall 1980; Wells and Doe 2014). Many effluents were tested for acute toxicity over the years, including wastewater from pulp mills, oil refineries, mines, chlor-alkali plants, fish processing plants, sewage treatment plants, battery plants, and textile plants.

Scientific results were presented initially as internal governmental reports and directed to a technical committee (the Intergovernmental Aquatic Toxicity Group) and an oversight management committee, with both committees addressing the details of proposed regulations and guidelines. Many of the reports were later published in the EPS regional technical report series. The information was used directly to establish compositional limits for effluents of various kinds; it was exchanged with similar programs across the country, especially for quality control. The scientific studies on effluents and the associated ancillary studies, for example, studies on reference toxicants and studies of effects of various physicochemical parameters, species, fish size, and water quality, on toxicity, were closely linked to the needs of managers and policy makers, as well as to the lawyers in EC who were drafting the relevant guidelines and regulations. The studies were initiated by both the science and management units of EC and guided by the new policies and management needs for water pollution control. This is a clear example of an effective science–policy interface operating to further the protection of Canadian watersheds and coastal waters.

16.4.2.2 Toxicity Studies of Forestry Pesticides (1976–1980)

The forestry industry in New Brunswick in the 1970s used various pesticides in its long battle against the spruce budworm, raising several environmental and human health concerns. Due to effects on wildlife, DDT had been terminated from most uses in Canada by the mid-1970s and was discontinued from all uses in 1985. However, replacement pesticides such as organophosphates and carbamates were being introduced, the active pesticide ingredient being in proprietary formulations. The federal government took the lead in the environmental research, conducted by both EC and DFO; it was considered a high priority due to concerns about effects on water quality and aquatic life. The research focused on the composition and toxicity of the different pesticide products being registered for use, and the toxicity of various formulations being applied. The parent pesticide was always applied in a solvent, and these varied in composition. There were also concerns about human health effects, as pesticides were often sprayed close to residential areas; this research was undertaken by Health Canada. The aquatic science studies conducted by EC provided information and advice to senior managers and policy makers (Lord et al. 1978; Wells et al. 1979; Wells and Doe 2014). Guidelines for use were considered for the newly introduced pesticides. One outcome of this research was that the organic solvent nonylphenol was identified as being in the organophosphate pesticide (Fenitrothion) formulations; because of its high toxicity to fish and its connection with a human illness called Reye's syndrome, this solvent was eventually replaced. These studies again showed a close working linkage between science, policy, and program management within the federal government, the policy of trying to find environmentally benign pesticides driving the science and its applications, and the scientific findings influencing new pesticide approval and use.

16.4.2.3 Sediment Toxicity Studies in Halifax Harbor, Nova Scotia, and Bay of Fundy (1991–1999)

Two events led to marine sediment toxicity studies in Nova Scotia and the Bay of Fundy. First, the Halifax municipal government decided to treat Metro's sewage that was entering into Halifax harbor largely untreated, hence abiding by Canada's imminent sewage guidelines; baseline data on the water and sediment quality of the harbor were needed to demonstrate the efficacy of the cleanup. The second event was the growing concern about the quality of intertidal sediments in the upper Bay of Fundy and their effects on shorebirds. Sediments were thought to be chemically contaminated and physically changed by land-based activities, especially agriculture, sewage, and coastal barriers (affecting hydrological cycles), hence having effects on food organisms, especially crustaceans, of the birds dependent on that environment. Both studies were stimulated by policy and management decisions, one by the Halifax Regional Municipality, and one by the federal government (EC—Canadian Wildlife Service).

Each study resulted in a description of the toxicity of surficial sediments using the microscale, bacteria-based toxicity test Microtox® (Cook and Wells 1996; Wells et al. 1998; Cook et al. 1999). The Halifax harbor study's information was used, with other information, in further discussions regarding the location and level of treatment of the sewage treatment plant(s) for Halifax. A decision was made in the late 1990s to build three advanced primary sewage treatment plants. The Bay of Fundy information contributed generally to an understanding of intertidal sediment quality in the upper Bay (Wells et al. 1997), especially near farmlands, and supported a continued effort to describe its quality and ecology in relation to shorebirds (Hamilton et al. 2006). Both applied studies were in response to management and policy decisions on marine environmental protection (sewage) and conservation (shorebirds). They contributed to a better understanding of the effects of chemical pollutants in coastal waters, an understanding critical to effective coastal management.

16.4.2.4 Gulfwatch: A Chemical Contaminants Monitoring Program

The Gulfwatch program is an international, intergovernmental, chemical contaminants monitoring program sponsored by the Gulf of Maine Council on the

Marine Environment. It began in 1991 as part of the Council's first action plan and has continued due to government agencies in both Canada and the United States recognizing the need to monitor aspects of ecosystem health in the Gulf of Maine and Bay of Fundy. The program measures trace chemicals in mussel (*Mytilus edulis*) tissues, following the standard U.S. Environmental Protection Agency toxic chemicals listing of metals, pesticides, polycyclic aromatic hydrocarbons, and polychlorinated biphenyls. Mussels are collected annually from a selection of over 50 intertidal stations around the Gulf of Maine and Bay of Fundy. The data are available at www.gulfofmaine.org/gulfwatch; many reports and papers have been published (e.g., Chase et al. 2001; Chamberlain and Wells 2014). The information is used by state and provincial authorities to advise on the use of coastal waters for molluscan shellfish fisheries (clams, scallops, mussels, oysters), on amenity use by the public for swimming and fishing, and on the general environmental quality of coastal waters of the Gulf of Maine (Chamberlain 2014; Harding 2013; Harding et al. 2005). A close linkage exists between this monitoring program, regional marine environmental policy, and legislation in two countries protecting water quality, molluscan shellfish, and human health. In this case, biological monitoring is considered a cornerstone of integrated coastal environmental management.

16.4.2.5 Other Studies

During my career with EC, I was involved in numerous other studies, some basic science, some directly applied. They were all motivated by the federal policy in Canada of preventing and controlling aquatic pollution from toxic chemicals. The research helped generate a new subdiscipline of aquatic toxicology (Wells et al. 1998), it contributed to the development and application of applied toxicity (biological assessment) protocols for environmental pollutants, used federally and provincially (Taylor et al. 2013), and it contributed to senior management's acceptance of life-cycle analysis and monitoring as a way of controlling the hazards of toxic chemicals (Côté and Wells 1991). Close collaboration with and direction from management and senior policy makers, as well as with the legal profession working for EC, was the overriding operating principle. A well-tuned, practical science–policy interface in the department, with information and advice moving in both directions, was a functional reality throughout my scientific career.

16.5 Discussion: Career-Based Perspective of the Interface between Science and Policy

Some principles and general features of the science–policy interface, or what in reality is the *science–information–management–policy interface*, are apparent from considering the previous eight personal case studies (see Table 16.1). As well, there is value in recalling the general experience of working as an applied scientist within a federal government department (EC) for 32 years. What were the nuggets of this experience, the particulars of the studies being well documented in the primary and gray literature? Are there some novel insights, helpful to the overall discussion of the role of information and information management in ICM and ICOM?

In broad terms, most of the science pursued by the Canadian federal government (the public service) scientist is conducted under policy guidance. That is, the research fits under, is funded by, and is justified by the mandate of the particular department, its mission, and relevant legislation. Although there are exceptions, for example the National Research Council of Canada until recently, it is generally not undirected, curiosity-based research, as is conducted in university departments and associated research institutes such as medical centers. Although some undirected research is indeed encouraged by progressive managers, it is largely mission-oriented research, linked directly to departmental mandates, legislative responsibilities, and governmental policies and priorities of the day. Most of the research and applied studies described here fit this description, and all in one way or the other contributed to Canada's efforts in water pollution control, a component of integrated coastal management (see Chapter 1 in this book). Information from both forms of research contributed to ICM and ICOM in Canadian waters.

The science–management–policy linkages may be formal or informal. If formal, they are based on the demands of already-established legislation, or the anticipated needs of planned legislation; the aquatic toxicity studies of industrial effluents (see Section 16.4.2) conducted for Fisheries Act applications are an example of this kind of science. The linkages may also be informal. Some scientific studies were conducted with decision makers involved, but with the implications for formal policies as yet unknown, for example, the toxicity studies of crude oil and pesticides. In both cases (formal and informal), the opportunity for unexpected discovery is present and may be profound, as shown by discovering the high sensitivity of decapod larvae to chemical contaminants in refinery effluents and the presence of a toxic solvent in the pesticide formulations. However, the ever-present goal of making an unplanned discovery is not the primary motivation for the work, even if it may be important to policy. In my experience, all parties to the research, the researchers, managers, and policy makers, usually recognized the implications of unexpected findings.

The science–management–policy linkages (or relationships), as they involve personnel and budgeted programs, are often layered in time and space within the working department. The linkages may be nonsequential and opportunistic, and they are generally unique to the problem at hand. As well, how the new scientific information might be utilized may be serendipitous, especially if the science is conducted with maximum awareness of the possibility for unforeseen results. Hence, the policy pull on the science may be tempered with new understanding of the problem at hand, and all actors must be flexible and accept compromise. Indeed, this may be an underlying principle of information use for ICM and ICOM; what may advance management of an environmental problem, for example, how to respond to an oil spill, or manage a fishery, may require a novel approach.

The scientific literature is the repository for the results of studies as described previously, conducted in response to a policy directive or leading to a new policy. Whether the information comes from literature that is *gray* or *primary* is generally of little interest to the policy maker or program manager; their concern is in the information's timeliness, reliability, and utility. Often, all the policy maker sees is a briefing note, expertly written and based on relevant internal studies and/or on externally published information. A program manager might see a report or paper itself and use its information appropriately to fit the problem at hand. Either way, in my experience, the science–policy interface functioned for a range of problems affecting ocean management.

16.6 Recent Activities on the Science–Policy Interface in Environment Canada

Various recent initiatives have occurred within EC and the related Canadian Council of Ministers of the Environment (CCME) on the topic of the science–policy interface. A series of science policy workshops were held in the early 2000s, apparently in response partly to the Walkerton, Ontario, incident of contaminated drinking water in 2000 that killed several people and sickened many others (Schaefer and Bielak 2005, 2006; Bielak et al. 2008). Workshops, a favored tool of policy makers to initiate activity on a priority issue, were held on several freshwater issues, including the effects of agricultural activity, groundwater, water reuse and recycling, water quality monitoring, and wastewater treatment. The workshops were considered useful for informing the decision-making process. One workshop also addressed water and climate change in this context (Government of Canada 2008). The final report of the workshop series gave a summary of lessons learned. A major lesson was that

> there is growing consensus that to ensure science better informs the decision-making process, researchers and policy/program managers need to understand and respect each other's way of working, culture and operational timelines. They must also interact more routinely. (Schaefer and Bielak 2005, p. 4)

According to Bielak et al. (2008),

> EC is both a significant environmental science performer and the responsible federal authority for policy and regulation development, program delivery, and enforcement in a range of environmental areas. This being so, the interface between science and policy is critically important in ensuring effective use of limited resources to deliver on an extensive mandate.

CCMF, and EC summarized the workshop initiative through the report, *Effectively Bridging the Gap: The Case for Science–Policy Workshops*. The report identifies two needs, "to customize and target science knowledge to the preferences of user audiences to improve uptake and utility in making decisions" and "to develop mechanisms for sustained interactions between the two groups (science producers and science users) to ensure both a push of science knowledge (science push) and opportunity for science to inform the research agenda (policy pull)" (EC 2010, p. 2).

Of direct relevance to the core discussion in this chapter (Section 16.4) are recent articles by Bielak (2013, p. 647) and Schaefer (2013, p. 1000). Bielak states, "Ultimately, science must be used to be useful. Science–policy linkages are key to ensuring that policy-relevant research reaches those who need to have it, to make decisions in timely fashion." Schaefer states

> There is a need to develop mechanisms or opportunities for sustained interaction between these two groups (scientist and user audience) not only to ensure a regular "push" of science knowledge but also to allow science users to inform/influence the research agenda ("policy pull").

These statements, seem unarguable and certainly were my experience. One wonders just how informed the writers were of the past history of EC, its operations across Canada since 1971, and the intimate tie between its science and the day-to-day application of its mandate, at least in the so-called *regions*, that is, those departmental units located far from the Ottawa and Ontario department bureaucracy.

As for EC's marine and ocean responsibilities, some programs have been jointly run between EC and other federal departments, and the science is intimately linked to legislation and policy, for example, shellfish monitoring by EC, DFO, and the Canadian Food and Inspection Agency, and oil spill emergency response by Transport Canada, DFO (including the Canadian Coast Guard), and EC. For other issues, for example, marine environmental quality (Wells and Rolston 1991), the science advice of DFO may be brought forward to EC for consideration, where appropriate. Another example is the report on marine ecosystem status and trends (CSAS 2010), which documents Canada's progress toward meeting the Convention on Biological Diversity biodiversity targets for 2010. Both departments, DFO and EC, were involved in these efforts. The report discusses the science–policy interface and concludes that improved collaboration among Canada's ecological research, monitoring and policy communities and institutions, focused on identifying and addressing policy-relevant questions, would enhance future assessments of status and trends related to the oceans.

16.7 Conclusion

Environment Canada is a federal department whose mandate from its beginning in 1971 has demanded a close science–policy linkage. Its broad mandate covers all aspects of water research and management, and it shares responsibilities for marine waters with numerous federal departments, as well as with other jurisdictions, including the provinces and municipalities. Hence, its scientific and policy contributions to ICOM are many, varied, interdisciplinary, and continuous. The extent of such contributions often depends as much on the scientific and technical practitioners as on any formal policies. From the inception of the department, the goals for the aquatic scientist have been to identify, understand, and manage water-related problems, as well as to protect and conserve aquatic ecosystems generally. Its scientists, though seldom schooled in public policy, are by and large enthusiastic contributors to the policy process when opportunities and demands exist—witness the contributions of Bielak and associates, noted previously and in Sections 16.4 and 16.5. Recognizing the land–sea connections, EC has a long history of contributing information essential to ICOM in Canadian waters. The examples in this perspective chapter are merely a glimpse into this contribution, and to an understanding of the dynamics of information generation and flow at the science–policy interface in coastal and ocean management.

References

Araujo, A. R. P., K. Momo, E. Gherardi-Goldstein, et al. 1987. Marine dispersant program for licensing and research in Sao Paulo state, Brazil. *Proceedings 1987 Oil Spill Conference (Prevention, Behavior, Control, Cleanup), Tenth Biennial*, 6–9 April 1987, Baltimore, MD. American Petroleum Institute Publication No. 4452, 289–292. Washington, DC: API.

Bielak, A. T. 2013. Knowledge translation and knowledge brokering in ecotoxicology. In *Encyclopedia of Aquatic Ecotoxicology*, edited by J.-F. Ferard and C. Blaise, 643–647. Dordrecht: Springer.

Bielak, A. T., A. Campbell, S. Pope, et al. 2008. From science communication to knowledge brokering: The shift from science push to policy pull. In *Communicating Science in Social Contexts. New Models, New Practices*, edited by D. Cheng, M. Claessens, N. R. J. Gascoigne, J. Metcalfe, B. Schiele, and S. Shi, 210–226. Dordrecht: Springer Science and Business Media.

Chamberlain, S. 2014. Developing and implementing a research framework to determine the use and influence of a long-term marine environmental monitoring program: A case study on Gulfwatch in Nova Scotia. Master's project report. Halifax, N.S.: Dalhousie University.

Chamberlain, S. and P. G. Wells. 2014. Gulfwatch bibliography. Unpublished manuscript. Halifax, NS: Dalhousie University.

Chase, M. E., S. H. Jones, P. Hennigar, et al. 2001. Gulfwatch: Monitoring spatial and temporal patterns of trace metal and organic contaminants in the Gulf of Maine (1991–1997) with the blue mussel, *Mytilus edulis* L. *Marine Pollution Bulletin* 42 (6): 491–505.

Cook, N. H., I. Blum, and P. G. Wells. 1999. Quality of Bay of Fundy sediments: Further insights using the Microtox7 solid phase assay. Poster abstract. In *Understanding Change in the Bay of Fundy Ecosystem. Proceedings of the 3rd Bay of Fundy Science Workshop*, Mount Allison University, Sackville, New Brunswick, 22–24 April 1999, edited by J. Ollerhead, P. W. Hicklin, P. G. Wells, and K. Ramsey. Environment Canada—Atlantic Region, Occasional Report No. 12. Dartmouth, NS: Environment Canada.

Cook, N. H. and P. G. Wells. 1996. Toxicity of Halifax harbour sediments: An evaluation of the Microtox(R) solid phase test. *Water Quality Research Journal of Canada* 31 (4): 673–708.

Côté, R. P. and P. G. Wells, eds. 1991. *Controlling Chemical Hazards—Fundamentals of the Management of Toxic Chemicals*. London: Unwin Hyman.

CSAS (Canadian Science Advisory Secretariat). 2010. 2010 Canadian marine ecosystem status and trends report. Canadian Science Advisory Secretariat, Scientific Advisory Report 2010/030. Ottawa: Department of Fisheries and Oceans, National Capital Region.

Doe, K. G. and P. G. Wells. 1978. Acute aquatic toxicity and dispersing effectiveness of oil spill dispersants: Results of a Canadian oil dispersant testing program (1973–1977). In *Chemical Dispersants for the Control of Oil Spills, ASTM STP 659*, edited by L. T. McCarthy, Jr., G. P. Lindblom, and H. F. Walters, 50–65. Philadelphia: American Society for Testing and Materials.

Environment Canada. 1984. Guidelines on the use and acceptability of oil spill dispersants. Environment Canada, Environmental Protection Service, Report EPS-1-EP-84-1.

Environment Canada. 2010. Effectively Bridging the Gap: The Case for Science–Policy Workshops. Strengthening Science–Policy Links: Study Series. Ottawa: Environment Canada.

Environmental Contaminants Act, 1975. R.S., c. E-12 [repealed 1985].

Fisheries Act, 1985. R.S.C. c., F-14.

GESAMP (Joint Group of Experts on Scientific Aspects of Marine Environmental Protection). 1993. Impact of oil and related chemicals and wastes on the marine environment. GESAMP Reports and Studies No. 50. London: IMO.

GESAMP (Joint Group of Experts on Scientific Aspects of Marine Environmental Protection). 2007. Estimates of oil entering the marine environment from sea-based activities. GESAMP Reports and Studies No. 75. London: IMO.

Government of Canada. 2008. *The Science–Policy Interface—Water and Climate Change and the Energy–Water Nexus*. Ottawa: Government of Canada, Policy Research Initiative.

Government Organization Act, 1979, 1978–79, c. 13.

Hamilton, D. J., A. W. Diamond, and P. G. Wells. 2006. Shorebirds, snails and the amphipod *Corophium volutator*, in the upper Bay of Fundy: Top-down versus bottom-up factors, and the influence of compensatory interactions on mudflat ecology. *Hydrobiologia* 567: 285–306.

Harding, G. and C. Burbidge. 2013. Toxic chemical contaminants. State of the Gulf of Maine report. Gulf of Maine Council on the Marine Environment. http://www.gulfofmaine.org/2/wp-content/uploads/2014/03/toxic-chemical-contaminants-theme-paper.pdf.

Harding, G., S. Jones, P. G. Wells, et al. 2005. Blue mussels: Canaries of the sea. In Bedford Institute of Oceanography, 2004 in Review, Annual Report, edited by J. Ryan, 50–52. Dartmouth, NS: Fisheries and Oceans Canada, Bedford Institute of Oceanography.

London Convention (1996 Protocol to the Convention on the Prevention of Marine Pollution by Dumping Wastes and Other Matter, 1972, as amended in 2006). Can TS 2006 No 5.

Lord, D. A., R. A. F. Matheson, L. Stuart, et al. 1978. Environmental monitoring of the 1976 spruce budworm spray program in New Brunswick, Canada. Fisheries and Environment Canada, EPS, Surveillance Report EPS-5-AR-78-3.

MacDonald, B. H., R. E. Cordes, and P. G. Wells. 2004. Grey literature in the life of GESAMP, an international marine scientific advisory body. *Publishing Research Quarterly* 20 (1): 25–41.

MARPOL 73/78 (International Convention for the Prevention of Pollution from Ships, 1973; Protocol of 1978 relating to the International Convention for the Prevention of Pollution from Ships of 1973). 17 February 1978, 1340 UNTS 61.

Ocean Dumping Control Act, 1975. R.S.C., c. 0-2 [repealed 1985].

Oceans Act, 1996. S.C., c. 31.

Pessah, E. and G. M. Cornwall. 1980. Use of toxicity tests in regulating the quality of industrial wastes in Canada. In *Aquatic Toxicology, Proceedings of the Third Annual Symposium on Aquatic Toxicology*, edited by J. G. Eaton, P. R. Parish, and A. C. Hendricks, ASTM STP 707, 130–141. Philadelphia: ASTM.

Ricketts, P. and P. Harrison. 2007. Coastal and ocean management in Canada: Moving into the 21st century. *Coastal Management* 35: 5–22.

Ricketts, P. and L. Hildebrand. 2011. Coastal and ocean management in Canada: Progress or paralysis? *Coastal Management* 39 (1): 4–19.

Schaefer, K. A. 2013. Science–policy linkages in ecotoxicology. In *Encyclopedia of Aquatic Ecotoxicology*, edited by J.-F. Ferard and C. Blaise, 997–1001. Dordrecht: Springer.

Schaefer, K. A. and A. T. Bielak. 2005. CCME linking water science to policy workshop. Final report: An overview and lessons learned. CCME Linking Water to Policy Workshop Series Report No. 7. Winnipeg, MB: Canadian Council of Ministers of the Environment. https://www.ec.gc.ca/inre-nwri/.../water_wkshp_smryrpt_2004_e.pdf.

Schaefer, K. A. and A. T. Bielak. 2006. Linking water science to policy: Results from a series of national workshops on water. *Environmental Monitoring and Assessment* 113: 431–442.

Stewart, C. 2010. Implementation status of the marine protected areas: Aspects of the federal government's HOTO initiatives. Unpublished internal document, World Wildlife Fund, Office of the Auditor General of Canada, January 8.

Taylor, L. N., K. G. Doe, R. P. Scroggins, et al. 2013. Regulatory ecotoxicology testing in Canada–activities and influence of the inter-governmental ecotoxicological testing group. *Water Quality Research Journal of Canada*, 48: 14–29.

UNCLOS (United Nations Convention on the Law of the Sea). 10 December 1982, 1833 UNTS 3.

UNEP (United Nations Environment Programme). 1985. Protection of the marine environment against pollution from land-based sources. Montreal guidelines. *Environmental Policy and Law* 14 (2–3): 77–83.

Vandermeulen, H. 1998. The development of marine indicators for coastal zone management. *Ocean & Coastal Management* 39 (1–2): 63–71.

Vandermeulen, H. and D. Cobb. 2004. Marine environmental quality: A Canadian history and options for the future. *Ocean & Coastal Management* 47 (5–6): 243–256.

Wells, P. G. 1973. Acute toxicity bioassay with an oil effluent and larvae of *Homarus americanus*. In A toxicity study of an oil refinery effluent in Saint John, New Brunswick, edited by W. R. Parker and R. P. Côté. Environmental Protection Surveillance Report EPS-5-WP-73-1, 20–25. Dartmouth, NS: Environment Canada.

Wells, P. G. 1976. Effects of crude oil on American lobster (*Homarus americanus*) larvae in the laboratory. PhD diss., University of Guelph.

Wells, P. G. 1984. The toxicity of oil spill dispersants to marine organisms–A current perspective. In *Oil Spill Chemical Dispersants: Research, Experience and Recommendations. STP 840*, edited by T. E. Allen, 177–220. Philadelphia: American Society for Testing and Materials.

Wells, P. G. 1996. Measuring ocean health: Resetting a cornerstone of Canadian marine policy, science and management. In *Annexes of the Canadian Ocean Assessment: A Review of Canadian Ocean Policy and Practice*, edited by S. Coffen-Smout, 241–249. Halifax, NS: International Ocean Institute, Dalhousie University.

Wells, P. G. 2005. Assessing marine ecosystem health–Concepts and indicators, with reference to the Bay of Fundy and Gulf of Maine, Northwest Atlantic. In *Handbook of Ecological Indicators for Assessment of Ecosystem Health*, edited by S. E. Jorgensen, F-L. Xu, and R. Costanza, Chapter 17, 395–430. Boca Raton, FL: Taylor & Francis/CRC Press.

Wells, P. G. 2015. Curriculum Vitae (with Comprehensive List of Publications). Unpublished.

Wells, P. G., N. H. Cook, A. Nimmo, et al. 1997. Factors influencing estimates of sediment toxicity using the Microtox(R) solid-phase test: Insights gained from studies on Halifax harbour and Bay of Fundy sediments. In Proceedings of the 24th Annual Aquatic Toxicity Workshop: Oct. 20–22, 1997, Niagara Falls, ON. Canadian Technical Report Fisheries and Aquatic Sciences, No. 2192, 29–30. Ottawa: Fisheries and Oceans Canada.

Wells, P. G. and K. G. Doe. 2014. Applied aquatic toxicology: Protecting the waters of Atlantic Canada. In *Voyage of Discovery*, edited by D. N. Nettleship, D. C. Gordon, M. Lewis, and M. P. Latremouille, 315–321. Dartmouth, NS: BIO-Oceans Association, Bedford Institute of Oceanography.

Wells, P. G. and J. Gratwick, eds. 1988. *Canadian Conference on Marine Environmental Quality: Proceedings*. 29 February–3 March, 1988, Halifax, N.S. Halifax: International Institute for Transportation and Ocean Policy Studies, Dalhousie University.

Wells, P. G., L. Harding, J. Karau, et al. 1987. Marine environmental quality in Canada. In *Oceans '87 Proceedings*. Volume Five. Coastal and Estuarine Pollution, 1633–1636. Washington, DC: Marine Technology Society.

Wells, P. G., K. Lee, and C. Blaise, eds. 1998. *Microscale Testing in Aquatic Toxicology: Advances, Techniques and Practice*. Boca Raton, FL: CRC Press.

Wells, P. G., R. A. Matheson, D. A. Lord, et al. 1979. Environmental monitoring of the 1978 spruce budworm spray program in New Brunswick, Canada-field sampling and aquatic toxicity studies with fish. Environment Canada, Environmental Protection Service Surveillance Report EPS-5-AR-79-1. Dartmouth, NS: Environment Canada.

Wells, P. G. and S. J. Rolston, eds. 1991. *Health of Our Oceans. A Status Report on Canadian Marine Environmental Quality*. 2nd ed. Ottawa: Conservation and Protection, Environment Canada.

Wells, P. G. and J. B. Sprague. 1976. Effects of crude oil on American lobster (*Homarus americanus*) larvae in the laboratory. *Journal of the Fisheries Research Board of Canada* 33 (7): 1604–1614.

17

Bridging the Science–Policy Divide to
Promote Fisheries Knowledge for All:
The Case of the Food and Agriculture
Organization of the United Nations

Lahsen Ababouch, Marc Taconet, Julian Plummer,
Luca Garibaldi, and Stefania Vannuccini

CONTENTS

17.1 Introduction

Established in 1945 as an agency of the United Nations (UN), the Food and Agriculture Organization (FAO) leads international efforts to eradicate hunger, food insecurity, and malnutrition. Its vision is "[a] world free from hunger and malnutrition where food and agriculture contribute to improving the living standards of all, especially the poorest, in an economically, socially, and environmentally sustainable manner." Three global goals underpin this vision:

- Eradication of hunger, food insecurity, and malnutrition, progressively ensuring a world in which people at all times have sufficient, safe, and nutritious food that meets their dietary needs and food preferences for an active and healthy life
- Elimination of poverty and the driving forwards of economic and social progress for all, with increased food production, enhanced rural development, and sustainable livelihoods
- Sustainable management and use of natural resources, including land, water, air, climate, and genetic resources, for the benefit of present and future generations.

FAO is made up of 197 members.* Its governing body, the Conference, meets every two years to review FAO's work and to approve the Programme of Work and Budget for the next two-year period. It elects the Council (49 members serving three-year rotating terms), which acts as an interim governing body, and the director-general (four-year term), who heads the agency.

The members fund FAO through contributions set at the Conference. Members and other partners provide additional voluntary contributions in support of FAO's Programme of Work and, in particular, its implementation in the field. The Programme of Work and Budget is delivered in four broad and complementary ways:

- Putting information within reach
- Sharing policy expertise
- Providing a forum for members and other stakeholders to meet, consult, and decide on international efforts to eradicate hunger

---

* They include 194 member nations, 1 member organization (the European Union), and 2 associate members (Faroe Islands and Tokelau).

- Bringing knowledge to the field, in particular in developing countries and countries in transition

FAO puts information within reach of governments, academia, civil society, and the donor community by collecting global statistics, studies, and analyses and by disseminating this information through its channels and in national, regional, and international fora. FAO also plays a lead role in developing and implementing methodologies and standards for data collection, validation, processing, and analysis and provides essential statistical capacity development to its members.

The FAO Fisheries and Aquaculture Department (the Department) is mandated to strengthen global governance and members' managerial and technical capacities and to lead consensus building toward improved conservation and use of aquatic resources. Its work centers on the "sustainable management and use of fisheries and aquaculture resources," embracing normative activities, for example, statistics, and operational activities. Fisheries statistics have a pivotal role in this work, and the trend information they provide is the basis for the formulation of fisheries policy and management (FAO 2003). The emphasis on evidence-based decision-making in governments and organizations at all levels puts a greater focus on statistics and their role in measuring and monitoring progress toward national and international development goals and targets.

FAO's work in statistics is gaining further visibility and importance within the framework of the post-2015 Sustainable Development Goals (SDGs) (www.sustainabledevelopment.un.org) through identifying indicators for monitoring progress toward goals and targets that fall within FAO's mandate, in particular, proposed Goal 2 (End hunger, achieve food security and improved nutrition, and promote sustainable agriculture) and proposed Goal 14 (Conserve and sustainably use the oceans, seas, and marine resources for sustainable development).

This chapter summarizes FAO's work in the area of fisheries and aquaculture statistics and information, analyses its links and interface with policy within the framework of global fisheries and aquaculture governance, and assesses the challenges and opportunities for statistics and the information needs in support of policy within the framework of the SDGs and FAO's Blue Growth Initiative (BGI).

17.2 Statistics and Information in Support of Global Fisheries and Aquaculture Policy and Management

As a knowledge organization, FAO provides a wide array of statistics and information products and services. These have evolved over time to reflect the needs of FAO's members, governing bodies, and partners in their quest to assess the contribution of fisheries and aquaculture to food security and poverty alleviation and the issues affecting the sector's performance. Statistics and information are vital for policy-making and sectoral planning at all levels. In particular, they are crucial for effective fisheries management, which, in turn, is essential for sustaining fishery resources and protecting ecosystems, biodiversity, food supply, and livelihoods.

FAO is the only source of global fishery statistics, and its members have to report statistics to it. Over five decades, FAO's Coordinating Working Party on Fishery Statistics has developed and maintained global norms, standards, and classifications, for example, for species, commodities, vessels, and gear, and for fishery statistics in close coordination with regional fishery bodies (RFBs) and other agencies (FAO 2005–2015). FAO provides the secretariat for the Fisheries and Resources Monitoring System (FIRMS)—a partnership involving most RFBs—to provide access to high-quality information on the global monitoring and management of marine fishery resources. Fishery statistics and information collated, analyzed, and disseminated by FAO are also key to the preparation of reviews such as the biennial flagship publication *The State of World Fisheries and Aquaculture* (SOFIA).

Particularly challenging is the collecting of statistics and information on small-scale coastal fisheries (with their many small and often remote landing places, and many vessels and fishers). Landings from such fisheries are often sold directly to consumers or merchants, or consumed by the fishing communities (subsistence fisheries). Recognizing the need to improve the quality of fishery data and statistics in general, and for small-scale fisheries in particular, FAO members adopted the Strategy for Improving Information on Status and Trends of Capture Fisheries (Strategy–STF) in 2003 (FAO 2003; further details provided in Section 17.3.2), which was subsequently endorsed by the UN General Assembly. It aims to provide a framework, strategy, and plan for the improvement of knowledge and understanding of fishery status and trends as a basis for policy-making and management for the conservation and sustainable use of fishery resources within ecosystems. Before the adoption of the Strategy–STF, FAO had also developed widely used guidelines and field manuals, such as the Guidelines for the Routine Collection of Capture Fishery Data (FAO 1999) and a manual on sample-based fishery surveys (Stamatopoulos 2002).

A significant achievement in capacity building for improving national fishery statistics has been seen for China, the world's largest fish-producing country. With FAO support, China's authorities introduced a sampling scheme that replaced an inefficient system of complete enumeration. This resulted in a major revision to the national fishery statistics published by China and reported to FAO and several RFBs.

17.3 Review of Selected FAO Data and Information Products

This chapter illustrates how FAO's data and information products support the policy process of global fisheries. Together with its premier information and advocacy document, SOFIA (see following sections and Box 17.1), this chapter reviews four other FAO information and data products to illustrate how FAO communicates with its audiences at the science–policy interface. For each, it examines the information flow and exploitation of information products by users and draws conclusions regarding the impact on target audiences, including at the policy level (see Figure 17.1).

17.3.1 SOFIA and Global Fisheries Policy

The Department produces SOFIA, its flagship publication, every two years to coincide with the sessions of the FAO Committee on Fisheries (COFI— www.fao.org/cofi/en/). Here, policy drives the demand for data, information, and knowledge, which, in turn, inform fisheries policy in an iterative process (see Box 17.1). Following the general approach of previous editions, the 2014 edition of SOFIA (FAO 2014d) presents a global review of fisheries and aquaculture, including trends and statistics. It also highlights issues debated worldwide and profiles future scenarios to provide readers with the most current global view and perspectives on fisheries and aquaculture. SOFIA is available in the six official languages of FAO (Arabic, Chinese, English, French, Russian, and Spanish) in both print and online versions. As originally requested by COFI, SOFIA presents policy-oriented knowledge and informs COFI as it works through the policy cycle (Box 17.1). The publication consists of four main parts in sequence:

1. Overview—It reports on socioeconomic and natural resources trends as well as policy implementation in the fisheries and aquaculture sector, using FAO's established indicators and monitoring systems.

2. Issues—These emerge from debate brought to COFI's attention, or from the analysis of FAO's official statistics, information products (e.g., FIRMS, FBS; see Figure 17.1), or partners (e.g. RFBs). Significant issues debated worldwide warrant specific and comprehensive reviews that may be used to support the adoption by COFI of global action plans; such action plans might include, for example, requirements for indicators and related monitoring systems to be set up by FAO.

3. Highlights—These present recent FAO studies on actions taken to address the issues. Thus, what was an issue in one edition in SOFIA may be reported on as a highlight in a later edition.

BOX 17.1   COFI AND SOFIA

The Committee on Fisheries (COFI), a subsidiary body of the FAO Council, was established by the FAO Conference at its Thirteenth Session in 1965. COFI constitutes the only global intergovernmental forum where major international fisheries and aquaculture issues are examined and recommendations addressed to governments, regional fishery bodies, NGOs, fishworkers, FAO, and the international community, on a worldwide basis. COFI has also been used as a forum for the negotiation of global binding and nonbinding instruments.

The Thirty-First Session of COFI considered the role and influence of the publication *The State of World Fisheries and Aquaculture* (SOFIA) in supporting the work of decision makers in general and FAO in particular. Key information in SOFIA 2014 was presented in a document to COFI concerning the current status, recent trends, and prospects in the fisheries and aquaculture sector as well as on the role and influence of SOFIA (FAO 2014e). Reporting on the COFI session, a subsequent document prepared for the FAO Conference stated: "The Committee expressed praise and support for the State of World Fisheries and Aquaculture (SOFIA) 2014 publication" (FAO 2014c).

This second document also illustrates the interplay in the science–policy interface, with policy makers (COFI) requesting scientific information and then using that information (provided also through SOFIA) to influence policy and request further information:

> The Committee welcomed the new categorization of the status of marine stocks, as requested by the 30th Session of COFI. Most Members were encouraged by the results in SOFIA 2014. Some expressed cautious optimism with regard to the stock status and others remained concerned. They also emphasized the need for further measures to rebuild the stocks. There were requests to include more detailed information on the status of specific stocks, including data on fleet capacity and addition of socioeconomic data, with a strong call to add regional information and perspectives. Members called for disaggregating data in future editions, suggested that specific topics be emphasized or added, and requested that assessments on some stocks and species be updated and corrected. Specifically, some Members requested that the ratio of stocks fished sustainably/unsustainably be expressed in terms of volume (catch in tonnes) and value (FAO 2014c).

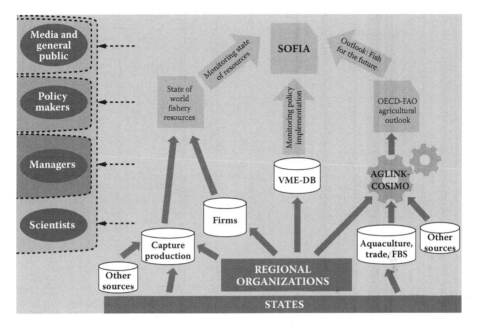

FIGURE 17.1
Information flow of some of FAO's fisheries and aquaculture data and information products.
SOFIA = *The State of World Fisheries and Aquaculture*; VME–DB = Vulnerable Marine Ecosystem
Database; FIRMS = Fisheries and Resources Monitoring System; FBS = food balance sheets.

4. Outlook—It presents considerations on emerging trends and issues and, hence, informs and advises on where FAO may need to take action as the policy cycle begins again.

17.3.1.1 Intended Outcomes

To take the example of SOFIA, its intended outcomes can be classed as immediate, intermediate, and long term. Immediate outcomes relate to user satisfaction as the dissemination and presentation of SOFIA facilitate discoverability, reading, and understanding of the issues/key messages raised. The outcomes also relate to SOFIA as a product or service perceived as offering technically sound, timely, and appropriate information and analyses on the status and trends of fisheries and aquaculture around the world.

Intermediate outcomes include the impact of SOFIA in terms of (1) informing the stakeholders, with target audiences being made aware of the latest trends and emerging issues in fisheries and aquaculture; (2) enhancing programs and practices, as its data inform the formulation of global, regional, and national programs on sustainable fisheries management and as international, national, and donor-funded projects adopt evidence-based practices proposed in SOFIA; (3) increasing responsible investment, as its analyses inform business and industry investment decisions;, and (4) enhancing education and research, as its data and analyses inform, enrich, and redirect future research, inform curriculum development, and help to update the content of training and educational programs.

Long-term outcomes relate to SOFIA's contribution to the sustainable management of fisheries and natural resources through better-informed policies, programs/practices, research/education, and investments, as well as to positive responses in indicators of performance related to achieving the global goals of improving food security and reducing poverty.

17.3.1.2 Questions Relating to Preparation and Impact of SOFIA

Referring to the issues and criteria discussed in Chapter 2 of this book, questions that could be asked about SOFIA include:

- Does it provide information that is useful, credible, and legitimate?
- Does it communicate information at the science–policy interface?
- Does it have an impact on policy?
- How do the preparers of SOFIA obtain, process, and communicate information?

The fourth question is addressed in more detail in Section 17.3.1.3, while the response to the first three questions is given in Section 17.3.1.5.

17.3.1.3 SOFIA Part 1: World Review of Fisheries and Aquaculture

As shown in Box 17.1, through SOFIA, policy decisions can be based on evidence and influenced by science. The process begins with SOFIA's inputs. These can be grouped into human resources, which consist of FAO staff and other experts from academia, research institutes, and nongovernmental organizations (NGOs), among other bodies, and institutional resources, which consist of key papers, national surveys, databases, models, networks, partners, infrastructure, and financial resources. To inform policy makers on the global state of fishery resources, FAO summarizes in SOFIA the findings of regular publications such as the *FAO Yearbook—Fishery and Aquaculture Statistics* (FAO 2014a), the *Review of the State of World Fishery Resources: Marine Fisheries* (FAO 2011), and other main series publications, for example, technical papers.

The global indicators on the state of fishery resources discussed by the *Review of the State of World Fishery Resources: Marine Fisheries* are based on stock assessment information, mostly collected by RFBs, and on FAO's global capture production statistics for those fishing areas where no stock assessment information is available. Figures on global trends in the state of world marine stocks are among the most cited and highly debated information produced by the FAO Fisheries and Aquaculture Department. In the latest published version of the *Review of the State of World Marine Fishery Resources* (FAO 2011), the number of categories used to classify the state of the stocks was reduced from six to three (overexploited, fully exploited, non-fully exploited), which were later renamed in SOFIA 2014 as overfished, fully fished, and underfished (FAO 2014d).

In the framework of the information work done by FAO on deep-sea fisheries (Bensch et al. 2009), the recently launched database on vulnerable marine ecosystems (Vulnerable Marine Ecosystem Database [VME–DB]; www.fao. org/in-action/vulnerable-marine-ecosystems/en/) is an example of a monitoring system that informs global users on policy implementation at regulatory level, fosters collaboration on best practices, and, in turn, contributes to conservation of fishery resources and biodiversity.

Similarly, for the global capture production statistics, the collected and processed statistical data are reviewed, checked, organized, and analyzed before being made accessible through databases such as FAOSTAT (faostat3.fao.org/home/E), FishStatJ (www.fao.org/fishery/statistics/software/fishstatj/en), and AGLINK–COSIMO (http://www.agri-outlook.org/database/). Further details are discussed in Sections 17.3.2 and 17.3.3.

Final publications are peer reviewed and approved by senior management. Dissemination strategies are then developed and implemented, for example, the official launch of SOFIA at COFI, press releases, websites, conferences, and events, to ensure the information reaches its target audiences and achieves the intended outcomes. Feedback from FAO members and target audiences is always sought and welcomed, as the science–policy interface needs to be a two-way interaction.

17.3.1.4 SOFIA Parts 2–4: Emerging Issues, Highlights, and Outlook

Part 2 of SOFIA reports on current and emerging issues of importance to the fisheries and aquaculture sector. In this way, FAO serves its members by raising awareness and providing objective and authoritative information. Part 3 of SOFIA highlights recent and forthcoming significant FAO publications on fisheries and aquaculture, some of which have been produced in response to issues raised in earlier editions of SOFIA. Figure 17.2 lists some of the issues covered and some of the studies highlighted in recent editions of SOFIA.

Part 4 of SOFIA is the Outlook section, which in the 2014 edition, focused on meeting future fish demand. It examined projected fish supply and demand and discussed assumptions used in the models as well as issues that may threaten the sector's ability to meet future fish demand and challenges. It provided the results of two main outlook studies, one based on the FAO Fish Model developed in collaboration with the Organisation for Economic Co-Operation and Development (OECD)—the work of this partnership was

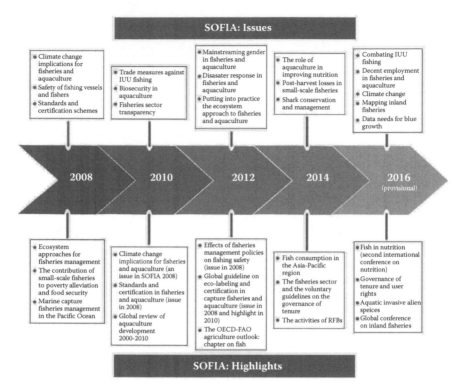

FIGURE 17.2
Key "issues" and "highlights" in recent editions of SOFIA.

highlighted in SOFIA 2012—and the other on the International Model for Policy Analysis of Agricultural Commodities (IMPACT) of the International Food Policy Research Institute (IFPRI). Model-based projections are intended to become a standard feature in future editions of the Outlook section.

17.3.1.5 Impact Assessment: FAO and SOFIA

FAO is fully cognizant of the need to assess the impact of its work from "on the ground" to the highest levels of international policy-making. For its publications (and the data, information, and knowledge they contain), it needs to know who is using them and how they are affecting policy decisions. Thus, in attempting to answer these questions, the extent to which SOFIA is achieving its intended outcomes has recently been the subject of a specific assessment conducted by FAO's Office of Evaluation within a broader overall evaluation of FAO publications. This assessment has captured evidence on the relevance, quality, and effectiveness of SOFIA through the following tools: (1) desk review; (2) interviews with key informants (about 200 key informants in 12 countries selected in consultation with FAO staff responded to a client survey; a member country survey administered through the permanent representatives to FAO received responses from 38 governments); (3) a user (readership) survey, based on an analysis of 252 completed questionnaires; and (4) web and cybermetric analyses. In terms of web traffic, data from FAO's Document Repository show SOFIA 2014 averaging 16,870 views per month (July 2014–April 2015), with 27,797 views in the most recent month for which data are available (April 2015). The data analysis has focused on determining SOFIA's contributions as per the outcomes listed Section 17.3.1.1. The main output still to come will be a report to inform the assessment of SOFIA's contributions (FAO 2015a). However, permission has been given to disclose some of the preliminary findings here, and Figure 17.3 is based on data from the user survey. The interviews with key informants have confirmed these findings. However, some pointed out that although SOFIA is very relevant to policy makers, not all readers have the level of expertise required to make full use of it.

The assessment discovered that among policy makers, SOFIA is perceived as a critical source of global trends and statistics and, to a lesser degree, of knowledge about fisheries in a variety of topics and contexts, which supports decision-making and policy decisions. One of SOFIA's key contributions relates to the interface between global and national statistics. In particular, SOFIA has helped to improve sectoral statistics over the years, which, in turn, are reused in a range of analytical products and research. Among program managers, about 64% of survey

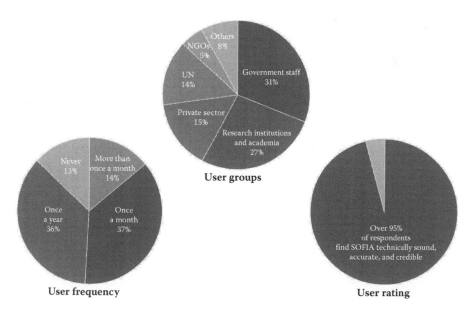

FIGURE 17.3
SOFIA user survey results.

respondents indicate that SOFIA reports have had a moderate to high guiding influence on the development/improvement of sectoral strategies or programs.

The assessment further finds that SOFIA is regularly quoted in workshops, scientific papers, and press articles by civil society and media outlets, as well as by partner international organizations. The assessment report concludes that the utility and uptake of SOFIA at the sectoral level and among different sets of users appear to be substantial, with uptake being evidenced in particular by policy makers, program managers, researchers, and academia.

This recent assessment follows a citation and content analysis of over 10,000 citations by Vanja Avdić of Dalhousie University in 2013 (Avdić 2013). She concludes that "SOFIA is the only marine assessment of its kind and is highly regarded" and that "it regularly provides both crucially important statistics on the states of fisheries and aquaculture and also all-important comparison of trends over long time-periods." However, she also comments that "major global assessments, such as SOFIA, are available to inform management or policy decisions, but whether the information is used is poorly understood." This comment links back to a question raised in Chapter 2 of this book: "are the reports being noticed and read and influencing decisions where it matters most?"

To respond to these questions, in addition to the evidence from the recent formal assessment, one can also cite a recent publication from the Dietary Guidelines Advisory Committee (DGAC) of the United States as an indicator of SOFIA's relevance and impact at the science–policy interface. Its 2015 report to the secretary of health and human services and the secretary of agriculture states: "Lastly, to address Question 4 on the worldwide capacity to produce enough nutritious seafood, the Committee used the FAO's report on the *State of World Fisheries and Aquaculture*, 2012. This was considered the most current and comprehensive source on this topic" (DGAC 2015, Part D, chapters 5, 7). In its conclusions to Question 4, the report states: "The DGAC concurs with the FAO report that consistent evidence demonstrates that capture fisheries increasingly managed in a sustainable way have remained stable over several decades" (Part D, Chapter 5, 21). It further states: "The DGAC endorses the FAO report that capture fisheries production plateaued around 1990 while aquaculture has increased since that time to meet increasing demand." Again, on the same page, it states: "The UN FAO report on *The State of World Fisheries and Agriculture* issued in 2012 formed the basis of the DGAC's evidence review on this topic. The FAO report addresses a wide variety of issues affecting capture fisheries and aquaculture, including economics, infrastructure, and labour and government policies." The DGAC report also reproduces a figure from SOFIA 2012 and provides a reference and web link to the publication. Thus, the comments in the DGAC report can be seen as confirming the authoritativeness and relevance of SOFIA at the science–policy interface.

17.3.2 Knowledge Products on the State of World Fishery Resources

17.3.2.1 Global Capture Fisheries Statistics Database

In the postwar reconstruction decades of the 1950s and 1960s, society's focus was on production development and related sector growth and employment concerns. These socioeconomic concerns were also the prevailing drivers when the Global Capture Production statistics database was instituted with the progressive establishment by FAO of the standards set for this database, mostly through the work of FAO's Coordinating Working Party on Fishery Statistics.

National correspondents from FAO members are mandated to provide annual catch statistics. The reported data are carefully checked, and when the figures are questionable, the national correspondent is consulted for clarifications. If a country does not report its catches or if those provided are questionable, FAO estimates the missing data and marks them in the database with an "F." The overall quality of capture statistics in the database is mostly dependent on the accuracy and reliability of the data collected nationally and provided to FAO. However, to improve completeness and coverage of information in the database, in some cases, FAO complements or replaces national data when better data are available.

The main sources of additional information are the RFBs that compile catch statistics for (1) tuna and shark species (Inter-American Tropical Tuna Commission, International Commission for the Conservation of Atlantic Tunas, Indian Ocean Tuna Commission, Western and Central Pacific Fisheries Commission); (2) non-tuna species in the high seas areas (South East Atlantic Fisheries Organisation, South Pacific Regional Fisheries Management Organisation); (3) all species in an entire ocean (for the Southern Ocean: Commission for the Conservation of Antarctic Marine Living Resources); or (4) a group of species at the global level, that is, marine mammals (International Whaling Commission). In addition to verifying whether distant-water fishing nations have reported their distant catches, FAO cross-checks statistics submitted by the flag states with information on foreign catches in their exclusive economic zone made available by coastal countries and territories, for example, Guinea-Bissau and Mauritania. If unreported catches are detected, they are added to the FAO database.

Despite efforts made by FAO to ensure the best possible coverage, the data submitted or not reported were considered inadequate in relation to the relative importance of capture fishery for more than half of the countries (Garibaldi 2012). Trend studies based on data included in the FAO database also may be biased by erroneous reporting (either under- or overreporting), incomplete identification of species, and changes in the national data collection system. The capture database is usually closed at the end of February, and in March, its updated version can be accessed at the Department's website through the FishStatJ (www.fao.org/fishery/statistics/software/fishstatj/en) stand-alone software and data set package or the online query panel (www.fao.org/fishery/topic/16140/en); both also include FAO's aquaculture production and fisheries commodities databases.

Data and information compiled and disseminated by FAO are widely used by the global community interested in fishery issues. Indicators of interest are the number of accesses to relevant web pages and the number of enquiries on fishery statistics received by FAO. In 2014, the entry page (www.fao.org/fishery/statistics/global-capture-production/en) for the FAO capture database received 15,000 unique page views, and the FishStatJ web page was viewed at least once about 36,000 times, both with peaks in March–April, when the updated annual version of the database was released.

To receive and address the enquiries on various aspects of the fishery statistics compiled by FAO, a dedicated e-mail account (fish-statistics-inquiries@fao.org) is available to the users, including scientists, NGOs, and the general public. This account receives an average of two enquiries per working day, to which FAO is committed to reply within a few days. More than one-third of the enquiries received (for a total of almost 200 e-mails per year) ask for meta-information on the data in the FAO fishery statistics databases, with the rest related to software and other technical issues. Through the dedicated account and e-mail correspondence, four FAO fishery statisticians regularly communicate with the national correspondents on data in the FAO databases, the quality and coverage of national data, and possible improvements in national data collection systems.

The FAO capture database is used largely by fishery scientists for their research work. A citation analysis performed in 2011 (Garibaldi 2012) showed that, although the database is cited in different ways, and often the data are used but the source is not quoted, between 1996 and mid-June 2011, it had been listed in the bibliographies of 622 articles in refereed journals. According to this result, the FAO capture database belongs to the restricted group of items that have been cited more than 500 times, representing 0.11% of the total of 20 million items published between 1900 and 2005 that have been cited at least once (Garfield 2005).

Given the extended time series available in the database (it currently includes data for 64 years, from 1950 to 2013), it is mostly used for studies on catch trends, which also have been characterized by trophic levels (Pauly et al. 1998; Caddy and Garibaldi 2000; Branch et al. 2010), catches in oceanic waters and large marine ecosystems (LMEs) (Garibaldi and Limongelli 2003), and quantification of depleted species to be restored (Garibaldi and Caddy 2004).

17.3.2.2 Moving beyond the Original Intended Use

In 1996, Grainger and Garcia (1996) made the first attempt to demonstrate that information in the capture database can also be used for interpreting developments in the world's fisheries. They calculated the percentage of major marine resources in the undeveloped, developing, mature, and senescent phases. The same approach was later applied also to national (Cuba) and regional (Eastern Central Atlantic) analyses of these major development phases (Baisre 2000; Garibaldi and Grainger 2004).

A scientific debate developed on the question of whether, in the absence of more detailed information from stock assessment, catch data can also be used to infer the status of the fished resources. Studies expressed a range of different views: (1) catch data in the FAO database can provide indications on stock status (e.g., Froese and Kesner-Reyes 2002; Froese et al. 2012; Niwa 2006; Kleisner et al. 2013; Vasconcellos and Cochrane 2005); (2) catch data should be complemented with other information, such as life history (e.g., Costello et al. 2012; Thorson et al.); (3) stock assessments provide far better information with which to estimate abundance (e.g., Agnew et al. 2013; Branch et al. 2011; Carruthers et al. 2012); and (4) the use of catch data is misleading (e.g., Cook 2013; Daan et al. 2011; de Mutsert et al. 2008).

In February 2013, *Nature* consulted Daniel Pauly, who has produced many studies based on catch data, and Ray Hilborn and Trevor Branch, for whom catch data alone are not a reliable indicator of stock status (Pauly et al. 2013). The latter position is embraceable from a theoretical point of view, but long time series of abundance assessment are available only for a fraction of the global fisheries, with coverage restricted to fisheries operated by developed countries and mostly focusing on highly commercial species.

17.3.2.3 FIRMS: State of Stocks and Fisheries

In addition to the mechanism that produces the global stock trends in the state or world marine stocks that are published in the *Review of the State of World Fishery Resources* (FAO 2011) and SOFIA, in the context of the Strategy–STF (FAO 2003), FAO also formed in 2004 an information-sharing partnership called the Fishery and Resources Monitoring System. Its establishment was prompted by the controversy described in Section 17.3.2.2 on the global assessment of the state of world fishery resources, the diagnosis that commercial fisheries are predominant in global catch reporting, and the need to collect a broader range of indicators to address socioeconomic and environmental issues. Given this context, the Strategy–STF proposed to proceed with an inventory-based approach to expand the coverage of individual stock status information, to enable better visibility of small-scale and subsistence fisheries, and to expand the thematic scope of reporting. It also proposed information-sharing partnerships as a mechanism for sustaining reporting in the long term.

FIRMS was launched to foster reporting on the status and trends of individual stocks and fisheries at the regional and national levels, to facilitate access of the general public to authoritative information on status and trends, and to support regional and global reviews in their regular updates of the status of fishery resources, in ways that enable the provenance of source statements to be tracked. As an information product, FIRMS is a website providing a single point of access to a collection of web-based fact sheets that present status and trends reports on stocks and fisheries according to a structured standard layout. The fact sheets present status information at different scales, from FAO's reports on the state of regional resources provided by FAO major areas down to subregional and local stocks. The sources of information are systematically provided. These reports are submitted by FIRMS' partners and published under their full control and ownership. Specific web-based services extract summaries on the status of stocks for user-defined combinations of species and/or areas.

While expanding data coverage is critical to its objectives, FIRMS has also succeeded in extending its geographic coverage. The partnership now includes 14 intergovernmental organizations, which together represent 17 RFBs. The Atlantic Ocean and Indian Ocean are quite well covered, while the Pacific Ocean is not. Partnerships with regional organizations for the detailed coverage of the Caribbean and the Bay of Bengal are under development. FIRMS contained, as at February 2015, a published inventory of 1,082 stocks and 322 fisheries. Status reports exist for about 600 stocks and 200 fisheries (FAO 2015b).

However, because of its current constituency, FIRMS has insufficient monitoring coverage in areas with no RFBs, in particular the waters of the exclusive economic zones of South America, North America, Australia and New Zealand, Southeast Asia, and Asia (Japan, China, the Republic of Korea, and the Russian Federation), where stocks are monitored by coastal states. In this respect, FIRMS is exploring the possibility of setting up NatFIRMS as a separate framework closely linked to FIRMS. NatFIRMS would include national fisheries institutions in the areas not covered by RFBs.

The success of FIRMS should also be measured against its capacity to respond to users' needs. In 2014, FIRMS registered 43,662 unique page views (4,466 for the home page alone), which compares well with the Global Capture Production page, totaling 15,001 unique page views in the same year for all languages. FIRMS has four categories of target users and related information products: (1) FIRMS partner RFBs and their member states; (2) governmental fisheries agencies for states involved in producing FIRMS fisheries status reports through partner RFBs; (3) global or regional marine science networks reporting on the state of the environment and ecosystems, for example, FAO's reporting on the state of world fishery resources, and other UN agencies and processes such as LMEs; and (4) NGOs and international NGOs promoting sustainable fisheries with consumers and industry, for example, Ocean Trust and Sustainable Fisheries Partnerships, both of which have shown keen interest in FIRMS.

While the first two categories of users are satisfied with FIRMS, the other two are not, because of limited data coverage. For the last category, the status-report formats are still inadequate with respect to audience expectations. These NGOs focus on science-based guidance regarding stock status; acceptance by management authorities of the impact of management on sustainability, education, and assistance of the seafood supply chain in their efforts at improving fisheries sustainability; and enhancement of the general public's recognition of such facts. They are interested in FIRMS as a reliable source of information on both resources status and fisheries management, but they point out the need for more information on the resource or fishery fact sheets in relation to sustainability.

FIRMS acknowledges the needs expressed by these interest groups for management frameworks and measurement. FIRMS envisions strengthening the capacity of indicators and scales to rate management performance, involving the performance reviews of regional fisheries management organizations (RFMOs).

17.3.2.4 Toward a Global Record of Stocks and Fisheries

Recently, opportunities have arisen for collaboration between the FAO mechanism that produces the global trends in the state of world marine stocks (see Section 17.3.2.3), FIRMS, and other projects compiling regional and global information on stock assessment toward a global record of stocks and fisheries. This global record will federate knowledge on status/trends of stocks and fisheries across various global sources (FIRMS information-sharing partnership among RFBs, the University of Washington's Ramm Legacy Stock Assessment database, the FishBase database of the FishBase Information and Research Group, and the Sustainable Fisheries Partnership's FishSource). This global record, which will rely on unique identifiers for stocks and fisheries, is expected to offer key services to stakeholders involved in "regional/global state of stocks indicators," for example, a dynamic "state of world marine resources," as well as public and private actors involved in eco-labeling, traceability, and sustainable fisheries. Thanks to its authoritative partnership and the standards it has set for global reporting on the status of stocks and fisheries, FIRMS is in a unique position to lead the development of such a global record.

17.3.3 Monitoring Policy Implementation: The Vulnerable Marine Ecosystem Database

The VME–DB offers a one-stop shop to global information on the actions taken by regional fisheries management organizations and arrangements (RFMO/As) to protect VMEs in the high seas (FAO 2015c). It illustrates the efforts of the international community involved in high seas fisheries in response to recognition by the UN General Assembly, in 2006, of the need to address impacts from deep-sea bottom fishing activities by identifying VMEs where these impacts are significant. It also demonstrates a commitment to responsible use of marine resources in the deep seas through funding the development of the VME–DB, which was provided after the adoption by COFI of the International Guidelines for the Management of Deep-sea Fisheries in the High Seas in 2009 (FAO 2009).

From the outset, the VME–DB has been developed as a tool to foster the linkage between science and policy. Its design is the result of a participatory process involving policy makers, scientists, industry representatives, and data managers. Over three years, FAO facilitated this process, including through two global meetings in Rome in 2011 (FAO 2013) and 2014 (FAO 2014b) and three regional workshops (in Japan, Mauritius, and Namibia). The process enabled contributors to become acquainted with the various concepts and practices around the world in an attempt to disseminate reliable and consistent information, although there is still room for more harmonized best practices at the global level.

The web product consists of the VME website and the database. Together, they host information that describes the actions taken by RFMO/As to protect VMEs in the areas beyond national jurisdiction (ABNJ), and they allow users to quickly discover the existing protocols and data each RFMO/A has developed for VMEs. This product boosts knowledge sharing and learning from other regions. While the VME–DB has been designed for decision makers, scientists, observers, and members of RFMO/As, the website is intended for the wider public (including academia, research, and education) interested in deep-sea fisheries, their interaction with deep-sea marine species, including marine habitats, and their management.

The VME website is a gateway to the VME–DB and includes comprehensive background information on the UN resolutions and other international guidelines and procedures involved in the VME process. The portal also links directly to the RFMO/A websites, which host additional information on each regulatory process. The VME–DB constitutes a unique repository and dissemination platform at the global level of all concrete measures implemented by the international fishery community in the high seas to prevent adverse impacts of deep-sea bottom fishing on vulnerable ecosystems. It has an interactive map viewer that can be filtered by year or by region to show geographic VME information, such as existing VME closures or bottom fishing areas, as well as links to more detailed information on the individual resolutions or protocols for that area. The application allows users to navigate the map across time and track the rapid progress made by RFMO/As since 2006 on identifying and protecting VMEs.

17.3.3.1 Toward Institutionalized Governance for Long-Term Sustainability

The sustainability of the VME–DB depends on an appropriate governance mechanism, data maintenance by the partners, and system maintenance by FAO. While FAO uploaded the portal and data using publicly available information provided by RFMO/As for the period 2006–2014, data entry is expected to migrate to the RFMO/A partners as they assume responsibility for data input. A web-based content management system powered by the iMarine platform has been set up for this purpose. Over time, the RFMO/As will continue to update the portal as new resolutions are made and as the identification and protection of VMEs in the ABNJ become more refined.

Future support and engagement by the regional bodies are necessary to gain approval within FAO for continued maintenance and development of the system. To increase ownership and returns on investment, that is, the data update, web services are offered to RFBs to embed the VME-Database Map viewer in their websites (SEAFO n.d.; NAFO n.d.). In the short to medium term, the governance framework managing the VME Database is likely to migrate to the FAO ABNJ Deep Seas project and partnership (FAO and RFBs). An option for the medium to long term is to anchor the VME Database as a component of the FIRMS partnership.

17.3.4 Fishery and Aquaculture Outlook

17.3.4.1 Understanding Future Trends and Addressing Forthcoming Challenges

Food security and nutrition represent a global challenge, as hunger and malnutrition remain among the most devastating problems facing the world. In the perspective of future population growth, this challenge is even more compelling. According to the UN, the current world population of about 7.3 billion is projected to reach 8.1 billion in 2025 and 9.7 billion in 2050, with most of the population growth occurring in developing regions (UN 2015). The fisheries and aquaculture sector can continue to play a prominent role in world food security, both directly, as a vital and nutritious source of food, and indirectly, as a contributor to economic growth and development by providing employment, livelihoods, and income to millions of people engaged in fish harvesting, culturing, processing, and trading. To continue fulfilling this role, the growing demand for fish requires sustainable growth in capture fisheries and aquaculture production. The need to develop specific projections for better understanding of the plausible outlook for fisheries and aquaculture and the challenges they might face has, therefore, attracted more international attention. Outlook studies can represent an important tool for international organizations, such as FAO, the OECD, and the World Bank, their members, and the international community to facilitate understanding of the impacts of changes in aquaculture and capture fisheries and of demand shifts and policy reforms, as well as to obtain relevant information for developing strategic responses to emerging challenges. Outlook projections can also help FAO, other international organizations, and donors to highlight work priorities and to develop a tailored strategy to support countries in addressing the major challenges facing the sector.

17.3.4.2 *Integrating Fish in Overall Agricultural Analysis*

In recent years, work has been undertaken to develop specific fish models in partnership with international organizations. However, it was considered important that this work should not be done in isolation, but integrated into an overall agricultural analysis aimed at achieving a more comprehensive and consistent examination of the medium- or long-run prospects for fish, together with those for food and agriculture. The two main outcomes are (1) the FAO Fish Model developed by FAO as a satellite to the OECD–FAO AGLINK–COSIMO Projection System,* with medium-term projections (10 years) annually included in the *OECD–FAO Agricultural Outlook* publication since 2011, and (2) the *Fish to 2030* publication (World Bank 2013), which shows the results of the IMPACT model of IFPRI. For both models, FAO fishery statistics on production, trade, and apparent consumption, the FAO Food Balance Sheets, represent the main data used. Based on key assumptions and uncertainties, these outlook models represent an important tool for providing insights on likely paths of development and constraints in supply and demand to determine regional vulnerabilities, changes in comparative advantage, price effects, and potential adaptation strategies in the fisheries and aquaculture sector.

17.3.4.3 *The* OECD–FAO Agricultural Outlook

The *OECD–FAO Agricultural Outlook* is an annual publication presenting projections and related market analysis for some 15 agricultural products (including fisheries and aquaculture) over a ten-year horizon. The projections are based on the AGLINK–COSIMO modeling system. The AGLINK model was developed by the OECD in the early 1990s and then enhanced through the development by FAO of the COSIMO (Commodity Simulation Model) component for a large number of developing countries. It brings together the commodity, policy, and country expertise of both organizations and input from collaborating members to provide an annual assessment of prospects for the coming decade for national, regional, and global agricultural commodity markets. It shows how these markets are influenced by economic developments and government policies, and it highlights some of the risks and uncertainties that may influence market outcomes. The capacity to capture interactions between commodities and between countries is a major strength of this model, allowing analysts to assess not only the direction but also the magnitude of market adjustments resulting from economic or policy changes. For many countries, agricultural policies are specifically modeled within AGLINK–COSIMO. This makes the model a powerful tool for forward-looking analysis of domestic and trade policies through the

---

* More information on the AGLINK–COSIMO modeling system and on the *OECD–FAO Agricultural Outlook* publication is available at http://www.agri-outlook.org/.

comparison of scenarios of alternative policy settings against the benchmark of the baseline projections.

FAO, with the collaboration and agreement of the OECD and FAO Secretariats for AGLINK–COSIMO, has recently built a dynamic, policy-specific, partial-equilibrium satellite model on fish and fishery products. This satellite model has followed the same general principles used to build the AGLINK–COSIMO modeling system to facilitate its eventual integration (at present, the fish model is not fully integrated in the overall AGLINK–COSIMO modeling system). The main results of the fish model, which are included in the "Fish and Seafood" chapter of the annual *OECD–FAO Agricultural Outlook* publication, provide insights on the most plausible scenario for a 10-year horizon in the fisheries and aquaculture sector. Under a certain set of assumptions, the results portray an outlook in terms of future production potential, projected demand for fisheries products, consumption, prices, and key factors that might influence future supply and demand. These trends guide FAO, OECD, and their members to plan the sustainable use and conservation of fisheries and aquaculture resources for economic growth, improved social welfare, and development.

The development of the model and the inclusion of the specific chapter on fisheries and aquaculture medium-term projections in the *OECD–FAO Agricultural Outlook* are very relevant. This report is all the more valuable as it provides in-depth knowledge and experience to policy makers, including through its discussion of the conceptual framework and the scenarios developed. Each year, the *OECD–FAO Agricultural Outlook* publication has a wide impact, with a high number of citations, and it is also one of the OECD's most frequently downloaded publication.

17.3.4.4 Fish to 2030

Fish to 2030 is the result of a collaboration between IFPRI, FAO, the University of Arkansas at Pine Bluff, and the World Bank (World Bank 2013). The report builds on the publication *Fish to 2020* (Delgado et al. 2003), which provided a comprehensive global overview of the food fish supply and demand balance. The *Fish to 2030* report employs IFPRI's IMPACT model to generate projections of global fish supply and demand up to 2030. The IMPACT model is a relatively straightforward partial equilibrium global agriculture sector model, covering the world in 115 model regions for a range of more than 40 agricultural commodities, to which fish and fish products were added for the *Fish to 2030* study (further information on the IMPACT model is available at http://www.ifpri.org/program/impact-model).

In the 1990s, IFPRI developed the IMPACT model to address a lack of long-term vision and consensus among policy makers and researchers about the actions necessary to feed the world in the future, reduce poverty, and protect the natural resource base. It serves as the basis for research on the linkage between the production of key food commodities and food demand and security at the national level in the context of scenarios of future change, with cutting-edge research results on rapidly evolving topics such as bioenergy, climate change, and diet/food preferences.

For the *Fish to 2030* report, the IMPACT model was calibrated and employed to evaluate different policies and alternative events and to illustrate the likely evolution of the global seafood economy. The results are structured according to a baseline scenario, considered the most plausible, and six alternative scenarios that investigate potential impacts of changes in the drivers of global fish markets under various assumptions. The overall publication is centered around three main topics: (1) the health of global capture fisheries, (2) the role of aquaculture in filling the global fish supply–demand gap and potentially reducing the pressure on capture fisheries, and (3) implications of changes in the global fish markets for fish consumption. Box 17.2 illustrates how the *Fish to 2030* report is used to steer international assistance to develop fisheries and aquaculture in sub-Saharan Africa.

17.4 Challenges and Opportunities

To make an impact through transformational change and effective implementation of sectoral policies, an organization such as FAO needs to address a diverse audience of stakeholders, from the general public to scientists, program managers, and high-level policy makers. Each level requires specific products and approaches with the potential for cross-fertilization, a need for regular alignment among the product layers, and interaction among the governance mechanisms driving each product. To improve monitoring of the use and impact of its data and knowledge products at the decision-making level and their communication to the target users, FAO has recognized the imperative need to adapt each product through regular monitoring and feedback to ensure continuous relevance to the evolving needs of the target users. In this respect,

- Partnerships are effective mechanisms to constantly adjust products to needs: considerations on return on investment prevail, and the established setup is regularly reviewed to reduce input costs and improve benefits. Benefits can materialize in the form of influence, communication, and visibility, or political will to promote transparency, science-based policy action, and contribution to global policy goals. Such concerns are conducive to aligning policy with information and partnership targets
- Full involvement of stakeholders throughout the life cycle is also a must: the stronger the involvement of stakeholders (sponsors,

BOX 17.2   *FISH TO 2030* IN SUB-SAHARAN AFRICA

The *Fish to 2030* report explores how the economics of fish supply and demand interact, and how increasing incomes and population growth may drive change. According to this report, the future for sub-Saharan Africa warrants particular concern. Under baseline assumptions, annual per capita fish supply in Africa is predicted to fall from the current 6.8 kg to 5.6 kg in 2030. In contrast, the global average will rise from 17.2 to 18.3 kg, and the guideline for fish consumption is about 14.5 kg per person per year. So, the simple fact is that, if fish is to be an adequate part of a balanced, nutritious diet for sub-Saharan Africans, even the current level is inadequate. A decline would have tragic consequences.

Furthermore, African production growth is projected to be only 4.5% compared with the world average of 23.6%, making sub-Saharan Africa more reliant on imports and vulnerable to shocks in global markets. FAO and other organizations have been alerting the international community and African leaders to this scenario and proposing recommendations to (1) sustain and enhance wild capture fisheries and (2) support aquaculture growth in Africa.

Indeed, the potential for aquaculture in Africa is enormous. Egypt, for example, has experienced spectacular growth in its aquaculture sector, and tilapia is now the least expensive animal-source food available and an important element of the diets of many Egyptians.

Furthermore, aquaculture as a household livelihood activity is practiced in several sub-Saharan countries, such as Nigeria and Ghana. However, its contribution to the continent's needs remains marginal. Deciding how best to support aquaculture growth will require systematic analysis of input and production value chains and of the enabling policy environment. These analyses need to identify the constraints and opportunities for the private sector, as well as where international development and government support is needed to remove roadblocks.

partners, development team, and beneficiaries) is at all stages of the information product's life cycle (including design and development), the higher up it can influence decision-making, because alignment is more likely see Chapter 9 of this volume). Moreover, stronger ownership stemming from such early and continuous involvement will boost outreach and the communication capacity as a result of joint action by all partners. The drawback is a slower process, and the frequently observed inability to respect this constraint might be a main cause of failure in influencing decision-making.

There is worldwide recognition that fisheries statistics are a necessary basis for sound fisheries management policies, yet this is an area that is under-funded and often neglected by donors. Adequate funding, in particular through voluntary contributions, is necessary to enable FAO and its partners to address the challenges of a modern global fisheries statistics and information system. Partnerships with various stakeholders are a cost-effective and reliable mechanism to address these challenges, although their potential has not been fully exploited.

As the world embarks on the post-2015 SDGs, the challenge of restoring the productive potential of oceans and wetlands will require new responsible and sustainable approaches to their economic development. A more environmentally, socially, and economically effective fish and seafood chain can contribute to sustainable growth and food security. It can also pave the way for reducing pressure on aquatic resources and deliver the potential for people employed in the sector to act not only as resource users but also as resource stewards. FAO recognizes the importance and need for the fisheries and aquaculture sector to grow sustainably to meet rising food demand and contribute to poverty alleviation, and the fact that zero growth is neither realistic nor desirable. In this context, FAO launched the BGI in 2013 in support of food security, poverty alleviation, and sustainable management of aquatic natural resource. An FAO working definition of *blue growth* is sustainable growth and development emanating from economic activities using living renewable resources of the oceans, wetlands, and coastal zones that minimize environmental degradation, biodiversity loss, and unsustainable use of renewable aquatic resources, and maximize economic and social benefits.

The concept of blue growth, which has also been referred to as "green economy in a blue world" (UNEP et al. 2012), "blue green economy" (Kelleher 2011), "blue growth, the new maritime blue economy of the EU," and "green growth in fisheries and aquaculture" (Asche 2011), is an emerging paradigm for the sustainable management of natural marine and freshwater resources. The term *blue growth* is preferred by many to *blue economy*, because there has been criticism in some development circles of the green economy concept, in particular its emphasis on zero or limited growth.

The blue growth concept featured prominently at the Rio+20 United Nations Conference on Sustainable Development in 2012 (www.uncsd2012. org). The Rio+20 outcomes have provided a strong catalyst for driving new efforts toward the implementation of previous and new commitments on oceans and wetlands to restore, use, and conserve natural aquatic resources. FAO's BGI aims to enable the catalyzing of policies, investment, and innovation that would underpin sustained growth and give rise to new economic opportunities in ecosystem goods and services. It aims to integrate key aspects of economic performance, such as poverty reduction, job creation, social inclusion, and community resilience, with those of environmental performance, such as climate change mitigation, ecosystems, and biodiversity restoration. It would mobilize financial and technical support, build local capacity for the design and implementation of blue growth strategies, and create action-oriented policy options and institutions tailored to the respective economic circumstances and constraints of FAO's members.

A key challenge for the promotion of the blue growth concept and approach is the collection of data and information, their sharing across a range of scientific domains, and the development of analytical methodologies on a range of criteria along its three dimensions of sustainability. Current methodologies on food security and fisheries and aquaculture economics will remain useful, although requiring some refinement. The new frontier will place greater emphasis on environmental aspects, such as fish stock restoration, fisheries, and aquaculture productivity in the context of ecosystem assessments, carbon footprints and sequestration, volume and types of certified fisheries and products, and mangrove and coral reef restoration, that can advance natural capital and ecosystem accounting in national economies. Such information and methodologies are needed to better account for the economic contributions of renewable aquatic resources and ecosystem services and to assess the long-term sustainability of national economies and business and investment models.

This growing momentum has led to the emergence of global initiatives to develop methodological guidance for demonstrating the value of ecosystem services as an input into policy and economic decision-making, for example, the UN System of Environmental-Economic Accounting (to which FAO brings its expertise on fisheries and aquaculture). Acknowledging that no organization in isolation can meet this demand, FAO is calling for a global partnership/alliance to forge a global data framework for blue growth. Such a framework would provide a mechanism to enable the collection and integrated use of data from diverse initiatives. This scenario can be achieved through collaborative data infrastructures (distributed e-infrastructure) with harmonization of concepts and references, improved data-sharing capacities, collaborative analysis through virtual research environments (VREs), and open data and information dissemination policies.

The iMarine initiative is an example of such a partnership (Taconet et al. 2014). This initiative offers a collaborative scientific platform able to network and connect a wide range of data management solutions in support of the ecosystem approach to fisheries. By enabling the pooling of data, software, methodologies, and expertise, this partnership expects to deliver cost-efficient solutions to the rising demand for information. Under the iMarine initiative, the recently launched BlueBRIDGE project will use VREs to address ambitious and wide-ranging objectives, thus consolidating its support for the ecosystem approach to fisheries while expanding it to other areas of blue growth. These VREs comprise (1) collaborative assessments of stock status, including computing-intensive capacities for ecological and food-web-based biological assessments; (2) production of a global record of stocks and fisheries to support regional and global policy-making for fishery resources, and responsible trade and consumption practices; (3) socioeconomic and environmental performance analysis of aquaculture farms and systems for empowering production companies in performance evaluation, benchmarking, and decision-making; (4) semiautomated recognition of spatial features from integrated remote sensing and geographic information system sources, to support spatial planning in aquaculture and habitat assessment and monitoring; and (5) mobile data collection to empower small-scale fishers in comanagement processes, such as identification of VMEs, onboard bycatch, and illegal, unreported, and unregulated sightings in small-scale fisheries.

More specifically in support of blue growth and food security, the Global Action Network was launched in March 2015 in Grenada. This network identified three action groups to address, respectively: (1) the fundamentals of Blue Growth and food security, facilitated by the Netherlands, Portugal, the FAO, and WorldFish; (2) the Investment Readiness Facility, facilitated by Grenada, the World Bank, and WorldFish; and (3) knowledge and technology, facilitated by Cabo Verde, the FAO, and Rare (an NGO). These three action groups were to work toward defining the fundamentals, scope, principles, guidelines, and indicators for best practices of blue growth, promotion of investment, and an exchange platform to monitor and evaluate progress and impact.

The BGI and the work of the three action groups in support of blue growth and food security are likely to play an important role in the post-2015 SDGs, in particular, Goals 2 and 14 (see Section 19.1). The pilot activities implemented under the BGI banner and the wide partnership it is building will generate important data, analyses, and information products that can help monitor progress toward achieving the goals and targets of global fisheries performance.

References

Agnew, D. J., N. L. Gutiérrez, and D. S. Butterworth. 2013. Fish catch data: Less than what meets the eyes. *Marine Policy* 42: 268–269.

Asche, F. 2011. *Green Growth in Fisheries and Aquaculture Production and Trade.* Paris: OECD.

Avdić, V. 2013. Measuring use and influence: An assessment of the FAO's flagship report The State of World Fisheries and Aquaculture. Master's project report, Dalhousie University.

Baisre, J. A. 2000. Chronicle of Cuban marine fisheries (1935–1995): Trend analysis and fisheries potential. FAO Fisheries Technical Paper No. 394. Rome: FAO. http://www.fao.org/docrep/X4529E/X4529E00.HTM.

Bensch, A., M. Gianni, D. Gréboval, et al. 2009. Worldwide review of bottom fisheries in the high seas. FAO Fisheries and Aquaculture Technical Paper No. 522, Rev. 1. Rome: FAO. http://www.fao.org/docrep/012/i1116e/i1116e00.htm.

Branch, T. A., O. P. Jensen, D. Ricard, et al. 2011. Contrasting global trends in marine fishery status obtained from catches and from stock assessments. *Conservation Biology* 25: 777–786.

Branch, T. A., R. Watson, E. A. Fulton, et al. 2010. The trophic fingerprint of marine fisheries. *Nature* 468: 431–435.

Caddy, J. F. and L. Garibaldi. 2000. Apparent changes in the trophic composition of the world marine harvest: The perspective from the FAO capture database. *Ocean & Coastal Management* 43: 615–655.

Carruthers, T. R., C. J. Walters, and M. K. McAllister. 2012. Evaluating methods that classify fisheries stock status using only catch data. *Fisheries Research* 120: 66–79.

Cook, R. M. 2013. A comment on "What catch data can tell us about the status of global fisheries (Froese et al., 2012)." *Marine Biology* 160 (7): 1761–1763.

Costello, C., D. Ovando, R. Hilborn, et al. 2012. Status and solutions for the world's unassessed fisheries. *Science* 338: 517–520.

Daan, N., H. Gislason, J. G. Pope, et al. 2011. Apocalypse in world fisheries? The reports of their death are greatly exaggerated. *ICES Journal of Marine Sciences* 68: 1375–1378.

de Mutsert, K., J. H. Cowan, Jr., T. E. Essington, et al. 2008. Reanalyses of Gulf of Mexico fisheries data: Landings can be misleading in assessments of fisheries and fisheries ecosystems. *Proceedings of the National Academy of Sciences* 105: 2740–2744.

Delgado, C. L., N. Wada, M. W. Rosegrant, et al. 2003. Fish to 2020: Supply and demand in changing global markets. WorldFish Center Technical Report No. 62. Washington, DC: International Food Policy Research Institute; and Penang: WorldFish Center. http://ebrary.ifpri.org/utils/getfile/collection/p15738coll2/id/87521/filename/87522.pdf.

DGAC (Dietary Guidelines Advisory Committee). 2015. Scientific report of the 2015 Dietary Guidelines Advisory Committee. Advisory report to the Secretary of Health and Human Services and the Secretary of Agriculture. Accessed March 9, 2015. http://www.health.gov/dietaryguidelines/2015-scientific-report/PDFs/Scientific-Report-of-the-2015-Dietary-Guidelines-Advisory-Committee.pdf.

FAO (Food and Agriculture Organization). 1999. Guidelines for the routine collection of capture fishery data. Prepared at the FAO/DANIDA Expert Consultation, Bangkok, Thailand, May 18–30, 1998. FAO Fisheries Technical Paper No. 382. Rome: FAO. http://www.fao.org/docrep/003/x2465e/x2465e00.htm.

FAO (Food and Agriculture Organization). 2003. Strategy for improving information on status and trends of capture fisheries. Rome: FAO. http://www.fao.org/docrep/006/Y4859T/Y4859T00.HTM.

FAO (Food and Agriculture Organization). 2005–2015. Fisheries and aquaculture topics. History of the coordinating working party. Topics fact sheets. In FAO Fisheries and Aquaculture Department [online]. Rome. Updated May 27, 2005. Accessed April 28, 2015. http://www.fao.org/fishery/cwp/history/en.

FAO (Food and Agriculture Organization). 2009. *International Guidelines for the Management of Deep-Sea Fisheries in the High Seas*. Rome: FAO. http://www.fao.org/docrep/011/i0816t/i0816t00.HTM.

FAO (Food and Agriculture Organization). 2011. Review of the state of world marine fishery resources. FAO Fisheries and Aquaculture Technical Paper No. 569. Rome: FAO. http://www.fao.org/docrep/015/i2389e/i2389e00.htm.

FAO (Food and Agriculture Organization). 2013. Report of the FAO Workshop for the Development of a Global Database for Vulnerable Marine Ecosystems, Rome, December 7–9, 2011. FAO Fisheries and Aquaculture Report No. 1018. Rome: FAO. http://www.fao.org/docrep/017/i3109e/i3109e00.htm.

FAO (Food and Agriculture Organization). 2014a. *FAO Yearbook. Fishery and Aquaculture Statistics. 2012.* Rome: FAO. http://www.fao.org/3/a-i3740t.pdf.

FAO (Food and Agriculture Organization). 2014b. Report of the FAO Workshop on the Global Database for Vulnerable Marine Ecosystems, Rome, May 7–9, 2014. FAO Fisheries and Aquaculture Report No. 1093. Rome: FAO. http://www.fao.org/3/a-i4209e.pdf.

FAO (Food and Agriculture Organization). 2014c. Report of the 31st Session of the Committee on Fisheries, Rome, June 9–13, 2014. Conference, Thirty-ninth Session, Rome, June 6–13, 2015. C 2015/23. http://www.fao.org/3/a-ML770e.pdf.

FAO (Food and Agriculture Organization). 2014d. *The State of World Fisheries and Aquaculture 2014.* Rome: FAO. http://www.fao.org/3/a-i3720e/index.html.

FAO (Food and Agriculture Organization). 2014e. State of world fisheries and aquaculture and progress in the implementation of the Code of Conduct for Responsible Fisheries and Related Instruments. Committee on Fisheries, Thirty-first Session, Rome, June 9–13, 2014. COFI/2014/2/Rev.1. http://www.fao.org/3/a-mk055e.pdf.

FAO (Food and Agriculture Organization). FAO 2015a. *Case Study: The State of World Fisheries and Aquaculture (SOFIA).* FAO Office of Evaluation. Rome: FAO.

FAO (Food and Agriculture Organization). 2015b. Report. FIRMS Steering Committee Meeting, Ninth Session, Swakopmund, Namibia, February 23–24 and 27, 2015. FIRMS FSC9/2015/Report. Accessed April 28, 2015. ftp://ftp.fao.org/FI/DOCUMENT/FIGIS_FIRMS/2015/FSC9_Report.pdf.

FAO (Food and Agriculture Organization). 2015c. Vulnerable Marine Ecosystems. Rome: FAO. http://www.fao.org/in-action/vulnerable-marine-ecosystems/en.

Froese, R. and K. Kesner-Reyes. 2002. Impact of Fishing on the Abundance of Marine Species. ICES Document CM 2002/L:12. Copenhagen: International Council for the Exploration of the Sea.

Froese, R., D. Zeller, K. Kleisner, et al. 2012. What catch data can tell us about status of global fisheries. *Marine Biology* 159: 1283–1292.

Garfield, E. 2005. *The Agony and the Ecstasy—The History and Meaning of the Journal Impact Factor.* Chicago: International Congress on Peer Review and Biomedical Publication.

Garibaldi, L. 2012. The FAO global capture production database: A six-decade effort to catch the trend. *Marine Policy* 36: 760–768.

Garibaldi, L. and J. F. Caddy. 2004. Depleted marine resources: An approach to quantification based on the FAO capture database. FAO Fisheries Circular No. 1011. Rome: FAO. http://www.fao.org/3/a-j3957e.pdf.

Garibaldi, L. and R. Grainger. 2004. Chronicles of catches from marine fisheries in the eastern central Atlantic for 1950–2000. In *Marine Fisheries, Ecosystems and Societies in West Africa: Half a Century of Change. International Symposium, Dakar, Senegal, 24–28 June 2002*, edited by P. Chavance, M. Bâ, D. Gascuel, J. M. Vakily, and D. Pauly. ACP-EU Fisheries Research Report, vol. 15, 99–112. Brussels: Office for Official Publications of the European Communities.

Garibaldi, L. and L. Limongelli. 2003. Trends in oceanic captures and clustering of large marine ecosystems: Two studies based on the FAO capture database. FAO Fisheries Technical Paper No. 435. Rome: FAO. http://www.fao.org/docrep/005/Y4449E/Y4449E00.HTM.

Grainger, R. J. R. and S. M. Garcia. 1996. Chronicles of marine fishery landings (1950–1994): Trend analysis and fisheries potential. FAO Fisheries Technical Paper No. 359. Rome: FAO. http://www.fao.org/docrep/003/W3244E/w3244e00.htm.

Kelleher, K. 2011. *Green Growth and Fisheries Issues*. TAD/FI(2011)5. Paris: OECD.

Kleisner, K., R. Froese, D. Zeller, et al. 2013. Using global catch data for inferences on the world's marine fisheries. *Fish and Fisheries* 14: 293–311.

NAFO (Northwest Atlantic Fisheries Organization). n.d. FAO VME database. NAFO. http://nafo.int/data/frames/data.html: under menu item "Geospatial Data/FAO VME-DB."

Niwa, H.-S. 2006. Exploitation dynamics of fish stocks. *Ecological Informatics* 1: 87–99.

Pauly, D., V. Christensen, J. Dalsgaard, et al. 1998. Fishing down marine food webs. *Science* 279: 860–863.

Pauly, D., R. Hilborn, and T. Branch. 2013. Fisheries: Does catch reflect abundance? *Nature* 494: 303–306.

SEAFO (South East Atlantic Fisheries Organisation). n.d. VME Map. SEAFO. http://www.seafo.org/Science/VME-Map.

Stamatopoulos, C. 2002. Sample-based fishery surveys: A technical handbook. FAO Fisheries Technical Paper No. 425. Rome: FAO. http://www.fao.org/docrep/004/y2790e/y2790e00.htm.

Taconet, M., A. Ellenbroek, D. Castelli, et al. 2014. Sustaining iMarine: A public partnership led business model. EU-FP7 iMarine Project Report. Accessed September 19, 2015. ftp://ftp.fao.org/FI/DOCUMENT/FIGIS_FIRMS/2015/Inf11e.pdf.

Thorson, J. T., T. A. Branch, and O. P. Jensen. 2012. Using model-based inference to evaluate global fisheries data from landings, location, and life history data. *Canadian Journal of Fisheries and Aquatic Sciences* 69: 645–655.

UN (United Nations). 2015. *World Population Prospects: The 2015 Revision. Medium Variant*. New York: Population Division of the Department of Economic and Social Affairs of the United Nations Secretariat. Accessed September 19, 2015. http://esa.un.org/unpd/wpp/index.htm.

UNEP, FAO, IMO, et al. 2012. *Green Economy in a Blue World*. Nairobi: UNEP.

Vasconcellos, M. and K. Cochrane. 2005. Overview of world status of data-limited fisheries: Inferences from landings statistics. In *Fisheries Assessment and Management in Data-Limited Situations*, edited by G. H. Kruse, V. F. Gallucci, D. E. Hay, R. I. Perry, R. M. Peterman, T. C. Shirley, P. D. Spencer, B. Wilson, and D. Woodby, 1–20. Fairbanks: Alaska Sea Grant College Program, University of Alaska.

World Bank. 2013. Fish to 2030: Prospects for fisheries and aquaculture. World Bank Report 83177-GLB. Washington, DC: World Bank. Accessed April 29, 2015. https://openknowledge.worldbank.org/bitstream/handle/10986/17579/831770WP0P11260ES003000Fish0to02030.pdf?sequence=1.

18

Informing and Improving Fisheries Management Outcomes: An Atlantic Canadian Large Pelagics Case Study by the Ecology Action Centre

Susanna D. Fuller, Kathryn E. Schleit,
Heather J. Grant, and Shannon Arnold

CONTENTS

18.1 Introduction

The use of scientific information in advocating for marine conservation and implementation of sustainable fisheries policies and practices has long been the focus of many conservation and environmental nongovernmental organizations (ENGOs) (Garcia et al. 2014). In Canada, environmental organizations have increasingly been included as *stakeholders* in various integrated management processes, including the failed Eastern Scotian Shelf Integrated Management process (Dutka et al. 2010; see also Chapter 12 in this volume) and the ongoing Pacific North Coast Integrated Management Area (www.pncima. org). Fisheries management processes have also begun to include environmental organizations, for example, the decade-old Marine Conservation Caucus on the west coast and the more recent Canadian federal Department of Fisheries Oceans (DFO) and ENGO Forum, established in 2012 with DFO Maritimes Region and local conservation organizations. In addition to these process-oriented initiatives, other examples of collaboration exist between ENGOs and marine industries to address specific marine conservation concerns. The redirection of shipping lanes in the Bay of Fundy in 2003 to protect the North Atlantic right whale was a successful initiative developed between the World Wide Fund for Nature (WWF); the Center for Coastal Studies in Provincetown, Massachusetts; Transport Canada (Canadian federal government department); and Irving Oil Limited, which has resulted in fewer ship–whale strikes in Canadian waters. This decision is seen as one of the contributing factors to a rebounding population of the endangered whale (CTV News 2015; Van Der Hoop et al. 2013), as noted in research on the subject conducted by the New England Aquarium (2015). On the west coast of Canada, the 2012 British Columbia Trawl Agreement stemmed from a collaboration between the David Suzuki Foundation, the Living Oceans Society, and the fishing trawl industry and resulted in a globally unique approach to protecting cold-water coral and sponge habitats through a combination of area closures and fleet-wide bycatch limits (Wallace et al. 2015). ENGOs can play an important role at the science–policy interface and can be effective intermediaries in the negotiation and implementation of novel approaches to marine conservation.

The Canadian government has been criticized for its failure to protect marine species at risk through existing legal frameworks and related policy mechanisms (McDevitt-Irwin et al. 2015; Hutchings and Fiesta-Blanchet 2009; Mooers et al. 2007). In Canada, DFO is tasked with implementing both the Fisheries Act and the Oceans Act. In addition, Canada's Species At Risk Act (SARA), implemented by DFO and Environment Canada, includes the protection of marine species. Together, these three pieces of legislation—which at a high level provide for fisheries management, protection of the ocean through marine protected areas, and for legal protection of species at risk—should, if implemented appropriately, ensure that Canada is a global leader in marine environmental protection from the perspective of living marine resources. Despite these legal tools, considerable evidence shows that Canada is failing its living marine resources, as well as the Canadian public, through its inability to protect marine biodiversity (Hutchings et al. 2012) and by not establishing marine protected areas in a timely manner (CPAWS 2015a,b). While scientific information is critical for achieving policy implementation, it is also important that this information be brought to the associated policy fora, and that the public is adequately engaged so that decisions, which are ultimately political in nature, are made with the contribution of civil society.

Governments at multiple levels are often caught in the push–pull of balancing economic gain with environmental protection, as well as allocating adequate financial resources for enforcement of environmental protection policies, including those focused on oceans and coastal management. ENGOs bring a unique set of skills to the table with the ability to integrate scientific information and advocacy for the application of that information through relevant policy implementation. In addition, ENGOs are able to engage the general public and communicate complicated scientific information and government policy in efforts to create agency involvement across a broad audience.

The example of large pelagics management in Atlantic Canada provides an informative case study about information use at the science–policy interface. The way that data are shared and interpreted has led to a situation where a swordfish fishery, which catches three sharks and two turtles for every swordfish, obtained Marine Stewardship Council (MSC) certification (MSC 2012a, 2012b). As well, western Atlantic bluefin tuna, assessed as endangered in 2011 by the Committee for the Status of Endangered Wildlife in Canada (COSEWIC) (COSEWIC 2011), are currently fished at levels higher than in 2011 (ICCAT 2014) and with few additional management measures in place since their assessment (McDevitt-Irwin et al. 2015). This case study describes the science and policy decisions that enabled the current status of these fisheries and related management and certification schemes to arise. The study further examines the role of an ENGO in engaging at the science–policy interface, with mixed results in achieving management measures to reduce human impact on large pelagic marine species, including swordfish, tuna, and sharks.

First, we provide background on the Ecology Action Centre (EAC), an environmental not-for-profit organization based in Atlantic Canada, and on its involvement in efforts to reduce shark bycatch and improve the management of Atlantic bluefin tuna within Canadian waters. We then describe the EAC's engagement in related decision-making fora, including acting as a stakeholder in the MSC eco-certification scheme and regional, national, and international fisheries management frameworks. We describe policy engagement processes as well as public awareness initiatives and conclude with recommendations for reducing conflict between fisheries management and conservation of species at risk in Canada. Finally, we outline ongoing challenges in transparency and decision-making in management processes related to the conservation of large pelagic fisheries.

18.2 Overview of the Ecology Action Centre

Founded in 1971, the Nova Scotia-based Ecology Action Centre is the oldest environmental not-for-profit organization in Nova Scotia and the largest in

Atlantic Canada. The EAC works in seven major "action areas," including energy, transportation, forestry and wilderness protection, coastal and water resources, local food, built environment, and marine and fisheries conservation. The EAC is membership based and relies on a combination of grassroots initiatives, research, public education, and policy advocacy to achieve its objectives. The Marine Program was founded in 1994 and works toward the sustainability of the ocean environment as well as promoting healthy coastal communities, with the vision that

> Canada's oceans are healthy and our coastal communities thrive. This is achieved through sound conservation-based management, equitable policy, and resilient markets that incentivize sustainable fishing practices, while ensuring that Canadians have access to fresh, fair fish. (EAC 2015)

The EAC relies on the best available scientific information provided by government and academic research to advocate for marine protection. The EAC is unique in its approach in using a wide variety of tools, from data collection to legal action to public awareness and policy advocacy, to achieve its goals. Because the EAC has a mandate of sustainable jobs and a sustainable environment, the marine work includes positive outcomes for coastal communities as well as the marine environment. Fisheries resources are managed nationally and internationally, and, as such, the EAC's marine work engages across these jurisdictions, in addition to working at the local community level to advocate for and incentivize improvements in fisheries management, fishing methods, and marine protection. The EAC also engages in market-based approaches as a stakeholder in Marine Stewardship Certification and the promotion of sustainably caught fish that result in a triple bottom line economy, that is, producing social, environmental, and economic benefits for coastal communities.

The impetus behind the founding of the EAC's Marine Program was the collapse of the Atlantic Canadian groundfish fishery in the early 1990s. Among its first projects was raising awareness of the existence of cold-water corals in Atlantic Canada (Breeze 1997), primarily through interviewing fishermen and combining their anecdotal information with existing scientific information. The EAC then collaborated on hosting the First International Symposium on Cold-Water Corals in 2000, which brought together international scientists and started a baseline of information on cold-water corals globally (Willison et al. 2001). At the same time that the EAC was working on increasing knowledge of cold-water corals, it was documenting impacts to seafloor ecosystems as observed by fishermen (Fuller and Cameron 1998) and engaging in policy advocacy to reduce the impact of destructive fishing practices, namely bottom trawling, on fragile seafloor ecosystems. Since that time, the EAC has increased its engagement in fisheries management and marine conservation stakeholder processes, including regional advisory councils, international fisheries management processes through participation at regional fisheries management organizations (RFMO), such as the Northwest Atlantic Fisheries Organization (NAFO) and the International Commission for the Conservation of Atlantic Tunas (ICCAT) meetings, and communication activities designed to raise public awareness about fisheries management in Canada. While focused in Atlantic Canada, the EAC's Marine Program is an active participant in national and international coalitions of ENGOs—namely, the Deep Sea Conservation Coalition and the High Seas Alliance, working together to achieve marine protection both within Canada and on the high seas.

18.3 EAC Focus on Swordfish, Shark, and Tuna Fisheries in Atlantic Canada

The EAC first began working on large pelagic fisheries in 1995, following reports from fishermen of high levels of juvenile bycatch and nontarget catch in the pelagic longline fishery targeting Atlantic swordfish (*Xiphias gladius*) (Fitzgerald 2000). Impacts of pelagic or surface longlines have been well documented, particularly on nontarget and threatened species (Lewison et al. 2004; Lewison and Crowder 2007; Mandelman et al. 2008; Gilman et al. 2008). When we began this work, commercial longline fisheries for swordfish and porbeagle shark existed, with a small-directed longline fishery for bluefin tuna. Because the EAC had become known for supporting low-impact methods of fishing through advocating for restrictions on bottom trawling, local fishermen felt that the EAC might be able to assist in raising awareness about and changing policies to mitigate the impacts of the pelagic longline fishery. This method of fishing was in stark contrast to the targeted, low-impact harpoon fishery that has no bycatch or impact on the ocean floor (Fuller et al. 2008).

As part of building a case for the implementation of low-impact fishing gear and rebuilding of large pelagic fish populations, in the late 1990s, the EAC conducted a study entitled "The Decline of the Cape Breton Swordfishery: An Exploration of the Past and Recommendations for the Future of the Nova Scotia Fishery" (Fitzgerald 2000). This report focused on collecting data from active fishermen on trends and patterns observed over time in the inshore swordfish fishery and documented scientific information available from the Department of Fisheries and Oceans (DFO) on landings of swordfish over time, as well as information on bycatch of the longline fishery. Observations from fishermen provided significant insight into the problems with the longline fishery, including shark bycatch and catch of juvenile fish:

> You could get a thousand sharks, one on every hook. In the last few years, they have been finning ... that is a bad thing because someday

someone's going to want them. It is like taking the tongues out of the buffalo. (Respondent #36)

We were getting a pile of fish, that was the hell of it. When you were longlining, you got fish that big, 70–80 small fish some days. It was a damn shame to get those babies. (Respondent #25) (Fitzgerald 2000)

The report provided 10 recommendations to improve the swordfish stocks and the fishery, including, but not limited to, improved data collection on bycatch, reducing the allocation to the longline fishery and increasing the harpoon quota as a method of reducing both juvenile catch and bycatch, implementing closed areas, and enforcing existing conservation measures agreed to at ICCAT, the international body where catch allocations are decided (Fitzgerald 2000). In addition to the high levels of bycatch in the longline fishery, this capital-intensive method results in more wealth concentration. This means that coastal communities do not benefit as well or as equally from the swordfish resource when they are caught with longline. Conversely, with the harpoon method, individual fishermen own licenses and are able to maximize the return from swordfish. This report led the EAC to increase engagement in advocacy for conservation and bycatch reduction measures, as well as outreach to the public and fisheries managers.

As a part of a follow-up to this work, the EAC, together with the U.S.-based Marine Conservation Institute (MCI) and the British Columbia-based Living Oceans Society, published a report in 2008 entitled How We Fish Matters: Addressing the Ecological Impacts of Canadian Fishing Gear (Fuller et al. 2008). This report reviewed the impacts of all fishing gear used in Canada and provided specific recommendations, including the development of a national bycatch policy. A previous report by the MCI documented the impacts of fishing gear in the United States (Chuenpagdee et al. 2003); however, the EAC recognized that different methods of fishing as well as different environmental conditions and regulatory frameworks warranted a similar study in Canada. The report development included a workshop that brought together fisheries managers, fishermen, First Nations peoples, and conservation organizations to review and assess the impacts of various fishing gears and rank the level of their impact (Fuller et al. 2008). Pelagic longlines were determined to have a high impact on sharks and large pelagic species, medium/high impact on seabirds, and medium impact on marine mammals in Canadian fisheries.

The results of this work were communicated broadly to regulators and fishing industry associations, as well as through traditional and social media. The assessment of collateral impacts of fishing gear created a longer-term policy advocacy agenda for the EAC's conservation work on Canadian fisheries. As part of carrying forward the report recommendations, the EAC began to become involved more actively in the Atlantic Large Pelagic Advisory Committee (ALPAC), as well as communicate with organizations advocating for conservation through ICCAT, the regional fisheries management

TABLE 18.1

COSEWIC Status of Bycatch Species in the Atlantic Canadian Swordfish Fishery

Common Name	Scientific Name	COSEWIC Status	Date of COSEWIC Assessment	SARA Listing	Years under Consideration for Listing
Atlantic bluefin tuna	*Thunnus thynnus*	Endangered	May 2011	None	2.5
Basking shark	*Cetorhinus maximus*	Special Concern	Nov. 2009	None	3.5
Blue shark	*Prionace glauca*	Special Concern	April 2006	None	7.5
Shortfin mako shark	*Isurus oxyrinchus*	Threatened	April 2006	None	7.5
Porbeagle shark	*Lamna nasus*	Endangered	Reassessed May 2014	None	11 since initial assessment, 9 since listing decision
Leatherback turtle	*Dermochelys coriacea*	Endangered	May 2001	Schedule 1	2
Loggerhead turtle	*Caretta caretta*	Endangered	April 2010	None	5

organization responsible for the management of highly migratory species in the Atlantic Ocean.

In the years between the initial examination of the decline of the swordfish fishery and the assessment of the impacts of Canadian fishing gear, several studies documented the impact of longline fisheries on shark populations in the Northwest Atlantic (Baum et al. 2003; Baum et al. 2005). Species caught in the Atlantic Canadian longline fishery increasingly were being reviewed by COSEWIC and subsequently being considered for listing under Canada's Species at Risk Act (Table 18.1).

The increasing vulnerability of shark species as indicated by the COSEWIC assessments and academic publications, the assessment of Atlantic bluefin tuna as endangered, and the failure of fisheries management bodies to adequately address fishing impacts on these species, set the stage for specific policy advocacy and public engagement through traditional and social media avenues. In 2009, the Atlantic Canadian Longline Swordfish fishery also entered into certification by the MSC (MSC 2009), resulting in it being the first surface longline fishery globally to seek the eco-certification label. In the following sections, we describe specific information-based campaigns and advocacy within policy fora conducted by the EAC between 2010 and 2015 to achieve conservation measures to protect large pelagic species in Atlantic Canada.

We begin with chronicling the EAC's input into the MSC certification, beginning with the initial assessment by the MSC in 2009, through to the annual surveillance audit in August 2015. We then focus on species-specific campaigns for the protection of sharks, with a focus on the directed Canadian fishery for porbeagle shark and management of the endangered Atlantic bluefin tuna. We discuss the EAC's role in bringing data to the table, raising public awareness as well as advocating for protective measures in government decision-making bodies, including ALPAC and ICCAT, and through Canadian law, including the Species at Risk Act. Through action, education, and advocacy, the EAC has focused on achieving the following:

- Fishing quotas and total allowable catches (TAC) for swordfish and tuna fisheries that are in line with robust precautionary scientific advice, particularly in the face of scientific and environmental uncertainty related to stock rebuilding
- Improvements in fishing behavior and fishing gear substitution to reduce bycatch of sharks and turtles in the Atlantic Canadian swordfish longline fishery
- Alternatives to quota increases that can allow for maintenance of a fishery, as well as rebuilding of populations, primarily through the support of a recreational fishery for Atlantic bluefin tuna
- Harmonization of Canada's requirement under the Species at Risk Act to protect marine species at risk, with the conflicting decisions made through fisheries management of fisheries for species at risk
- Inclusion of civil society and ENGO participation in science and policy processes to increase transparency and accountability of decision-making
- Ensuring that third-party eco-certification schemes are upholding best practices of fisheries management and are using the best available scientific information when making decisions on certifications

18.4 Marine Stewardship Council Certification

The rise of eco-certification schemes in commercial fisheries can ostensibly be linked to the desire for environmentally sustainable food production. The largest of these schemes, the Marine Stewardship Council, was formed as a partnership in 1996 by WWF and the Unilever Corporation to improve environmental practices in large-scale commercial fisheries. MSC assesses fisheries across three principles and has developed an elaborate and reasonably comprehensive scoring guidance for these principles. The three principles are (1) sustainable fish stocks, (2) minimizing environmental impact, and (3) effective management (MSC 2015). The MSC program allows for fisheries to become certified with conditions, meaning that the certification is one that seeks continuous improvement rather than the existing practices of a fishery at the time of certification. The MSC itself does not conduct the certification process; instead, the fishing industry employs certification bodies to conduct the assessment, and all stakeholder input is submitted to, managed by, and responded to by the certification body.

The actual conservation benefits of the MSC, in terms of leading to concrete measures that have improved marine resources, have been questioned (Ward 2008; Jacquet et al. 2010; Galil et al. 2013). However, some have found concrete improvements as a result of certification (Martin et al. 2012). Where there is capacity and detailed knowledge of a particular fishery, often by environmental organizations, and in some cases by other fishing entities, the certifications have been challenged through official objections (Christian et al. 2013).

18.4.1 EAC Stakeholder Engagement

Because of our background in research and policy advocacy for lower-impact fishing methods wherever possible, the EAC engaged heavily in the MSC process for longline swordfish. The certification process was further complicated by the fact that the harpoon fishery and the longline fishery entered the assessment process simultaneously, largely as a cost saving measure, as the information under Principle 1 (stock status) and Principle 3 (management measures) was more or less the same, with the difference between the fisheries occurring under Principle 2 (minimizing environmental impact). Throughout our engagement, it has been necessary for us to obtain access to the best information possible regarding the fishery and seek the expertise of both government and academic scientists. Our opposition to the certification was based on the following information, a combination of the quantification of fishery impacts and lack of associated fisheries management measures (EAC 2011):

- This fishery has the highest bycatch to target catch ratio of any fishery in Atlantic Canada at over 50% of the catch. Of these discarded animals, 90% by weight are species considered at risk including loggerhead and leatherback turtles, and blue, shortfin mako, and porbeagle sharks.
- For turtles, there are currently no bycatch limits, no mandated gear, fishing methods, or bait configurations, and no spatial/temporal closures.
- For sharks, there are currently no spatial/temporal closures, no gear/ bait restrictions, no limits on blue shark catch, a bycatch limit on porbeagle that does not include discards and does not close the fishery if a catch is over the limit, and a guideline tonnage for shortfin mako that is not biologically based and does not include discards (Wang 2013).
- Mandated observer coverage agreed at ICCAT is only 5%, too low to characterize bycatch reliably, that is, spatially and temporally

(Anderson and Small 2013), and is not spatially representative of the fishery.

- Data on the bycatch species are characterized as insufficient by the regulator. There are clear gaps in knowledge of discard levels and impacts on these species' status and recovery.

The EAC engaged in the MSC process from the outset, with recommendations for independent science representation on the assessment team, participation in initial site visits by the selected assessment team, and provision of additional information to be considered as part of the Public Comment Draft Report (PCRD). We contributed data based on the best available science, and described where the PCRD failed to address these significant conservation concerns. We worked together with other Canadian environmental organizations, including the WWF and the David Suzuki Foundation, submitting joint comments with the latter and communicating our submissions to the former. Finally, when it became clear that the fishery would be recommended for certification, the EAC launched an official objection to the certification in September 2011. Several other submissions by ENGOs were made, including from the WWF, the Turtle Island Research Network, and Greenpeace. Despite the objection and specific recommendations to amend the scoring, the objection was overturned through an independent adjudication process and the swordfish fishery was certified in April 2012 (MSC 2012a), three years after the certification process began. The EAC continued its engagement as a stakeholder during the annual audit process, having engaged in the third such audit in August 2015. Despite our repeated concerns that the fishery continues to adversely impact threatened and endangered bycatch species, there remain large gaps in both science and management, including the lack of science capacity at DFO, as of summer 2015, to address outstanding gaps in shark assessments as well as continued low observer coverage that does not allow for robust estimates of bycatch. At the same time that the EAC was heavily involved as a stakeholder in this certification process, we attended ALPAC, ICCAT, and related science meetings where stock assessments and recovery potential analyses are conducted. ALPAC began a process of developing an Ecosystem Working Group, where issues of bycatch and other research programs could be addressed; however, this working group last met during the annual ALPAC meeting in 2010 but has not been reconvened since that time.

18.5 Shark Conservation as a Vehicle for Change

In an effort to raise awareness of the problematic certification of the Atlantic Canadian swordfish fishery and the declining status of sharks, the EAC launched a public campaign drawing attention to the impact on sharks and using them as a vehicle for further public engagement. We recognized that sharks are a charismatic group of animals, for which there was already growing public concern following the exposure of shark finning and the shark fin trade in the media. We decided to focus a campaign specifically on porbeagle sharks, because they were the most in danger of extinction of all of the sharks commonly caught by the longline fishery. The campaign had two main goals: a closure of the directed fishery for porbeagle sharks through domestic policy decisions in Canada and a ban on the retention of these sharks within tuna fisheries at the international level with ICCAT. The campaign illustrates how highlighting data on the status of a species, using data to question current policies, and repeatedly bringing these facts to the decision-making table can influence public policy.

COSEWIC assessed porbeagle sharks as endangered in 2004 (COSEWIC 2004). It was denied listing under the Species at Risk Act in 2006, a decision that mainly cited socioeconomic reasons (Species at Risk Registry 2006). In addition, according to ICCAT's Standing Committee on Research and Statistics, porbeagle sharks are one of the most vulnerable in the ICCAT area, and they have also been assessed as endangered by the International Union for the Conservation of Nature (IUCN). In March 2013, porbeagle sharks were included in Appendix II of the Convention on International Trade in Endangered Species of Wild Fauna and Flora (CITES), which requires countries to obtain export permits in order to trade products derived from listed species. An Appendix II listing acknowledges that the species may become threatened with extinction if trade is not carefully controlled (www.cites.org/eng/app/index.php). Further, the porbeagle was reassessed as endangered in 2014 by COSEWIC, which found that the significant threats to the species had not diminished, and that there was no notable recovery of the population (COSEWIC 2014). Ten years after the original assessment, the species is now awaiting a listing decision on SARA yet again. In the Northwest Atlantic, fishing mortality adds decades to the already slow recovery trajectory, pushing it to upwards of 100 years (DFO 2005) for this vulnerable shark. Estimates of dead discards, postrelease mortality, or environmental changes over the possible century of recovery time are not accounted for, and there may be unregulated and unreported high seas catch not represented in abundance models.

18.5.1 Closure of the Directed Fishery

Canada was the last country to have a fishery that targeted porbeagle sharks. While Canadian landings for porbeagle have declined over the last decade, the catch was over 1000 tonnes during the 1990s. The EAC was concerned that if the market for this shark resumed, and Canada continued to allow the exploitation of this species with no hard cap limit, there was a real danger that catches would rise once again. The campaign highlighted the endangered status of porbeagle sharks as assessed by Canadian scientists through COSEWIC and though the international community by the IUCN and ICCAT scientists. Data from the DFO Recovery Population Assessment (RPA) (DFO 2005) were used to point out that recovery timelines are in the order of decades to a century, and that this is unacceptable for an endangered species. The Government of Canada continued to point out that, following a declining population trend for the last 50 years, there were indications that porbeagle numbers have stabilized in the last few years. While we agree that this trend should be celebrated, it should not be used as a justification to continue fishing. These "signs of recovery" only mean that the shark is holding steady at 20% of its 1960s population, which was likely lower than virgin biomass levels. Porbeagle is still considered to be in the "critical zone" according to Canada's own Precautionary Approach Policy (DFO 2013).

DFO continued to argue that the department needed the directed fishery for data collection and a continued time series. Canada had some of the world's leading nonlethal porbeagle shark-tagging research, which has contributed immensely to our understanding of this species. We maintained that further fisheries-independent research on porbeagle should be supported by the government and our international partners, but data collection should not be an excuse for maintaining a directed fishery for an endangered species with little market value. The directed porbeagle fishery was also still considered an "exploratory" fishery under Canadian fisheries management since it is not yet a self-sustaining fishery economically. In fact, only a handful of fishermen participated in the fishery, and there have been almost no landings by the fleet in recent years. Using these facts, we were able to bring reality to the statement that the directed fishery was of high economic importance to Canadians.

In addition to bringing these facts to the forefront at domestic and international management meetings, our public petition was perhaps the most successful. Working with Change.org, we were able to secure over 21,000 signatures that we delivered to the federal Minister of Fisheries and Oceans (Friends of Hector 2015). Additionally, we received coverage in the national publication *Ipolitics* (Zilio 2012), which was effective in communicating the porbeagle shark issue to elected officials in Ottawa through a data driven infographic.

In 2013, the Canadian government announced the suspension of the directed fishery for porbeagle shark. By moving the conversation forward from the directed fishery, we then were able to focus on our larger goal of achieving a ban on the retention of porbeagle sharks at an international level.

18.5.2 Bycatch Fishery and Advocating for Retention Bans

Porbeagle sharks are frequently caught as bycatch in the longline swordfish fishery, which comes under the jurisdiction of ICCAT. Thus, the EAC's campaign on bycatch was focused mostly at the international level. Canada has consistently blocked the adoption of international bans on the retention of porbeagle at ICCAT, asserting that it has a well-managed porbeagle fishery based on sound science, and that the population is on track to recover (despite the estimated 100 plus year recovery time). The EAC has maintained that, while Canada's research has contributed immensely to ICCAT's understanding of the porbeagle shark population, Canada's management of porbeagle sharks is high risk, rather than precautionary, given the lack of any target timeline for recovery and insufficient monitoring and enforcement to prevent further population decline. Furthermore, Canada's refusal to support the retention ban results in the failure of ICCAT to protect the species in other jurisdictions where management is even more flawed, as proposals require consensus in order to pass. For several years, Canada was the only country to block the retention ban proposal; however, in more recent years other countries have stepped in to support the ban.

While retention bans are not always the most effective way to protect a species from overfishing, we decided that in this case, we would support a ban. Our rationale was that banning the retention of the species would remove any incentive for fishermen to target porbeagle sharks opportunistically when the markets are good. It would also mean that any landing of the species in the Atlantic would be illegal, which makes management extremely straightforward, and is important for a species that is subject to illegal, unreported, and unregulated fishing. Since observer coverage is often too low to provide meaningful enforcement in this fishery, a retention ban is the most simple and straightforward way to protect the species from fishing-related mortality.

Canada claimed that it was managing porbeagle shark bycatch well, with a total allowable catch in place. The agreed 185 tonne TAC that Canada has set for porbeagle is not actually an enforced limit. In the past, Canada's fishery management officials have confirmed that they will not stop other fleets from catching the species once the quota has been reached. Further, the success of the Canadian recovery program is contingent on proper accounting of all catches, including those taken by high seas fleets. With the current management model so risky already, any unaccounted for mortality may slow the projected recovery time considerably. There has been mounting concern regarding unaccounted discards of porbeagle sharks as bycatch in several fishing fleets in Atlantic Canada, particularly juvenile discards in an area identified as a porbeagle nursery ground.

Increased observer coverage is needed to fully assess total mortality on this vulnerable species and to ensure that signs of further decline are noted in a timely fashion. As yet, no limit reference points or timeline targets for recovery have been set, nor any harvest control rules laid out for various porbeagle shark population scenarios, as is required under Canada's Precautionary Approach Policy for species considered to be in the "critical zone." Additionally, there are no maximum length restrictions or mandatory spatial closures to protect important life stages for porbeagle sharks, like pupping and mating.

The desire to influence these decisions led the EAC to apply to be an official observer at ICCAT. In 2015, the EAC remains the only Canadian ENGO to attend the ICCAT annual meetings, and has been attending these meetings since 2010. As an observer, the EAC is able to attend meetings and sit in plenary sessions.

Observers do not participate in decision-making and do not put forward official recommendations or vote. Therefore, our ability to influence decision-making at ICCAT is through side meetings with governments and submitted written statements and interventions made during the meetings.

Unfortunately Canada, as well as some other countries, continues to block efforts to ban the retention of porbeagle sharks. Nonetheless, progress has been made at ICCAT and in other fora. There is growing support from other countries for a ban on porbeagle sharks. Further, Canada did table a proposal at the 2014 ICCAT meeting that would allow for only the retention of dead sharks, which shows a desire to move the conversation forward. While member governments and the EAC did not feel that Canada's proposal went far enough to protect porbeagle sharks, it was the first time that Canada has shown any proactive behavior on the species at ICCAT.

The Appendix II listing of porbeagle sharks on CITES in 2013, which came into force in 2014, means that those wishing to export the species must obtain a permit from the Canadian government. These permits are issued on the basis that trade at these levels will not jeopardize the future population. We have heard anecdotally that the market has already been impacted by raised awareness around the endangered status of porbeagle sharks, and that many traders have not deemed it worth the effort to seek permits to trade. Hopefully, this development will lead to a decline in catch levels of porbeagle as well, and open doors to further conversations about the retention of the species (Box 18.1).

BOX 18.1   FINS ATTACHED

Canada's mysterious defense of the 5% rule

Shark finning is considered one of the most egregious and wasteful practices, as sharks are caught, fins removed, and then the rest of the shark is disposed of at sea. While many countries have adopted fins attached regulations, Canada is one of the few developing fishing nations that has refused to adopt fins attached and prefers adhere to the "5% rule" which means that when sharks are landed and weighed the total weight of the fins must add up to 5% of the total weight of sharks landed. However loopholes exist with the 5% rule that mean illegal shark fins are still being landed. Requiring sharks to be landed with their fins naturally attached at the first point of landing is the most straightforward way of enforcing the finning ban and will greatly improve species-specific data collection for sharks (Biery and Pauly 2012). Despite defending the 5% practice as comparable to fins attached and more efficient for the fishing industry as storing sharks with their fins attached is seen as impractical, Canada has yet to produce any data on the 5% rule and its supposed comparable measures to demonstrate that they are adequately working, despite multiple requests.

18.6 Atlantic Bluefin Tuna

Canada's decision-making with regard to Atlantic bluefin tuna in the last few years, and the resulting campaign activities by the EAC, is illustrative of the challenges in large pelagic management and the dichotomies at the science–policy interface. DFO has often found itself in conflict between a desire to allocate fisheries resources and significant pressure by the fishing industry to increase quotas and the legal obligation to protect species at risk under the Species at Risk Act. The management of the western stock of Atlantic bluefin tuna offers an interesting case study into this conflict, and illustrates how the Canadian government can be at odds with itself when it comes to the protection of marine fish populations.

In 2011, Atlantic bluefin tuna (*Thunnus thynnus*) was assessed by COSEWIC as endangered (COSEWIC 2011). The western stock of Atlantic bluefin tuna is currently only at 55% of its 1970s levels, a time when the population was already in decline and with overfishing cited as a leading cause of this decline. In the four years that this species has been under consideration for Species at Risk listing, it has received no new protective measures or management plans from DFO. The Integrated Fisheries Management Plan (IFMP), the public guiding document on fisheries management for Atlantic bluefin tuna, has not been updated since 2008—a draft updated version was circulated in 2015. Most egregiously, Canada agreed to a quota increase at the ICCAT meeting in 2014 despite the fact that fishing mortality is the main threat to the species, and no decision has been released by the federal fisheries minister on the SARA listing decision. In fact, Canada started advocating for a quota increase in 2012, after COSWEIC had just assessed the species as endangered.

Due to the highly migratory nature of bluefin tuna, the species cannot be effectively managed by one country alone. The amount of Atlantic tuna (including bluefin) that each of ICCAT's contracting parties can catch is negotiated during annual or biannual ICCAT meetings. As a result of the decline in the western Atlantic bluefin tuna population, in 1998 ICCAT adopted a 20-year rebuilding plan for the species after efforts up to that point had failed to rebuild the population and instead led to further declines. Now, 15 years into this plan, the population is still severely depleted. Given the commercial importance of the charismatic species to Atlantic Canada and its population decline, the EAC launched a campaign on bluefin tuna conservation in 2013. The bulk of the public and outreach part of this work took place in Canada, while influence at the decision-making level occurs at ICCAT.

18.6.1 Canada's Role in Bluefin Tuna Management

Western Atlantic bluefin tuna can live to be 40 years old, and it takes these iconic fish up to 10 or more years to mature (ICCAT 2012). Since they take longer to reach sexual maturity than most fish species, there is a high probability that they will be caught before they are able to spawn. This makes them especially vulnerable to overfishing. The Canadian bluefin tuna fishery, particularly in the Gulf of St. Lawrence, not only targets the giants— the largest and most biologically important members of the population—but it is the only country in the world to do so. Therefore, Canada holds a critical role in determining the fate of western bluefin. Canada was among the first contracting parties when the treaty that established ICCAT was signed in 1966 and, until recently, had been considered a leader in precautionary bluefin tuna management. Canada remains an influential party within ICCAT decisions and holds 23% of the western Atlantic bluefin tuna quota (as well as quota transfers from other countries).

18.6.2 EAC's Efforts to Raise Awareness and Influence Policy on Bluefin Tuna Management

The EAC bluefin tuna campaign focuses on the dichotomy between Canada on the one hand reviewing Atlantic bluefin tuna to be listed on the Species at Risk Act and, on the other, advocating for quota increases. Since this is an internationally managed species, the EAC's engagement with the government has been mainly through ICCAT and regional advisory bodies. An important role for the EAC has also been communicating complicated and lengthy scientific documents on the status of Atlantic bluefin tuna.

The EAC is a member of ALPAC, the regional body that meets annually and provides recommendations to the federal fisheries minister on the management of tuna and swordfish. At these meetings, DFO presents science data collection plans on large pelagics. As an official member of ALPAC, the EAC has the opportunity to sit at the same table as industry and Aboriginal members.

As noted above, the EAC remains the only Canadian ENGO to attend the ICCAT annual meetings, and has been attending these meetings since 2010, originally to advocate for greater protection for sharks. While being at the decision-making table is necessary for advocacy campaign success, an important role that we play as an ENGO is the ability to link the public to the issues. The EAC continues to highlight the status of bluefin tuna through traditional and social media. In 2014, as a way to refute the argument that the only way for fishermen to make more money from bluefin tuna is through a quota raise, we commissioned a study on the newly emerging live release tuna fishery. While the tuna fishery in Canada began as a recreational fishery in the first part of the twentieth century, in recent decades it has been exclusively a commercial fishery. However, the Gulf of St. Lawrence has seen the development of a live release recreational fishery since 2010. In the last four years, the live release fishery has grown consistently, and data suggest that it has the potential to bring in almost six times as much revenue to fishers as the commercial fishery. This fishery has a low level of tuna mortality—roughly 5%—and therefore has a much lower impact on the population (EAC 2014).

In their advice for the 2015 fishing year, ICCAT scientists said that catches of less than 2250 tonnes had a 50% chance of keeping the population where it was or potentially increasing it. However, they went on to say that keeping the quota at 1750 metric tonnes would allow the population to increase more quickly and help to decrease some of the uncertainty with stock assessment models. In order to communicate this message to the public, the EAC placed an advertisement in a newspaper, the *Ottawa Citizen*, that distilled the science down to these facts (Figure 18.1). We were disappointed that, despite this advice, ICCAT parties agreed to raise the quota to 2000 tonnes, as they felt that a slightly more optimistic outlook than previous years justified a quota increase.

FIGURE 18.1
Public interpretation of the scientific risk assessment regarding the rebuilding of the Western Atlantic bluefin tuna stocks presented at ICCAT in 2014.

Throughout the campaign, the EAC continues to be the voice of long-term sustainable management of the stock. It is critical that DFO renew its commitment to using a precautionary, science-based approach to western Atlantic bluefin tuna management. Bluefin is an important part of eastern Canada's history, culture, and economy, along with playing an important role as an apex predator in Canada's waters. While rebuilding may require short-term moderation, it will ensure the long-term sustainability of the western bluefin population and all that depend on it. Thus, fishery managers must only authorize removals at a level that promotes a full recovery.

18.6.3 Public-Facing Aspects of EAC's Campaigns: Hector the Shark and Tina Tuna

The EAC has long recognized that building public support and awareness around issues are important tools that can be used to put pressure on governments and managers to make policy changes. One of the challenges in this task can be finding ways to distill complex, scientific information into interesting and easily digestible content for the general public. In order to gain the attention of the public and to provide a fun, casual, and interesting platform for the dissemination of information, the EAC has employed mascots or "spokes-fish" to act as mouthpieces for the public-facing aspects of its large pelagics campaign.

Following the launch of the initial campaign that was concerned with the MSC certification of the Atlantic Canadian longline swordfish fishery, the EAC developed Hector the Blue Shark, a cartoon blue shark that was concerned with the wasteful practices of the pelagic fleet. A blue shark was chosen to represent these concerns because of all the species caught in the Atlantic Canadian longline swordfish fishery, blue sharks were caught in the largest numbers, with an estimated 100,000 blue sharks caught every year in the fishery (Campana et al. 2005), 35,000 of which do not survive (Campana et al. 2008, 2009). While blue sharks are not as threatened (designated as special concern by COSEWIC in 2006) as the porbeagle shark, this advocacy campaign was seen as an opportunity to protect a species *before* its population continued its trajectory. Hector the Blue Shark was set up with a website, a Facebook page, and a Twitter profile in order to post updates on the MSC certification process and to build support for the EAC's objections against it. A photo campaign was also launched, which encouraged supporters to upload photos of themselves with Hector to show their concern and become a "Friend of Hector."

Hector's Facebook page currently has close to 4000 "likes" from around the world and has regular engagement on the information he shares. On Twitter, Hector has about 2000 followers and is frequently asked to weigh in on issues relating to the conservation of Canadian sharks by the scientific community, other advocacy groups, and the general public. Tweets receive anywhere from 3000 views per month in slower campaign periods, to 15,000 during more active periods such as the ICCAT annual meeting.

With the launch of the EAC's bluefin tuna campaign in more recent years, we decided that Hector the Blue Shark needed a new friend, and so Tina the Bluefin Tuna was created. She was also set up with her own website, and Facebook and Twitter accounts. We use Tina to disseminate press releases and news items related to tuna management around the world. During the lead up to ICCAT, Tina encourages her followers to contact the federal minister and the Department of Fisheries and Oceans and ask them to be precautionary in their management of tuna. In past years, she has encouraged her "Bluefinatics" to bring attention to the fact that the federal government has been "two-faced on tuna" by advocating for quota increases when the species is still waiting for a listing decision under the Species at Risk Act. Tina has nearly 400 followers on Twitter and over 300 on Facebook. Tina's tweets receive anywhere between 2,000 views per month in slower campaign periods, to 15,000 during the time leading up to and including the ICCAT annual meeting.

18.7 Conclusion

Despite the successes along the way, there remain significant challenges to obtaining and using scientific information to achieve marine conservation and fisheries management outcomes in Canada that are consistent with implementing national and international laws and policies that have been created to mitigate human impacts on our marine ecosystems. The key challenges include:

- *Data availability*: Fisheries observer data and logbook data are the only source available where bycatch is recorded on a regular basis. These data are often difficult to obtain and there is no consistent standard for recording across regions (i.e., the Maritimes Region and Newfoundland Region in Atlantic Canada do not share observer databases). Observer data are often considered proprietary and not available to all stakeholders, although academic institutions have had reasonable success with data-sharing arrangements (Cosandey-Godin 2015).

- *Transparency in decision-making* is a continuous challenge. It is common in meeting reports from both national and international management processes that the record of discussion is not presented, but only a record of the final determination or decision. In international meetings, Canada has never opened up its delegations to NGOs, although the fishing industry and representative associations, arguably nongovernment, are always present on the delegation. In fact, Canada regularly appoints ICCAT commissioners, whose role is to "represent Canada"; however, they are members of the fishing industry who also represent their own fishing interests and not necessarily the views of the Canadian public (DFO 2009). Additionally, at regional meetings of ALPAC there are frequently industry-only meetings that do not include all stakeholders.

- *Reductions in scientific capacity* through budget reductions have resulted in less data being collected (e.g., De Souza 2013). For example, the August 2012 MSC certification audit of the Atlantic Canadian swordfish longline fishery includes conditions for research on shark populations, when DFO no longer has a dedicated scientist with this area of expertise in Atlantic Canada. DFO has also moved to multiyear stock assessments, which means that less information is available. At the same time, there have been reductions in collaborative research efforts with organizations such as the Fishermen & Scientists Research Society (FSRS). In addition, significant concerns have been expressed about the muzzling of scientists in various government departments, leading to a reduction of information available to the public (CBC 2015).

- *Clientism that favors industry participation in decision-making*: While the EAC fully supports the concepts and actions of community-based comanagement of fisheries, we do not support decision-making where only the fishing industry is engaged, particularly when Canada has not committed to ensuring that quota setting does not compromise the ability of a fish population to recover from depleted levels. We see the need for an increased stakeholder base, as well as more collaborative decision-making across various interest sectors. As the EAC is committed to seeing coastal communities flourish and wild fish populations recover and thrive, we see many areas of mutual interest where decision-making can benefit both fishermen and the marine environment.

The EAC has managed to use a wide variety of information—from observations and anecdotes from fishermen, to scientific information generated by government and nongovernment scientists—to effectively engage in national and international policy forms. By combining credible science and traditional knowledge with innovative public engagement campaigns, and creating a platform for individuals to take action and have a sense of urgency in changing how Canada manages and protects large pelagic species, changes in public policy and species management have been achieved. At the same time, significant challenges remain in seeking conservation improvements through the MSC eco-certification process. As the marine conservation movement evolves, one of the difficult aspects is ensuring that Canadian policies are being implemented by government agencies, and that third-party certifications are not seen as the major implementation mechanism. The complexities of assessing, compiling, and interpreting scientific information, effectively communicating with policy makers, engaging in multistakeholder management advisory fora—sometimes as a legitimate member and other times as an observer—requires a diverse skill set. It is important to be able to comprehend and interpret the science as well as have a general knowledge of the specific scientific field. Knowing the science is only half the battle; the science must then be communicated strategically. Relationships must be established so that policy makers and regulators take the interjections and information from an environmental organization seriously. Knowledge of the policy framework, assessments of implementation, and the details around how and when policies are implemented is also necessary. Finally, as these processes are unfolding, it is necessary to engage the general public and the media, both traditional and social, so that policy makers are held accountable to greater interests than those of the specific fishing industry.

Environmental nongovernmental organizations can be effective actors at the science–policy interface, and, in this case, the EAC has played a critical role in making public scientific information regarding the impact of the Atlantic Canadian large pelagic fishery on nontarget species, bringing that information to relevant policy fora both nationally and internationally, engaging in the eco-certification process, and bringing critical information to the Canadian public. In all cases, the EAC has brought an important voice to the table, advocating for implementation of existing conservation-based legislation and policies in Canada.

References

Anderson, O. R. J. and C. L. Small. 2013. Review of tuna regional fisheries management organization longline scientific observer programmes. *ICCAT Collective Volume of Scientific Papers* 69 (5): 2220–2232. http://www.iccat.es/Documents/CVSP/CV069_2013/n_5/CV069052220.pdf.

Baum, J., D. Kehler, and R. Myers. 2005. Robust estimates of decline for pelagic shark populations in the Northwest Atlantic and Gulf of Mexico. *Fisheries-Bethesda* 30 (10): 27.

Baum, J. K., R. A. Myers, D. G. Kehler, et al. 2003. Collapse and conservation of shark populations in the Northwest Atlantic. *Science* 299 (5605): 389–392.

Biery, L., and D. Pauly. 2012. A global review of species-specific shark-fin-to-body-mass ratios and relevant legislation. *Journal of Fish Biology* 80: 1643–1677.

Breeze, H. 1997. *Distribution and Status of Deep Sea Corals off Nova Scotia*. Marine Issues Committee Publication No 1. Halifax: Ecology Action Centre.

Campana, S. E., W. Joyce, L. Marks, et al. 2008. The rise and fall (again) of the por-beagle shark population in the Northwest Atlantic. In *Sharks of the Open Ocean: Biology, Fisheries and Conservation*, edited by M. Camhi, E. K. Pikitch, and E. A. Babcock, Chapter 35, 445–461. Oxford: Blackwell Science.

Campana, S. E., W. Joyce, and M. J. Manning. 2009. Bycatch and discard mortality in commercially caught blue sharks *Prionace glauca* assessed using archival satel-lite pop-up tags. *Marine Ecology Progress Series* 387: 241–253.

Campana, S. E., L. Marks, W. Joyce, et al. 2005. Catch, by-catch and indices of popu-lation status of blue shark (*Prionace glauca*) in the Canadian Atlantic. *ICCAT Collective Volume of Scientific Papers* 58 (3): 891–934.

CBC. 2015. Steve Campana, Canadian biologist, "disgusted" with government muz-zling. *CBC News*, 19 May. Accessed 23 September 2015. http://www.cbc.ca/news/canada/nova-scotia/steve-campana-canadian-biologist-disgusted-with-government-muzzling-1.3078587.

Christian, C., D. Ainley, M. Bailey, et al. 2013. A review of formal objections to Marine Stewardship Council fisheries certifications. *Biological Conservation* 161: 10–17.

Chuenpagdee, R., L. E. Morgan, S. Maxwell, et al. 2003. Shifting gears: Assessing collateral impacts of fishing methods in U.S. waters. *Frontiers in Ecology and the Environment* 1 (10): 517–524.

CITES (Convention on International Trade in Endangered Species of Wild Fauna and Flora). 3 March 1973. 993 UNTS 243.

Cosandey-Godin, A. 2015. Elasmobranch bycatch in the Canadian Northwest Atlantic and Arctic adjacent seas: Composition, biogeography, and mitigation. PhD diss. Dalhousie University.

COSEWIC (Committee on the Status of Endangered Wildlife in Canada). 2004. Status assessment porbeagle shark. http://www.sararegistry.gc.ca/document/dspText_e.cfm?ocid=1014.

COSEWIC (Committee on the Status of Endangered Wildlife in Canada). 2011. COSEWIC assessment on the Atlantic bluefin tuna *Thunnus thynnus* in Canada. http://www.cosewic.gc.ca/eng/sct1/searchdetail_e.cfm?id=1148&StartRow=1&boxStatus=All&boxTaxonomic=All&location=All&change=All&board=All&commonName=tuna&scienceName=&returnFlag=0&Page=1.

COSEWIC (Committee on the Status of Endangered Wildlife in Canada). 2014. Assessment and status report on the porbeagle shark *Lamna nasus* in Canada. http://www.registrelep-sararegistry.gc.ca/document/default_e.cfm?documentID=465.

CPAWS (Canadian Parks and Wilderness Society). 2015a. Dare to be deep: Are Canada's marine protected areas really "protected"? Annual report on Canada's progress in protecting our ocean. Ottawa: CPAWS. http://cpaws.org/uploads/CPAWS_DareDeep2015_v10singleLR.pdf.

CPAWS (Canadian Parks and Wilderness Society). 2015b. Protecting Canada: Is it in our nature? How Canada can achieve its international commitment to pro-tect our land and freshwater. Accessed July 2015. http://cpaws.org/uploads/CPAWS_Parks_Report_2015-Single_Page.pdf.

CTV News. 2015. App, shipping lane changes keep right whales safe in the Bay of Fundy. *CTV Atlantic*, 10 June. http://atlantic.ctvnews.ca/app-shipping-lane-changes-keep-right-whales-safe-in-bay-of-fundy-1.2415146.

De Souza, M. 2013. Harper government cutting more than $100 million related to protection of water. Canada.com. Accessed 23 September 2015. http://o.canada.

com/news/harper-government-cutting-more-than-100-million-related-to-protection-of-water.

DFO (Department of Fisheries and Oceans). 2005. Recovery Assessment Report on NAFO Subareas 3–6 Porbeagle Shark. DFO Canadian Science Advisory Secretariat Science Advisory Report 2005/043. http://www.dfo-mpo.gc.ca/csas/Csas/status/2005/SAR-AS2005_043_E.pdf.

DFO (Department of Fisheries and Oceans). 2009. Minister Shea appoints commissioner to the International Commission for the Conservation of Atlantic Tuna. News release. http://news.gc.ca/web/article-en.do?m=/index&nid=450799.

DFO (Department of Fisheries and Oceans). 2013. Overview of the precautionary approach framework guidelines. http://www.dfo-mpo.gc.ca/fm-gp/peches-fisheries/fish-ren-peche/sff-cpd/precautionary-precaution-back-fiche-eng.htm.

Dutka, S., R. Hunka, and J. McNeely. 2010. ESSIM: Eastern Scotian Shelf Integrated Management Plan. A case study of a successful IMCAM plan (ESSIM Plan) lacking leadership for implementation. Truro Heights, NS: Maritime Aboriginal Peoples Council and Maritime Aboriginal Aquatic Resources Secretariat. http://www.mapcmaars.ca/theblog/archive/essimstudy.pdf.

EAC (Ecology Action Centre). 2011. Notice of objection: Submission of objection to the Marine Stewardship Council certification of the Canadian Atlantic longline swordfish assessment process. https://www.msc.org/track-a-fishery/fisheries-in-the-program/certified/north-west-atlantic/north_west_atlantic_canada_longline_swordfish/assessment-downloads-1/MSC_Objection_CAN_LL_SWO_FINAL.pdf.

EAC (Ecology Action Centre). 2014. Reeling in revenue: Opportunities to increase the value of Atlantic bluefin and support recovery through the live-release fishery. https://www.ecologyaction.ca/reelinginrevenue.

EAC 2015. Marine Issues. https://www.ecologyaction.ca/marine.

Fisheries Act, 1985, RSC, chF-14.

Fitzgerald, G. 2000. *The Decline of the Cape Breton Swordfishery: An Exploration of the Past and Recommendations for the Future of the Nova Scotia Fishery.* Halifax: Ecology Action Centre.

Friends of Hector. 2015. Accessed 25 September 2015. http://friendsofhector.org/porbeagle1.

Fuller, S. D. and P. E. Cameron. 1998. *Marine Benthic Seascapes: Fishermen's Perspectives.* Ecology Action Centre Marine Issues Committee Publication. Halifax: Ecology Action Centre.

Fuller, S. D., C. Picco, J. Ford, et al. 2008. How we fish matters: Addressing the ecological impacts of Canadian fishing gear. Halifax: Ecology Action Centre, Living Oceans Society, and Marine Conservation Institute.

Galil, B. S., P. Genovesi, H. Ojaveer, et al. 2013. Mislabeled: Eco-labeling an invasive alien shellfish fishery. *Biological Invasions* 15 (11): 2363–2365.

Garcia, S. M., J. C. Rice, and A. T. Charles, eds. 2014. *Governance of Marine Fisheries and Biodiversity Conservation: Interaction and Co-Evolution.* Hoboken, NJ: Wiley Blackwell.

Gilman, E., S. Clarke, N. Brothers, et al. 2008. Shark interactions in pelagic longline fisheries. *Marine Policy* 32 (1): 1–18.

Hutchings, J. A., I. M. Côté, J. J. Dodson, et al. 2012. Sustaining Canadian marine biodiversity: Responding to the challenges posed by climate change, fisheries, and aquaculture. Expert panel report prepared for the Royal Society of

Canada. Ottawa: Royal Society of Canada. http://rsc-src.ca/sites/default/files/pdf/RSCMarineBiodiversity2012_ENFINAL.pdf.

Hutchings, J. A., and M. Festa-Bianchet. 2009. Canadian species at risk (2006–2008), with particular emphasis on fishes. *Environmental Reviews* 17: 53–65.

ICCAT (International Commission for the Conservation of Atlantic Tunas). 2012. Report of the 2012 Atlantic Bluefin Tuna Stock Assessment Session, Madrid, Spain— 4–11 September 11, 2012. ICCAT Doc. No. SCI-033/2012. http://www.iccat.int/Documents/Meetings/Docs/2012_BFT_ASSESS.pdf.

ICCAT (International Commission for the Conservation of Atlantic Tunas). 2014. Recommendation by ICCAT amending the supplemental recommendation by ICCAT concerning the Western Atlantic Bluefin Tuna Rebuilding Program. Madrid: ICCAT. http://www.iccat.int/Documents/Recs/compendiopdf-e/2014-05-e.pdf.

Jacquet, J., D. Pauly, D. Ainley, S, et al. 2010. Seafood stewardship in crisis. *Nature* 467 (7311): 28–29.

Lewison, R. L. and L. B. Crowder. 2007. Putting longline bycatch of sea turtles into perspective. *Conservation Biology* 21 (1): 79–86.

Lewison, R. L., L. B. Crowder, A. J. Read, et al. 2004. Understanding impacts of fisheries bycatch on marine megafauna. *Trends in Ecology & Evolution* 19 (11): 598–604.

Mandelman, J. W., P. W. Cooper, T. B. Werner, et al. 2008. Shark bycatch and depredation in the U.S. Atlantic pelagic longline fishery. *Reviews in Fish Biology and Fisheries* 18 (4): 427–442.

Martin, S. M., T. A. Cambridge, C. Grieve, et al. 2012. An evaluation of environmental changes within fisheries involved in the Marine Stewardship Council certification scheme. *Reviews in Fisheries Science* 20 (2): 61–69.

McDevitt-Irwin, J. M., S. D. Fuller, C. Grant, et al. 2015. Missing the safety net: Evidence for inconsistent and insufficient management of at-risk marine fishes in Canada. *Canadian Journal of Fisheries and Aquatic Sciences* 72 (11).

Mooers, A. O., L. R. Prugh, M. Festa-Bianchet, et al. 2007. Biases in legal listing under Canadian endangered species legislation. *Conservation Biology* 21 (3): 572–575.

MSC (Marine Stewardship Council). 2009. Northwest Atlantic Canada longline and harpoon swordfish fisheries. Full MSC certification assessment notification. https://www.msc.org/track-a-fishery/fisheries-in-the-program/certified/north-west-atlantic/north_west_atlantic_canada_longline_swordfish/assessment-downloads-1/13-03-2009-Swordfish-full-assessment-announcement.pdf.

MSC (Marine Stewardship Council). 2012a. *North Atlantic Swordfish* (Xiphias gladius) *Canadian Pelagic Longline Fishery*. Volume 1. Public Certification Report. https://www.msc.org/track-a-fishery/fisheries-in-the-program/certified/north-west-atlantic/north_west_atlantic_canada_longline_swordfish/assessment-downloads-1/PCR.pdf.

MSC (Marine Stewardship Council). 2012b. North West Atlantic Canada longline swordfish. Accessed 30 July 2015. https://www.msc.org/track-a-fishery/fisheries-in-the-program/certified/north-west-atlantic/north_west_atlantic_canada_longline_swordfish.

MSC (Marine Stewardship Council). 2015. MSC fisheries standard. https://www.msc.org/about-us/standards/fisheries-standard/msc-environmental-standard-for-sustainable-fishing.

New England Aquarium. 2015. Right whale research. Accessed 23 September 2015. http://www.neaq.org/conservation_and_research/projects/endangered_species_habitats/right_whale_research/.

Oceans Act, 1996, S.C., c 31.

Species at Risk Act, 2002, S.C., c 29.

Species at Risk Registry. 2006. Order giving notice of decisions not to add certain species to the list of endangered species. http://www.registrelep-sararegistry.gc.ca/default.asp?lang=En&n=ED2B8BBF-1.

Van Der Hoop, J. M., M. J. Moore, S. G. Barco, et al. 2013. Assessment of management to mitigate anthropogenic effects on large whales. *Conservation Biology* 27 (1): 121–133.

Wallace, S., B. Turris, J. Driscoll, et al. 2015. Canada's Pacific groundfish trawl habitat agreement: A global first in an ecosystem approach to bottom trawl impacts. *Marine Policy* 60: 240–248.

Wang, R. J. 2013. Analyzing bycatch mitigation in the MSC certified Canadian Northwest Atlantic longline swordfish fishery. Master's project report. Dalhousie University.

Ward, T. J. 2008. Barriers to biodiversity conservation in marine fishery certification. *Fish and Fisheries* 9 (2): 169–177.

Willison, J. H. M., J. Hall, S. E. Gass, et al. 2001. *Proceedings of the First International Symposium on Deep-Sea Corals.* Halifax: Ecology Action Centre.

Zilio, M. 2012. "Canada's shark" still swimming without Canada's protection. *iPolitics.* http://ipolitics.ca/2012/10/30/canadas-shark-still-cant-get-canadas-protection/.

Section IV

The Way Forward

19

Does Information Matter in ICOM?
Critical Issues and the Path Forward

Elizabeth M. De Santo, Suzuette S. Soomai,
Peter G. Wells, and Bertrum H. MacDonald

CONTENTS

19.1 Introduction

In identifying issues of critical importance regarding the role of information in integrated coastal and ocean management (ICOM), it is essential to recall the various dimensions and complexity of ICOM and the need for a solid scientific basis for decision-making and effective policies in coastal and ocean management. As discussed in Chapters 1 and 2, the concept and practice of ICOM is not new. It has been evolving in various ways for several decades. Many terms (e.g., CZM, ICZM, ICM, ICAM, ICOM, previously defined in Chapters 1 and 2) are used interchangeably to describe the range of management approaches that take into account the complex ecological (biophysical), socioeconomic, and governance characteristics of coastal and open ocean areas. In using the term ICOM, this volume acknowledges its broad acceptance by the international ocean management community (UNESCO 2006).

It is clear from the contributions to this book that scientific information is indispensable in ICOM. The complex nature of the coastal and ocean environment poses significant challenges to environmental and resource decision-making, highlighting the need for robust information at every stage of its management. Problems with the coastal and ocean environment have become so complex that they are described as *wicked* (Rittel and Webber 1973), as they are difficult to define and delineate, and simple solutions do not exist, as discussed by McNie et al. (Chapter 9, this volume) and Coffey and O'Toole (Chapter 3).

This volume also recognizes that the prevailing approaches of evidence-based (evidence-informed) decision-making and policy-making have been described recently by other researchers, for example, Young (2013). But, as noted by MacDonald et al. in Chapter 2 and by Coffey and O'Toole in Chapter 3, decision-making is often more complex than "imagined in the frequently expressed statement that 'science should inform decision-making'." Many factors come into play, including the use of different forms of knowledge, such as anecdotal and traditional information. Recognizing this milieu of factors provides "a richer understanding of the complex dynamics involved in marine and coastal management" (Coffey and O'Toole, Chapter 3). Furthermore, as shown by the chapter by Graham and Jones (Chapter 14), case studies from different disciplines—in this case the health sector—also provide insight into a range of the core principles pertaining to the acquisition and determination of the validity of data and information. These principles must be kept in mind when considering the role of information in ICOM.

Distinguishing between use and influence of information in ICOM is a challenge in itself. What does it mean to *use* information? We have been grappling with this seemingly simplistic question for some time. How information is utilized and with what effect varies with the individual, the context, and/or the issue involved. Many studies have focused on the information needs of and use by coastal and ocean managers, as shown by numerous examples in this book. To actually parse the difference between use and influence requires gaining access to managers, senior decision makers and policy makers, and other stakeholders, and this process has begun (e.g., the EIUI research program (2015), among others). A primary message is that developing an understanding of how scientific information is used by any of these groups is fundamental to building better policies and better management practices in ICOM.

Within the context of ICOM, both governmental and nongovernmental actors (also known as state and non-state actors) play key roles, depending on the scale of the management issue being addressed. As explained in Chapter 2 and illustrated in Figure 2.1, the actors can be differentiated into several categories: decision makers (including law makers and senior government officials), managers (reporting to decision makers, typically within government), scientists (in government, academia, and industry), stakeholders (e.g., industry, tourism, and recreational users), and the wider public in coastal communities. A distinction can be made between senior decision makers, managers, and scientists on the one hand, and diverse stakeholders (i.e., industry, and the wider public) on the other. The latter also include advocacy organizations representing particular interests, ranging from pro-industry organizations to advocacy organizations for nature conservation. The multitude of actors involved in coastal jurisdictions contributes to the complexity of information communication and decision-making, that is, to the science–policy interface, regardless of the issue being considered (e.g., environmental quality, fisheries, aquaculture, industrial development, habitat protection, or species protection).

The science–policy interface occurs in many places in decision-making and policy-making in ICOM, and at several scales, ranging from geographic to institutional, political, and temporal (as shown in Figure 1.1 and discussed in Chapter 2). Managers, decision makers, and others responsible for policy can be inundated with information from a range of sources, as is evident throughout this volume. The working environment surrounding decision-making is multifaceted. It is subject not only to external socioeconomic and political contexts, but also an interweaving of flows of information and pressures from legislative mandates, management programs directly involved in ICOM, and a range of affected stakeholders (e.g., interest groups). Figure 19.1

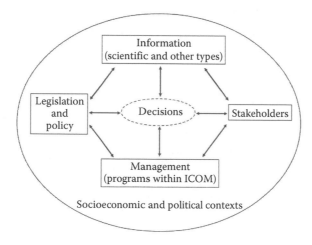

FIGURE 19.1
Information use at the science–policy interface in ICOM. Communication and information pathways occur throughout ICOM.

illustrates this conceptually; many information and communication pathways exist, to a greater or lesser extent depending on the context and case. But, as highlighted in the case studies discussed in this volume, the information and communication pathways between researchers and government, and government and the public at large, are often less transparent and efficient than intended.

The goal of this synthesis chapter is to draw together key themes and lessons from the concept chapters and case studies presented in the book. We begin with insights about the science–policy interface, and relevant aspects of information production, use, and influence in policy-making and decision-making. We discuss the political nature of decision-making and present some of the challenges and critical issues to be considered in future research. Above all, we attempt to clarify the role of information at the science–policy interface(s) in ICOM, recognizing that the topic is very broad. The numerous players in ICOM and marine affairs present unique and varied perspectives, and our collective efforts addressing the topic have undoubtedly left gaps. As Ouellette and Hardy (2010, p. 7) stressed, "integrated coastal management, in a sustainable manner, seems a formidable challenge yet it is particularly important that we … [act] strategically and efficiently with the best information and tools at hand." Ascher and Ascher (Chapter 7, p. 153) also emphasize this point very clearly: "Sound knowledge is crucial for good coastal policy and management decision-making." This statement is our credo and the reason we embarked on the voyage of preparing this book.

19.2 Themes and Critical Issues

The chapters on fundamental concepts and principles (Section II) and the case studies (Section III) set out the diverse roles of information in policy-making and decision-making. Most of the chapters refer directly to the needs of ICOM, and a few chapters draw on experiences in other issue areas, to introduce concepts important to evidence-based and evidence-informed policy-making. Many of the case studies have approached information use and influence from managerial and advisory levels in different institutions (e.g., see Gluckman and Allen, Chapter 10; Wells, Chapter 16). In examining how information is produced, how it influences behavior, and how it is used by the various actors, we gain insights into how these components of the ICOM information cycle work and are interconnected with one another.

19.2.1 Production of Information

Information important to ICOM is produced by government agencies, intergovernmental bodies, academia, and nongovernmental organizations

(NGOs), among other contributors, in various forms, primary and grey and, in recent decades, much of it (but not all) is available digitally. This production pattern is apparent in the various chapters in this book.

However, it should be recalled that information and knowledge are distinct entities and that information acquisition does not ensure that full knowledge is attained (Ascher and Ascher, Chapter 7). Information and knowledge are generated by a wide range of people and institutions, not all of whom are involved in the decision-making process. Certainly, this is true in marine science, where most scientists, especially in academia, can pursue basic or discovery science, regardless of its applications (use) in ICOM. Given time and confirmation, much new information becomes part of the body of knowledge. Knowledge generation occurs within particular social and governance contexts with varying levels of access and visibility. These points have also been highlighted by other researchers, including those in this volume.

McNie et al. (Chapter 9) and Ross and Breeze (Chapter 12) emphasize the production aspect in light of expected users, with the latter focusing on state of the environment (SOE) reports. They state that the process of generating scientific reports is as important, if not more important, than the reports themselves, given the interplay between credibility (soundness), legitimacy (fairness), and salience (relevance) as defined by Mitchell et al. (2006). This conclusion was also reached by Wells (2003) and Lexmond (2009). In addition, Gluckman and Allen (Chapter 10) describe operationalizing a research and scientific information standard for fisheries that includes government-led working groups and participatory workshops. The end product alone is not the only significant outcome. The production process can also be important as it often includes multiple stakeholders and multiple sources of information, and allows consensus building on a subject where information on an issue is controversial or incomplete.

It is well understood that information is produced in various formats. But the format of the information and the way in which it is communicated can affect its use in decision-making (Mitchell et al. 2006; Wells 2003; Ababouch et al., Chapter 17). For instance, decision makers in Environment Canada prefer and request summary information on various topics and issues to be presented in the form of briefing notes that summarize complex concepts succinctly, rather than the original, detailed technical papers (Wells, Chapter 16). How information is produced and presented to different audiences impacts the way it is perceived, and thus directly affects its use.

19.2.2 Use of Information

Stepping back to consider what makes science or knowledge most useful from a policy perspective, Mitchell et al. (2006) argued that the use of science depends on its perceived credibility, legitimacy, and salience (as defined in Section 19.2.1) by multiple stakeholders. Hartley (Chapter 8) draws on Wilson's (2009) emphasis on the importance of these factors (credibility, legitimacy, and salience) in the latter's evaluation of the use of scientific information by the International Council for the Exploration of the Sea (ICES). These factors are widely cited in political science scholarship on the science–policy interface. The neutral, authoritative, and politically distant view of science of "speaking truth to power" is frequently emphasized when discussing the role of scientists in the making of environmental policy (Gupta et al. 2012; Haas 1992). There is an assumption both in that literature and in several of the chapters in this volume that *useful* scientific information will inherently improve decision-making (McNie et al., Chapter 9; Gluckman and Allen, Chapter 10; Ross and Breeze, Chapter 12).

One might also note the institution or intent behind knowledge generation, as much internal government science is mandate specific and mission oriented. Formal processes for producing scientific advice that are embedded within organizations, considered to be the authority for coastal and ocean management (e.g., the Food and Agriculture Organization of the United Nations (FAO) and national fisheries departments), have been seen to provide credible and relevant information for decision-making (Gluckman and Allen, Chapter 10; Ababouch et al., Chapter 17; Soomai 2015).

Some operational units of organizations are more involved at the science–policy interface than others (Ross and Breeze, Chapter 12; Wells, Chapter 16; Ababouch et al., Chapter 17). As noted by Ababouch et al. (Chapter 17) and Ross and Breeze (Chapter 12), evidence-based policy-making emphasizes the importance of such an approach, and partnerships (e.g., among government, NGOs, and regional and international organizations) provide a mechanism for bringing data and information together from different sources. Partnerships with various stakeholders are also a cost-effective mechanism in information production with global austerity measures today (Ababouch et al., Chapter 17). Given the fact that major issues addressed in ICOM are intertwined with social and economic issues, for example, social well-being and food security, and that they are inherently complex, it is paramount that different sectors overcome the disparities in their technical languages and cultures (Rice, Chapter 4), and utilize both formal and informal linkages of communication (e.g., Hartley, Chapter 8 and Wells, Chapter 16).

How different units of organizations move information through to where it is needed is especially important. Networks, for example, are common in every community of practice. As Hartley points out (Chapter 8, p. 182), "Communication patterns in networks can develop mutual understanding, organizational learning, trust, and other features critical to the development of credibility and legitimacy of scientific information."

Determining how information is utilized is challenging since much work is occurring digitally today. With the increasing use of the Internet, issues related to accountability, traceability, archiving, and verification of data and information are of increasing importance. This development may be less of a problem for governments with central archives/registries of publications and communication activities. However, informal communication and related documentation are also common features of work places and decision-making, which is very difficult to document. There is a growing risk of the possible loss of information for situations and exchanges where no organized record is maintained, for example, NGO interactions or informal discussions in meetings with government and industry. In order to counter any potential bias, information should be gathered and combined from multiple sources and sectors, such as Fuller et al. (Chapter 18) noted with regard to the generation of fisheries statistics for international fisheries management.

Ascher and Ascher (Chapter 7) and Graham and Jones (Chapter 14) pointed out that misinformation can be disseminated deliberately or accidentally. How information is *misused* varies according to the context and the individuals involved (Nutley et al. 2007). For instance, *misuse* can be demonstrated by the selection of information or misinterpretation of information (Graham and Jones, Chapter 14). The impact of the *misuse* of information is of concern as a reduced perception of its credibility can result. With the current general overabundance of information, stakeholders may lack confidence in their ability to recognize credible information or may simply not be knowledgeable enough to filter out misinformation. As a consequence, a decision may be made without realizing that the information on which the decision is made is not credible, or a decision is not made at all because of a fear that available information lacks credibility (Ascher and Ascher, Chapter 7 and Graham and Jones, Chapter 14).

19.2.3 Influence of Information

An important point made by several of the case studies is that scientific information informs policy-making, but not necessarily policy outcomes, that is, knowledge alone should not be assumed to influence policies (Rice, Chapter 4; Gluckman and Allen, Chapter 10; DeSombre, Chapter 13; Graham and Jones, Chapter 14). This point underlies the recent literature on the distinction between evidence-based policy-making and evidence-informed policy-making. Information can be used to scope, as well as inform issues, a phenomenon known as *framing* (Niemeijer and de Groot 2008; Rice, Chapter 4; Ross and Breeze, Chapter 12; Graham and Jones, Chapter 14). Once the information is in the hands of managers and decision makers, however, other issues often come into play, including socioeconomic and political pressures. Consequently, the same information used by different organizations and in different contexts can lead to different outcomes (Rice, Chapter 4), prompting the question, whose information matters and who decides (De Santo 2010; Rice, Chapter 4; Gluckman and Allen, Chapter 10; Graham and Jones, Chapter 14; Wells, Chapter 16).

Typically, the audiences for information are varied, and they differ in how they search for and use it (Ascher and Ascher, Chapter 7; Ababouch et al., Chapter 17). Audiences, whether managers, special interest groups, or the public, apply a variety of filters in seeking information for decision-making, and their different underlying motivations need to be taken into account when examining how information influences behavior. Hence, the complexity of the governance dialogue, as described by Coffey and O'Toole (Chapter 3) and Rice (Chapter 4), accentuates the spectrum of information use practices and patterns due to multiple audiences, motivations, and information sources.

Trade-offs or compromises often come into play when multiple sources of information are available for decision-making (Toonen and Mol, Chapter 6; DeSombre, Chapter 13). DeSombre describes the *trickiness* of the compromises related to regulation when multiple stakeholder groups are involved and economic returns are paramount to the existence of the shipping industry. Such compromises are also evident in the pharmaceutical industry (Graham and Jones, Chapter 14). Trade-offs are also encountered when governmental and nongovernmental actors each fulfill governance roles. Toonen and Mol (Chapter 6) describe this development as polycentric or multiactor governance arrangements, which are notable in fisheries, shipping, and tourism. Independent bodies provide authoritative information via standardized labels, benchmarks, sustainability rankings, and certification schemes, which often act as a trade-off when government legislation and regulation do not result in the desired industry or public response to achieve environmental sustainability.

As shown in Figure 19.1, information flows in many pathways in ICOM. However, the trajectory of its influence may be long-term and not immediately apparent. For example, Rice (Chapter 4) explores the development of the regime for protecting marine biodiversity beyond national jurisdiction, a lengthy process that has evolved iteratively and less rapidly than many had hoped. In several cases of science–policy interactions, political factors have had a major impact on the timescale for decision-making and the effectiveness and influence of information. The work of the Intergovernmental Panel on Climate Change (see Chapter 2) is a case in point—decades passed before the scientific information on climate change reached the political stage and was given serious traction at the international level.

Environmental NGOs may be particularly influential in swaying both the public's view of marine issues and the decision-making process itself. Fuller et al. (Chapter 18) discuss the role of Nova Scotia's Ecology Action Centre (EAC) as a key communicator on fisheries issues by informing and engaging the public and by distilling complex science as well as legal and policy issues for wide consumption. Some organizations, such as EAC, also view themselves as allies of *responsible* industry, helping to inform them of policy developments, while also representing their voice in the advocacy arena. Environmental NGOs employ different approaches in a range of contexts, for example, using appealing *spokesfish* and web-based and social media information dissemination methods in public outreach, as well as more nuanced written statements and interventions at intergovernmental meetings, such as the International Commission for the Conservation of Atlantic Tunas

(ICCAT) (Fuller et al., Chapter 18). Another NGO profiled in this volume, the Atlantic Coastal Zone Information Steering Committee (ACZISC), fulfills its mandate by marshaling dissemination of coastal information and interfacing between scientists and policy makers, including widely distributing updates on ICOM and organizing numerous meetings for various sectors to interact informally (Sherin and Baccardax Westcott, Chapter 15).

19.2.4 The Role of Politics in Decision-Making

Several of the case studies presented in this volume underscore the political nature of decision-making (e.g., Rice, Chapter 4; Gluckman and Allen, Chapter 10; Graham and Jones, Chapter 14). Good governance that aims to implement best practices and adaptive management must maintain accountability and transparency. However, socioeconomic and political factors can eclipse the use and influence of information, as illustrated in Figure 19.1. Providing credible, relevant, and legitimate information does not guarantee that it will be used appropriately or, in some circumstances, at all (e.g., in policies on climate change in some countries). The information required to enact a decision may be readily available to decision makers, as in setting aside a coastal location as a marine protected area or marine sanctuary for endangered species, but opposition from affected stakeholders (e.g., the fishing industries) could preclude them from acting on it (Nursey-Bray et al. 2014). It is, therefore, important to strengthen existing pathways and/or add new ones to achieve transparent movement of information and understanding of the ways in which governance structures affect the flow of information (Rice, Chapter 4; Graham and Jones, Chapter 14; Wells, Chapter 16).

19.2.5 Science–Policy Interface: Actors, Uncertainty, and Trust

Important interrelated themes running throughout this book are the roles of actors at the science–policy interface, the way uncertainty (in information) is handled and communicated, and the related issue of trust. Toonen and Mol (Chapter 6) and Graham and Jones (Chapter 14) note the possibility and benefits of involving multiple stakeholder groups in decision-making and policy-making, including engaging civil society in knowledge/informational governance. However, while managers often aim to include many stakeholders in environmental planning, this effort can also lead to *stakeholder paralysis*, wherein no agreement is reached, as Hartley notes (Chapter 8), or consensus is difficult to realize among many stakeholders who hold differing values. Iterative communication can overcome these challenges by providing opportunities for questions to be answered and time to consider various viewpoints, and build understanding of the perspectives of other stakeholders (Sarkki et al. 2015). However, as Hartley (Chapter 8, p. 180) points out, "Few professional opportunities exist for scientists, managers and other ... stakeholders to engage in long-term, iterative dialogue in order to establish the salience of scientific information, credibility of the science, and the perceived legitimacy of the scientific process among non-scientist stakeholders."

The related concept of informational governance is examined in this book in light of stakeholder empowerment and the capacity of citizens to achieve behavioral change (e.g., of polluters) (Chapter 6). In order for such empowerment and change to occur, it is essential that stakeholders trust and understand the regulatory and decision-making process, as shown by DeSombre (Chapter 13) with regard to shipping, and by Graham and Jones (Chapter 14) with respect to regulatory processes involving the pharmaceutical industry. However, scientific uncertainty often affects the way in which information and evidence are communicated to and perceived by various audiences. Fisheries management recommendations for allowable catches, for example, can be seen as having a *yo-yo* effect, up one year, down the next, based on natural system fluctuations, as well as changes in measurement methods, which affects perceptions of trust by some stakeholders (Hartley, Chapter 8).

Scientific uncertainty, discussed here and in Chapters 2 and 11, is viewed differently in political (i.e., governance) versus scientific realms, as well as by the wider public. Scientific uncertainty is not a homogenous concept; different forms of uncertainty pose particular challenges for interpreting evidence and creating appropriate management measures (Ascough et al. 2008; IOM 2013; O'Riordan and Jordan 1995). Outsiders viewing "debates" over science may not realize that an abundance of data may exist. The media can inflate the importance of issues and/or the strength of counterarguments in order to create a sharp dichotomy, for example, with biased depictions of climate change denial (Carvalho 2007; Boykoff and Boykoff 2004). Consequently, the onus on all parties in the science–policy divide to provide and use the best information to make decisions is increased. All stakeholders, whether organizations directly involved in ICOM decisions or the general public, need to proactively seek out information and be critical in assessing its validity. Ascher and Ascher (Chapter 7, p. 171) outline reasons and strategies for motivating "stakeholders to search for, make sense of, and make decisions based on the knowledge needed for sound resource practices and stances toward conservation policies."

Communication and professional networks serve multiple functions, of which the dissemination of information is notably important. As Hartley outlines in Chapter 8, network analysis is a particularly informative method for illustrating the communication pathways between organizations, individuals, and relevant government authorities, and the method can help to increase understanding of issues among the public, while also building trust via transparency. Visualizations of networks may also highlight the role of bridgers, that is, individuals or organizations that connect other actors such as stakeholder groups with researchers and such groups with each other or with decision makers. As Hartley stated, and another recent study has shown (Wilson 2015), bridgers can be pivotal for effective dissemination of information or they can act as gatekeepers since they may selectively disseminate information either because they are overwhelmed by their role or because they distribute information in a discriminating manner.

Mol (2008, p. 152) notes that, "trust has changed in the information age," wherein institutions, procedures, and social/experts systems outrank individuals and leaders. Trust is an issue in the risk governance process (Quigley, Chapter 5) and relates to how information is communicated (Hartley, Chapter 8). Uncertainties surrounding risk assessment affect its transparency and can lead to different ways in which information is perceived by different groups (Quigley, Chapter 5).

19.3 Oceans of Information in ICOM: Future Challenges

The role of information in the various phases of ICOM has not been explicitly recognized or widely studied in ocean management circles (Chapters 1 and 2). Why is this? To date, few researchers (e.g., Mitchell et al. 2006) have focused their attention on the influence of information in global environmental assessments and decision-making more broadly. Information, with its myriad attributes and dimensions, is somewhat of an elephant in the room, unrecognized for its importance and influence in policy-making. Numerous aspects of its relationship to decision-making need further investigation. Understanding how science is used in policy development requires examining what makes reliable, valid, and compelling policy arguments from both managerial and senior decision-making perspectives (Ascher and Ascher, Chapter 7; McNie et al., Chapter 9; NRC 2012). Researchers and decision makers also need to appreciate how knowledge creation works, in order to optimize the opportunities and overcome the challenges for transitioning scientific information and knowledge into policy, especially good public policy pertaining to the coastal areas and oceans (Chapter 2; Coffey and O'Toole, Chapter 3).

Further exploration of the science–policy interface will help to shed light on two essential questions: (1) How do decision makers and practitioners determine what sources of information are credible and authoritative, and hence useful to their particular roles in ICOM? and (2) How can attention to scientific information in policy-making and decision-making related to the various program activities in ICOM be enhanced? Other challenges merit study, some of which are briefly described in this section.

19.3.1 Improving Reliability of Information

The potential biases of information used in various regulatory and policy development systems need to be better understood, particularly with regard to scientific uncertainty and the related matter of trust (Toonen and Mol,

Chapter 6; Gluckman and Allen, Chapter 10; DeSombre, Chapter 13; Graham and Jones, Chapter 14). Biases can have negative consequences. For instance, the reliability of evidence brought to bear in regulatory or policy decisions may be questioned with regard to its accuracy and precision and the ability for research conclusions to be reproduced in further studies. Questions about the reliability of information ultimately relate to the credibility of its source. Is the information right for the question being considered?

19.3.2 Enhancing Knowledge Sharing

Knowledge sharing, described as "a key part of managing the coast" (Nursey-Bray et al. 2014, p. 115), is another aspect of ICOM that merits further attention. For this aspect, solidifying understanding of all of the relationships among knowledge production, bridging organizations, and adaptive management is needed. With this understanding in hand, then drawing on the experience of knowledge translation and exchange (KTE) in other disciplines (see Chapter 2) could enhance information and knowledge sharing in ICOM. "Successful joint knowledge production" (van Tatenhove et al. 2015) must also meet the criteria of credibility, legitimacy, and relevance. How this is achieved is still being debated and "requires more systematic and empirical comparative research" on policy processes (van Tatenhove et al. 2015). Understanding the relationships between knowledge production and governance also needs to be improved.

19.3.3 Strengthening Information Management and Communication Skills

There is a continuing need to strengthen the various skills of ICOM practitioners, from scientists to program managers to policy makers. These skills include both data and information management and communication capabilities. Indicators, such as the existence and application of scientific advice, have been proposed to measure progress in ICOM (UNESCO 2006, pp. 137–139). ICOM practitioners must remain current by responding to the pace of change in the information age, which is particularly driven by advances in digital technology, including social media.

While this view is changing, marine scientists may still believe that their primary role is to conduct research and publish their findings (Rudd 2015). In contrast, the public may believe that researchers have an obligation to educate society about their discoveries (Sarewitz 2013). Policy makers and decision makers assert that researchers need to communicate their results effectively to decision makers and the public so that appropriately informed decisions can be made (Lalor and Hickey 2013; Gewin 2014; McNie 2007). For scientists involved in ICOM, finding a balance between effectively communicating their findings to the public and policy makers while safeguarding the credibility of their work within scientific circles may be difficult. Nonetheless, public outreach is essential if comprehensive evidence-based decisions are to be made.

Strategies to improve stakeholder information-seeking in the development of environmental knowledge could positively affect decision-making (Ascher and Ascher Chapter 7). Various communication methods to enhance the flow of information among the ICOM actors could be applied:

- A communication gap often exists between researchers and those in a position to make decisions or policy. Thus, asking questions about the communication processes will help to increase understanding of the enablers and barriers in the communication of research findings (Cossarini et al. 2014).

- Studies on the dissemination of information have found that managers and other information users place a high value on data visualization in decision support systems, as available in digital atlases (McLean 2014).

- Transparency is an important feature of effective decision-making processes. Openness and accountability in data and information delivery, coupled with access to the data and information by all actors in public policy decisions, will help to overcome distrust in decision-making processes (McNie et al., Chapter 9; Graham and Jones, Chapter 14).

- Collaboration and overlapping membership within and among organizations contribute to strong networks with higher degrees of interorganizational and intraorganizational communication and effective decision-making (Soomai 2015).

- The methods used to disseminate publications and establish awareness of marine information, such as state of the environment reports, often only reach the *interested public*. These individuals and groups typically respond to governmental and nongovernmental requests for input, they may be better able to contribute to policy development because they are already active in coastal and ocean conservation issues (Soomai et al. 2013). Additional communication effort to reach beyond established networks requires understanding the communication behaviors of many groups. For example, the rise in social media use by younger members of a population means that communication through social media channels is more likely to be effective in engaging younger age-groups in ICOM issues.

- ICOM practitioners will benefit from gaining experience in the different coastal and ocean management sectors to build appreciation for the perspectives of each sector. This understanding, including how and what information is used in the different sectors, will contribute to more efficient and effective decision-making that accounts for numerous points of view commonly expressed about ICOM issues.

19.3.4 Increasing the Visibility of Information about the Oceans

Publications prepared and released by major ocean organizations, such as the Global Ocean Commission (2014), "may be very influential if given coverage by the press, are noticed by politicians, policy makers and marine resource managers, and offer proposals for positive action" (Wilson and Wells 2014). This recommendation emphasizes the value of a promotion and advertising campaign on the completion and during distribution of any major ICOM-related report.

Raising awareness and use of research-based environmental information is often challenging due to the requirement of communicating it to diverse audiences and also engaging the public (stakeholders and individuals outside of established networks) in ICOM subjects. Soomai et al. (Chapter 11) showed that studies on the awareness, use, and influence of information in the processes of coastal and marine resource management can further understanding of problems with information flow at the science–policy interface.

Documenting and communicating the value of the oceans can increase visibility on ocean issues. While this is not a new concept, it is often difficult to achieve (Barbier et al. 2011; de Groot et al. 2010). A recent WWF report (Hoegh-Guldberg et al. 2015) calculated the value of the oceans at US$24 trillion. According to Cressey (2015), "this unique state of the oceans report with its blunt and proactive message will serve to bridge the science–policy gap in a way that scientific publications alone have not been able to do."

19.3.5 Encouraging Interdisciplinary Collaboration for Innovative ICOM Initiatives and Research

Identifying solution-oriented benefits of appropriate information management for ocean health is a necessity. Simply stated, if problems facing the ocean are appropriately managed as a result of generating and using accurate information, then fewer threats to resources, human health, and water quality, for example, should occur. The day-to-day practitioners in ICOM and marine information management specialists should continue to collaborate and consider ways to maximize the use of information in specific programs. As illustrated by the Gulf of Maine Council on the Marine Environment (GOMC), managers from Canada and the United States have formally collaborated on coastal programs for over 25 years (www.gulfofmaine.org). Similarly, member countries of global bodies like the Food and Agriculture Organization of the United Nations (FAO) collaborate in regional and global fisheries management (Ababouch et al., Chapter 17). The Marine Environmental Observation Prediction and Response Network (MEOPAR) is another example of scientists collaborating with national and international bodies, including industry, in multidisciplinary research projects (MEOPAR 2015). Such initiatives recognize the need for data management and information management expertise in research partnerships.

Not unlike the efforts of organizations such as the American Association for the Advancement of Science (AAAS) over many years, scientists are increasingly encouraging each other to become more engaged in the policy process in coastal and ocean management (Rudd 2015). While scientists are calling for new levels of engagement and collaboration between scientists and policy makers, this is dependent on scientists' willingness to overcome the fear of appearing to advocate and losing objectivity in the scientific community. Furthermore, engagement and collaboration between the two groups depends on their current and evolving frames of reference (i.e., their ideas, beliefs, and discourses). Priority-setting fora that include scientists and nonscientists are seriously needed (Neff 2011). This *science push* concept recognizes that good public policy should be evidence-based or evidence-informed.

Boundary organizations (as described in Chapter 2) bridge the science and policy realms and are becoming more noticeable. Such organizations create a neutral setting where producers and users of scientific information interact while maintaining accountability to science or to policy (Guston 2001). The recent literature describes *boundary chains* which join complementary boundary organizations and facilitate cooperation and cost sharing (Kirchhoff et al. 2015). Enhanced understanding of the role and characteristics of such organizations is needed.

19.3.6 Creating Policy Solutions and Considering Trade-Offs

Wesselink et al. (2013, p. 2) argued that, "policy solutions are created by the interweaving of expertise and politics." That expertise includes both technical knowledge and experience-based understanding of different agencies and actors involved in and affected by policy decisions. As Nursey-Bray et al. (2014) point out, the "influence of power" affects integration of knowledge into policy. This observation is clearly relevant to ICOM, given the multiple and overlapping political and economic jurisdictions found in coastal and ocean areas. Understanding the stages of the policy cycle can help to inform which information trade-offs are most appropriate, for example, communication about uncertainties can be important in early decision-making stages, but less relevant near the end of the policy cycle (Sarkki et al. 2014).

19.3.7 Understanding the Consequences of Inaction on Issues

Given increasing growth of coastal populations and the myriad issues confronting the ICOM practitioner, the task of achieving effective and long-lasting management approaches and solutions to the many problems confronting coastal zones may seem insurmountable. Inaction is not an option, however. Understanding the need for and then practicing effective marine information management makes the task of dealing with the issues more tractable and represents a real contribution to ICOM.

Reprioritizing activities in response to austerity is critical. As government budgets tighten, information is often undervalued. Governmental organizations may reduce producing publications or, once they are produced, often do not optimize their use. The general sense that information is widely ubiquitous can easily result in actions that undervalue its importance.

19.4 Conclusion

This book demonstrates that the perspectives of information management are important, and actually essential, in the multidisciplinary approaches of ICOM. These perspectives affect all of the various ICOM practitioners, including on-the-ground managers of programs in coastal and ocean management. Thirty-five authors, in their individual capacities and disciplinary outlooks, have brought their experience to bear on this subject. The current state and continuing deterioration of the health of the oceans call for a genuine change in how different players, from scientists to policy makers, work so as to break down disciplinary silos inhibiting the interaction of science, policy, and information management. Each of these fields has a responsibility to work together to address the serious coastal and ocean problems now requiring urgent action. Collaborative ocean management programs are already operational in many jurisdictions (Cicin-Sain and Knecht 1998; Massachusetts Ocean Partnership 2009) and more should be encouraged and supported.

Greater appreciation for the important role of information at the science–policy interface is needed. This appreciation will draw on more active participation by information managers in coastal and ocean management programs. Modern day information managers can work side by side with other ICOM practitioners in seeking solutions to the problems noted above. Mandates from government often drive the science—these mandates are policy specific, based on particular legislative initiatives and funding pulses. To date, the majority of ICOM programs have been driven by scientists and managers with limited involvement from information management. It is time to change this imbalance, as is evident from the contributions of the authors of this book. The outcome would establish a workable balance between the science and information push and the policy pull. To this end, the ocean-related information being produced can help improve the health of the oceans, achieving the purpose for which it was generated.

The ideas and lessons learned from the contributions assembled in this book give momentum to the study of the role of information in policy-making and decision-making. Our work is compatible with the directions given by the United Nations to countries (e.g., UNESCO 2006), and this volume represents the first step toward ongoing work in an underexamined field. Our message is simple: despite the continued push to expand our knowledge bases, we also need to understand how to use the existing information that we have to inform policy-making and decision-making aimed at seeking solutions to serious coastal and ocean problems. Given the current and future threats facing the oceans, including the widespread loss of biodiversity, impacts of climate change, and ensuring human livelihoods and access to safe seafood, among others, we cannot act soon enough to remedy the challenges at the science–policy interface(s). While this volume contributes to our understanding of this complex interaction, there is still much more to be done.

References

Ascough II, J. C., H. R. Maier, J. K. Ravalico, et al. 2008. Future research challenges for incorporation of uncertainty in environmental and ecological decision-making. *Ecological Modeling* 219: 383–399.

Barbier, E. B., S. D. Hacker, C. Kennedy, et al. 2011. The value of estuarine and coastal ecosystem services. *Ecological Monographs* 81 (2): 169–193.

Boykoff, M. and J. Boykoff. 2004. Balance as bias: Global warming and the US prestige press. *Global Environmental Change* 14: 125–136.

Carvalho, A. 2007. Ideological cultures and media discourses on scientific knowledge: Re-reading news on climate change. *Public Understanding of Science* 16: 223–243.

Cicin-Sain, B. and R. W. Knecht. 1998. *Integrated Coastal and Ocean Management: Concepts and Practices*. Washington, DC: Island Press.

Cossarini, D. M., B. H. MacDonald, and P. G. Wells. 2014. Communicating marine environmental information to decision makers: Enablers and barriers to use of publications (grey literature) of the Gulf of Maine Council on the Marine Environment. *Ocean & Coastal Management* 96: 163–172.

Cressey, D. 2015. Oceans are "Worth US$24 trillion." *Nature*. 23 April. http://www.nature.com/news/oceans-are-worth-us-24-trillion-1.17394.

de Groot, R. S., R. Alkemade, L. Braat, et al. 2010. Challenges in integrating the concept of ecosystem services and values in landscape planning, management and decision-making. *Ecological Complexity* 7 (3): 260–272.

De Santo, E. M. 2010. "Whose science?" Precaution and power-play in European environmental decision-making. *Marine Policy* 34 (3): 414–420.

EIUI. 2015. Environmental information: Use and influence (EIUI) research program. http://www.eiui.ca.

Gewin, V. 2014. Hello, governor. *Nature* 511: 402–404.

Global Ocean Commission. 2014. From decline to recovery: A rescue package for the global ocean. Oxford, UK: Global Ocean Commission. https://s3.amazonaws.com/missionocean_www_uploads/reports/GOC+Full+Report.pdf.

Gupta, A., S. Andresen, B. Siebenhüner, et al. 2012. Science networks. In *Global Environmental Governance Reconsidered*, edited by F. Bierman and P. Pattberg, 69–93. Cambridge, MA: MIT Press.

Guston, D. H. 2001. Boundary organizations in environmental policy and science: An introduction. *Science, Technology, and Human Values* 29: 87–112.

Haas, P. M. 1992. Epistemic communities and international policy coordination. *International Organization* 46 (1): 1–35.

Hoegh-Guldberg, O., D. Beal, T. Chaudry, et al. 2015. *Reviving the Ocean Economy: The Case for Action—2015.* Gland, Switzerland: WWF International.

IOM (Institute of Medicine). 2013. *Environmental Decisions in the Face of Uncertainty.* Washington, DC: National Academies Press.

Kirchhoff, C. J., R. Esselman, and D. Brown. 2015. Boundary organizations to boundary chains: Prospects for advancing climate science application. *Climate Risk Management* 9: 20–29.

Lalor, B. M. and G. M. Hickey. 2013. Environmental science and public policy in executive government: Insights from Australia and Canada. *Science and Public Policy* 40: 767–778.

Lexmond, S. 2009. Improving the effectiveness of environmental regimes: Consilience, science, and common sense. In *The Future of Ocean Regime-Building Essays in Tribute to Douglas M. Johnston*, edited by A. E. Chircop, T. L. McDorman, and S. J. Rolston, 487–513. Leiden, the Netherlands: Martinus Nijhoff.

Massachusetts Ocean Partnership. 2009. A review of ocean management and integrated resource management programs from around the world. http://www.seaplan.org/wp-content/uploads/ProgramSummaries.pdf.

McLean, S. 2014. Coastal web-atlases in policy and decision-making: An EIUI study. Blog, 25 March. http://eiui.ca/?p=2367.

McNie, E. C. 2007. Reconciling the supply of scientific information with user demands: An analysis of the problem and review of the literature. *Environmental Science & Policy* 10: 17–38.

MEOPAR. 2015. Marine environmental observation prediction and response network. http://meopar.ca/.

Mitchell, R. B., W. C. Clark, D. W. Cash, eds, et al. 2006. *Global Environmental Assessments. Information and Influence.* Cambridge, MA: MIT Press.

Mol, A. P. J. 2008. *Environmental Reform in the Information Age: The Contours of Informational Governance.* Cambridge: Cambridge University Press.

Neff, M. W. 2011. What research should be done and why? Four competing visions among ecologists. *Frontiers in Ecology and the Environment* 9 (8): 462–469.

Niemeijer, D. and R. S. de Groot. 2008. Framing environmental indicators: Moving from causal chains to causal networks. *Environment, Development and Sustainability* 10 (1): 89–106.

NRC (National Research Council). 2012. *Using Science as Evidence in Public Policy.* Washington, DC: National Academies Press.

Nursey-Bray, M. J., J. Vince, M. Scott, et al. 2014. Science into policy? Discourse, coastal management and knowledge. *Environmental Science & Policy* 38: 107–119.

Nutley, S. M., I. Walter, and H. T. O. Davies. 2007. *Using Evidence: How Research Can Inform Public Services.* Bristol: Policy Press.

O'Riordan, T. and A. Jordan. 1995. The precautionary principle in contemporary environmental politics. *Environmental Values* 4 (3): 191–212.

Ouellette, M. and M. Hardy. 2010. Integrated coastal zone management: Bridging the land-water divide. *Proceedings of the ICES Annual Science Conference*, 20–24 September 2010. ICES ASC 2010/B:10. http://www.ices.dk/sites/pub/CM%20Doccuments/CM-2010/B/B1010.pdf.

Rittel, H. and M. Webber. 1973. Dilemmas in a general theory of planning. *Policy Sciences* 4: 155–169.

Rudd, M. A. 2015. Scientists' framing of the ocean science–policy interface. *Global Environmental Change* 33: 44–60.

Sarewitz, D. 2013. Science's rightful place is in service to society. *Nature* 502: 595.

Sarkki, S., J. Niemelä, R. Tinch, et al. 2014. Balancing credibility, relevance and legitimacy: A critical assessment of trade-offs in science–policy interfaces. *Science & Public Policy (SPP)* 41 (2): 194–206.

Sarkki, S., R. Tinch, J. Niemela, et al. 2015. Adding "iterativity" to the credibility, relevance, legitimacy: A novel scheme to highlight dynamic aspects of science–policy interfaces. *Environmental Science & Policy* 55: 505–512.

Soomai, S. S. 2015. Elucidating the role of scientific information in decision-making for fisheries management. PhD diss., Dalhousie University.

Soomai, S. S., B. H. MacDonald, and P. G. Wells. 2013. Communicating environmental information to the stakeholders in coastal and marine policy-making: Case studies from Nova Scotia and the Gulf of Maine/Bay of Fundy region. *Marine Policy* 40: 176–186.

UNESCO (United Nations Educational, Scientific, and Cultural Organization). 2006. *A Handbook for Measuring the Progress and Outcomes of Integrated Coastal and Ocean Management.* IOC Manuals and Guides, 46; ICAM Dossier, 2. Paris, France: UNESCO.

van Tatenhove, J. P. M., H. A. C. Runhaar, and H. J. van der Windt. 2016. Organizing productive science–policy interactions for sustainable coastal management: Lessons from the Wadden Sea. *Environmental Science & Policy* 55 (3): 377–472.

Wells, P. G. 2003. State of the Marine Environment (SOME) reports—A need to evaluate their role in marine environmental protection and conservation. *Marine Pollution Bulletin* 46: 1219–1223.

Wesselink, A., K. S. Buchanan, Y. Georgiadou, et al. 2013. Technical knowledge, discursive spaces and politics at the science-policy interface. *Environmental Science & Policy* 30: 1–9.

Wilson, D. C. 2009. *The Paradoxes of Transparency: Science and the Ecosystem Approach to Fisheries Management in Europe.* Amsterdam: Amsterdam University Press.

Wilson, L. T. and P. G. Wells. 2014. Vital global ocean reports must be noticed to have influence. Blog, 16 July. http://eiui.ca/?p=2086.

Wilson, L. T. 2015. The communication of information in multi-sectoral networks: A case study of tidal power network(s) in the Bay of Fundy region of Atlantic Canada. Master's thesis, Dalhousie University.

Young, S. P. (ed.), 2013. *Evidence-Based Policy-Making in Canada.* Don Mills, ON: Oxford University Press Canada.

Index

Printed and bound by CPI Group (UK) Ltd, Croydon, CR0 4YY

22/10/2024

01777623-0019